Pathology of Laboratory Rodents and Rabbits

Pathology of Laboratory Rodents and Rabbits

DEAN H. PERCY • STEPHEN W. BARTHOLD

Iowa State University Press/Ames

DEAN H. PERCY, D.V.M., M.Sc., Ph.D., is professor and chair, Department of Pathology, Ontario Veterinary College, University of Guelph, Guelph, Ontario.
STEPHEN W. BARTHOLD, D.V.M., Ph.D., is professor, Section of Comparative Medicine, Yale School of Medicine, New Haven, Connecticut.

♾ Printed on acid-free paper in the United States of America

First edition, 1993

Library of Congress Cataloging-in-Publication Data

Percy, Dean H.
Pathology of laboratory rodents and rabbits / Dean H. Percy, Stephen W. Barthold.—1st ed.
p. cm.
Includes bibliographical references (p. 225) and index.
ISBN 0-8138-1309-3
1. Laboratory animals—Diseases. 2. Rodents—Diseases. 3. Rabbits—Diseases. 4. Rodents as laboratory animals. 5. Rabbits as laboratory animals. I. Barthold, Stephen W. II. Title.
SF996.5.P47 1993
636′.932—dc20 93–17809

CONTENTS

4. GERBIL 137

5. GUINEA PIG 146

6. RABBIT 179

PREFACE

Pathology of Laboratory Rodents and Rabbits

is designed to serve as a general reference text for veterinary pathologists requiring general information on diagnostic features of the diseases of commonly used laboratory animals. The emphasis has been placed on characteristic diagnostic features, differential diagnoses, and significance of the disease. This text is not intended to be a comprehensive and detailed source of information on all aspects of any particular spontaneous disease seen in these species. There are excellent books available that deal with individual diseases in particular species in more depth than in this text, and we encourage the reader to refer to these sources for additional information. A bibliography follows each main heading, and the key references from these lists are cited in the text. We have written the text with the hope that diagnostic pathologists and graduate students will find it a useful source of information for the specialty of laboratory animal pathology.

We are indebted to our mentors and colleagues for their guidance, assistance, and moral support over the past few years. Drs. T. J. Hulland, H. J. Olander, Carl Olson, and A. M. Jonas were instrumental in guiding us through the discipline of pathology during those formative years. Other colleagues have also provided us with the stimulus and the encouragement to attempt this project. Special thanks are due Drs. R. B. Miller, R. O. Jacoby, and D. G. Brownstein in this regard. We are also indebted to our colleagues who generously provided material for photographs required for the text. Special thanks to Ted Eaton, Department of Pathology, Ontario Veterinary College, for the excellent technical assistance in preparing the photographs, and to Jean Bagg, Ontario Veterinary College, for providing the meticulous technical assistance necessary for the preparation of this manuscript.

We acknowledge the financial assistance provided by the Department of Pathology, Ontario Veterinary College, and the Section of Comparative Medicine, Yale School of Medicine, in order to complete this project.

GENERAL BIBLIOGRAPHY

Baker, H.J., et al., eds. 1979. *The Laboratory Rat. I. Biology and Diseases*. New York: Academic.

Benirschke, K., et al., eds. 1978. *Pathology of Laboratory Animals*. Vols. I and II. New York: Springer-Verlag.

Bhatt, P.N., et al., eds. 1986. *Viral and Mycoplasmal Infections of Laboratory Rodents*. New York: Academic.

Boorman, G.A., et al., eds. 1990. *Pathology of the Fischer Rat*. New York: Academic.

Burek, J.D. 1978. *Pathology of Aging Rats*. Boca Raton, Fla: CRC.

Cheeke, P.R. 1987. *Rabbit Feeding and Nutrition*. New York: Academic.

Cheeke, P.R., et al. 1986. *Rabbit Production*. 6th ed. Danville, Ill.: Interstate Printers and Publishers.

Feldman, D.B., and Seely, J.C. 1988. *Necropsy Guide: Rodents and the Rabbit*. Boca Raton, Fla: CRC.

Flynn, R.J. 1973. *Parasites of Laboratory Animals*. Ames: Iowa State University Press.

Foster, H.L., et al., eds. 1982. *The Mouse in Biomedical Research. IV. Experimental Biology and Oncology*. New York: Academic.

Fox, J.G., et al., eds. 1984. *Laboratory Animal Medicine*. New York: Academic.

Frith, C.H., and Ward, J.M. 1988a. *Color Atlas of Neoplastic and Non-neoplastic Lesions in Aging Mice*. New York: Elsevier.

Frith, C.H., et al. 1985. *A Color Atlas of Hematopoietic Pathology of Mice*. Little Rock, Ark.: Toxiologic Pathology Associates.

Greaves, P., and Faccini, J.M. 1984. *Rat Histopathology: A Glossary for Use in Toxicity and Carcinogenicity Studies*. Amsterdam and New York: Elsevier.

Harkness, J.E., and Wagner, J.E. 1989. *The Biology and Medicine of Rabbits and Rodents*. Philadelphia: Lea and Febiger.

Jones, T.C., et al., eds. *Monographs on Pathology of Laboratory Animals*. New York: Springer-Verlag. Series sponsored by International Life Sciences Institute.

______. 1983. *Endocrine System*.

______. 1985. *Digestive System*.

______. 1985. *Respiratory System*.

______. 1986. *Urinary System*.

______. 1987. *Genital System*.

______. 1988. *Nervous System*.

______. 1989. *Integument and Mammary Glands*.

______. 1990. *Hematopoietic System*.

______. 1991. *Cardiovascular and Musculoskeletal Systems*.

______. 1991. *Eye and Ear*.

Lindsey, J.R., et al., eds. 1982. *The Mouse in Biomedical Research*. II *Diseases*. New York: Academic.

Mohr, U. et al., eds. 1993. *Pathobiology of the Aging Rat*. Vols. 1 and 2. Washington, D.C.: ILSI Press.

Owen, D.G. 1992. *Parasites of Laboratory Animals*. London: Laboratory Animals Ltd.

Richardson, V.C.G. 1992. *Diseases of Domestic Guinea Pigs*. London: Blackwell Scientific.

Sanderson, J.H., and Phillips, C.E. 1981. *An Atlas of Laboratory Animal Haematology*. Oxford, Engl.: Clarendon Press.

Schmidt, R.E., et al. 1983. *Pathology of Aging Syrian Hamsters*. Boca Raton, Fla: CRC.

Seaman, W.J. 1987. Postmortem Change in the Rat: A Histologic Characterization. Ames: Iowa State University Press.

Stinson, S.F., et al., eds. 1989. *Atlas of Tumor Pathology of the Fischer Rat*. Boca Raton, Fla: CRC.

Van Hoosier, G.L., and McPherson, C.W., eds. 1987. *Laboratory Hamsters*. New York: Academic.

Wagner, J.E., and Manning, P.J., eds. 1976. *The Biology of the Guinea Pig*. New York: Academic.

Weisbroth, S.H., et al., eds. 1974. *The Biology of the Laboratory Rabbit*. New York: Academic.

Pathology of Laboratory Rodents and Rabbits

1 MOUSE

The laboratory mouse poses unique but surmountable problems for the diagnostic pathologist. Genetic background is an important determinant of disease expression in any species, but in the mouse, genotype is a particularly salient consideration. Each inbred, isogenic, congenic, mutant, recombinant, and recently, transgenic mouse strain, as well as outbred stock, express different patterns of disease, some unique to the strain. Expression of infectious and spontaneous diseases and response to experimental variables are very much genotype-related in mice. Genetic differences in susceptibility within strains can be quite striking. BALB/c mice, for example, rarely develop amyloidosis, while it is quite common in other strains. Differences can also be very subtle. A/J mice are highly resistant to virulent mouse hepatitis virus (MHV) strain 3, while they are quite susceptible to less virulent MHV strain JHM. Thus, it is impossible to generalize about disease patterns with laboratory mice, but it is likewise impossible to know specifics of disease patterns in all mouse genotypes. The pathologist must become familiar with the general picture of mouse pathology but understand and expect the marked variability in prevalence and expression that occurs.

A second diagnostic problem is the degree of sensitivity required for infectious disease diagnosis in mice. Concern over infectious diseases in mice is at a different level from that for other, larger species of laboratory animals because of the nature of research performed upon mice. Infectious agents that would otherwise be overlooked or unknown in other species are significant to the mouse biologist because of their potential effect on the biological response of the mouse to research variables. A number of agents, such as lactate dehydrogenase-elevating virus, cause minimal recognizable disease, even at the microscopic level but can have significant effects upon biological response, particularly immune responsiveness. This book emphasizes diseases or lesions that are apt to be recognized by the diagnostic pathologist, but one must be cognizant that murine diagnostics rely heavily on other methods, such as serology, that detect subclinical but significant murine agents.

ANATOMIC FEATURES

Hematology

This subject is thoroughly reviewed by Bannerman (1983). Mouse leukocytes resemble those of other mammals. Lymphocytes are the predominant circulating leukocyte and circulating basophils are nonexistent, although tissue mast cells are plentiful. Mature male mice have significantly higher granulocyte counts than do female mice.

Gastrointestinal System

Incisive foramina, located posterior to the upper incisors, communicate between the roof of the mouth and the anterior nasal cavity. Incisors grow continuously, but cheek teeth do not. Rodents do not have tonsils. The submaxillary salivary glands in sexually mature males reveal increased secretory granules in the cytoplasm of serous cells (Figs. 1.1 and 1.2). The intestine is simple. Paneth cells occupy crypt bases in the small intestine (Fig. 1.3). These specialized enterocytes have prominent eosinophilic cytoplasmic granules, which are larger in mice than in rats. Most microflora-associated mice possess prominent gram-positive filamentous bacteria attached to ileal enterocytes (Fig. 1.4). Pregnant and lactating mice have noticeably thickened bowel walls due to physiological mucosal hyperplasia. The liver has variable lobation. Hepatocytes frequently display anisokaryosis, polykarya, karyomegaly, and cytoplasmic invagination into the nucleus. These features are present at all ages, but increase with age and disease (see Aging, Degenerative, and Miscellaneous Disorders). Hematopoeisis normally occurs in the infant liver (Fig.1.5) but wanes by weaning age, although islands of hematopoeitic cells can be found in hepatic sinusoids of older mice, particularly in disease states. Hepatocytes frequently contain cytoplasmic fat vacuoles. Certain strains, such as BALB mice, normally have diffuse hepatocellular microvesicular fatty change (Fig. 1.6), resulting in grossly pallid livers, compared with the mahogany-colored livers of other mouse strains.

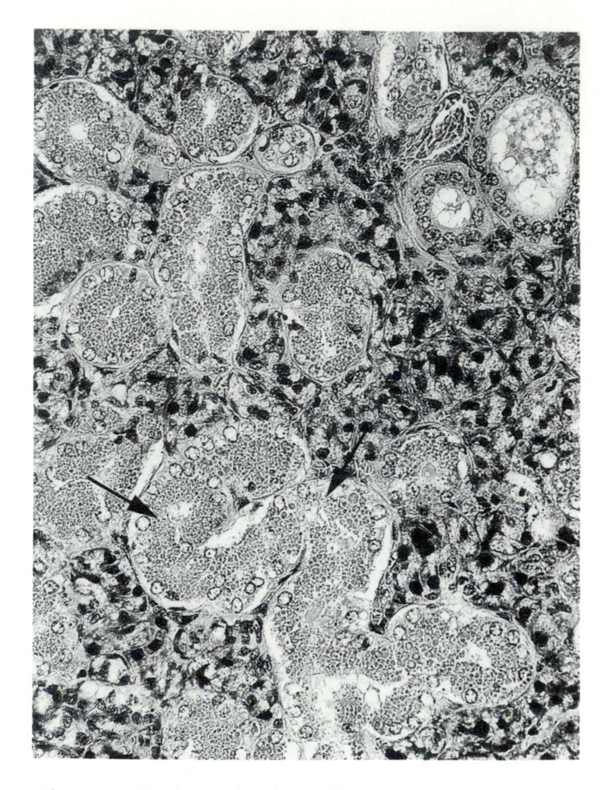

Fig. 1.1. Sections of submandibular (submaxillary) salivary glands from adult male mouse. Note the prominent secretory granules (*arrows*) in the cytoplasm of epithelial cells.

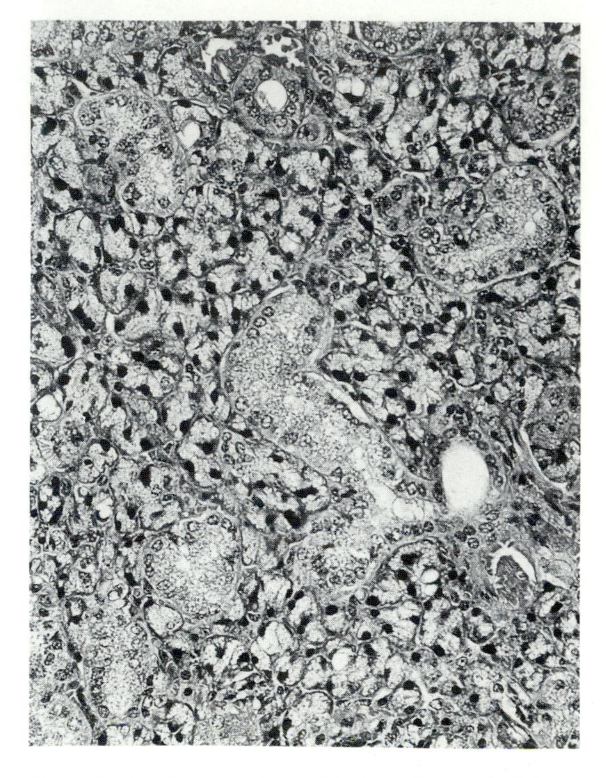

Fig. 1.2. Sections of submandibular salivary gland from sexually mature female. The secretory granules in ductal epithelial cells are less prominent, with fewer granules per unit area, compared with males.

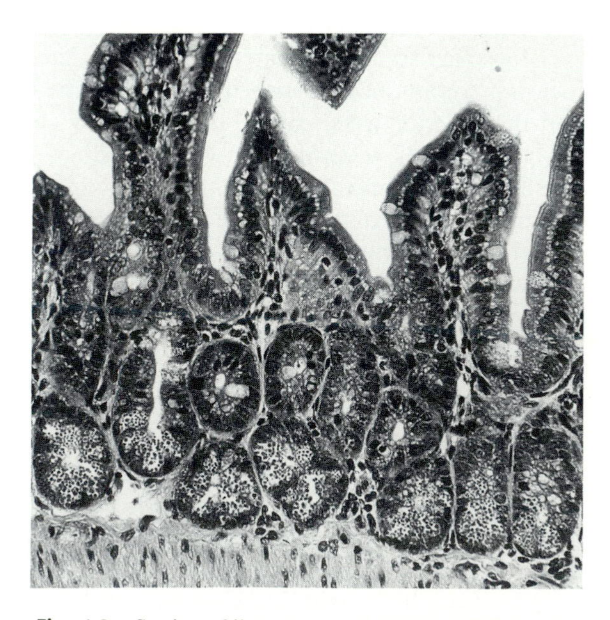

Fig. 1.3. Section of ileum from mouse illustrating the distinct cytoplasmic granules in the enterocytes lining the crypts (Paneth cells), a normal finding in mice.

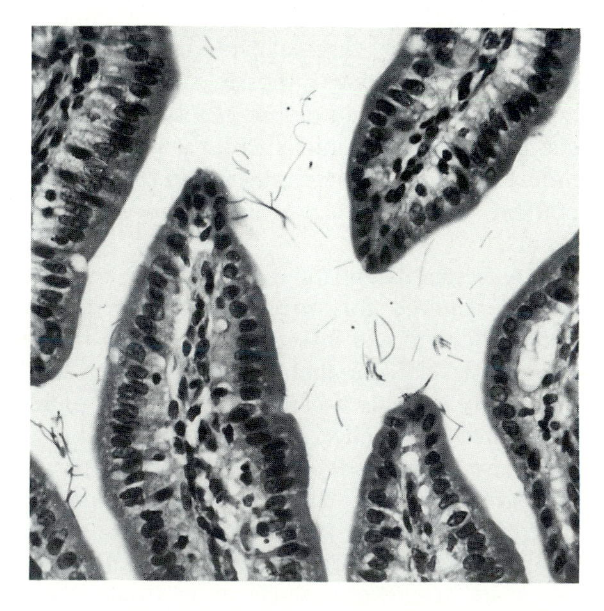

Fig. 1.4. Ileum from mouse demonstrating filamentous bacteria associated with the villi. These organisms are gram-positive and are considered to be a normal finding in this species.

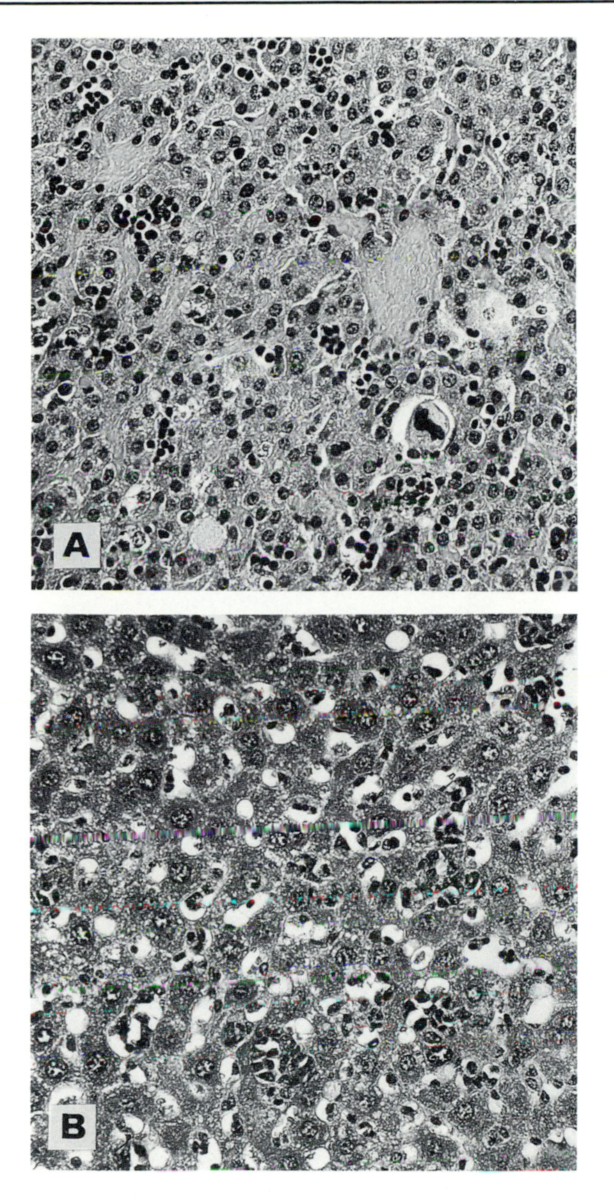

Fig. 1.5. Section of liver from newborn mouse (*A*) and adult mouse (*B*). There are numerous hematopoietic cells in the sinusoidal regions. This is an incidental finding, particularly in young mice.

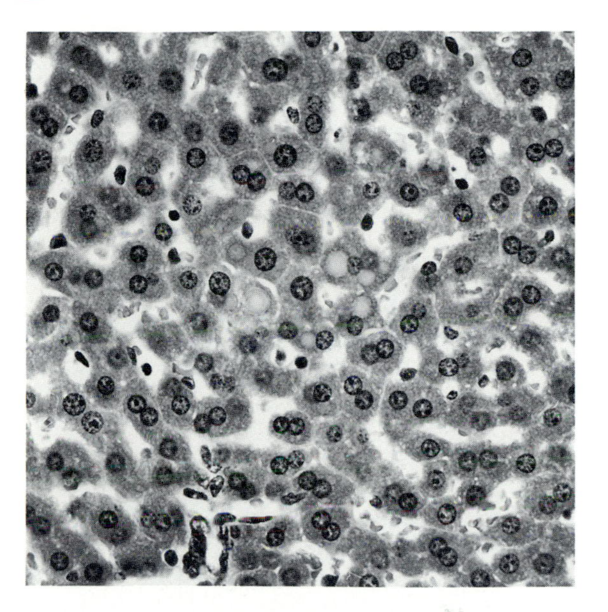

Fig. 1.6. Liver from adult mouse, BALB strain. There are prominent intracytoplasmic vacuoles in a few hepatocytes.

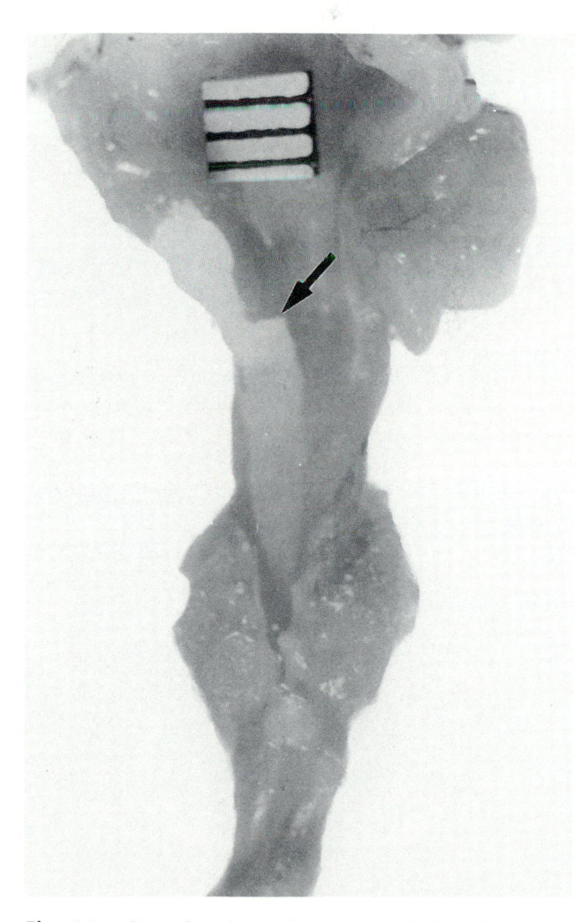

Fig. 1.7. Coagulum (*arrow*) present at the junction of the urinary bladder and urethra in a euthanized male mouse. They are expelled secretions from the accessory sex glands (copulatory plug). In the absence of any evidence of urinary obstruction, this is considered to be a terminal event and an incidental finding.

Genitourinary System

Female mice have a large clitoris, or genital papilla, with the urethral opening near its tip and anterior to the vaginal orifice. Males have redundant testes that readily retract into the abdominal cavity, particularly when they are picked up by the tail. Both sexes have well-developed preputial glands, and males have conspicuous accessory sex glands, including large seminal vesicles, coagulating glands, and prostate (Cook 1983). Ejaculation results in formation of a coagulum, or copulatory plug. This frequently occurs agonally; coagulum can be found in urinary bladder or urethra as a normal incidental finding and must not be misconstrued as a calculus or obstruction (Figs. 1.7 and 1.8). Sexual maturity in males results in expression of several sexual dimorphic features, including peripheral blood granulocytosis, salivary gland changes (mentioned previously), and renal changes. In adult males, the parietal layer of Bowman's capsule is lined by cuboidal epithelium, resembling tubular epithelium (Fig. 1.9). This is not absolute, since some glomeruli of male mice are surrounded by squamous epithelium and some glomeruli of female mice are surrounded by cuboidal epithelium. Proteinuria is also normal in mice, with highest levels in sexually mature male mice. Mice are endowed with relatively large numbers of glomeruli per unit area, compared with some species, such as the rat.

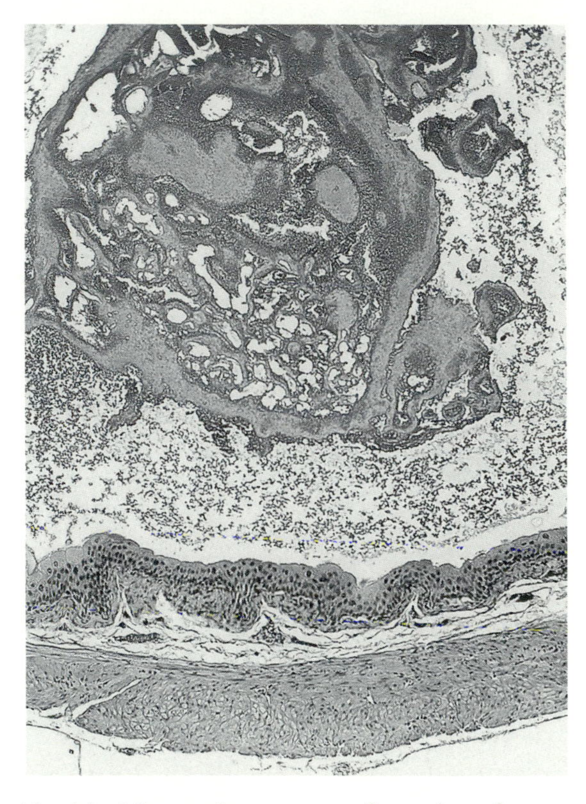

Fig. 1.8. Microscopic appearance of coagulum of accessory sex gland secretions in urinary bladder of male mouse.

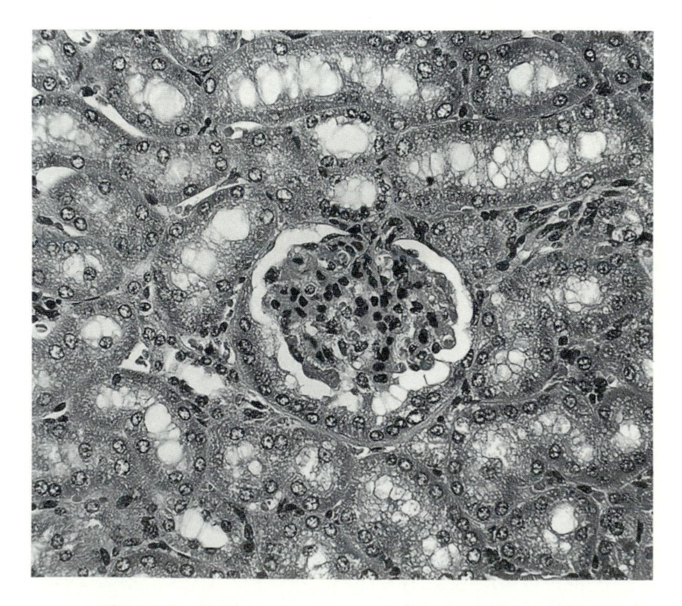

Fig. 1.9. Renal cortex from adult male mouse, illustrating the typical cuboidal epithelium lining the parietal surface of Bowman's capsule.

Skeletal System

Bones of mice, rats, and hamsters do not have haversian systems (Fig. 1.10), and ossification of epiphyseal plates with age is variable and incomplete, depending upon mouse genotype. Hematopoeisis remains active in long bones throughout life.

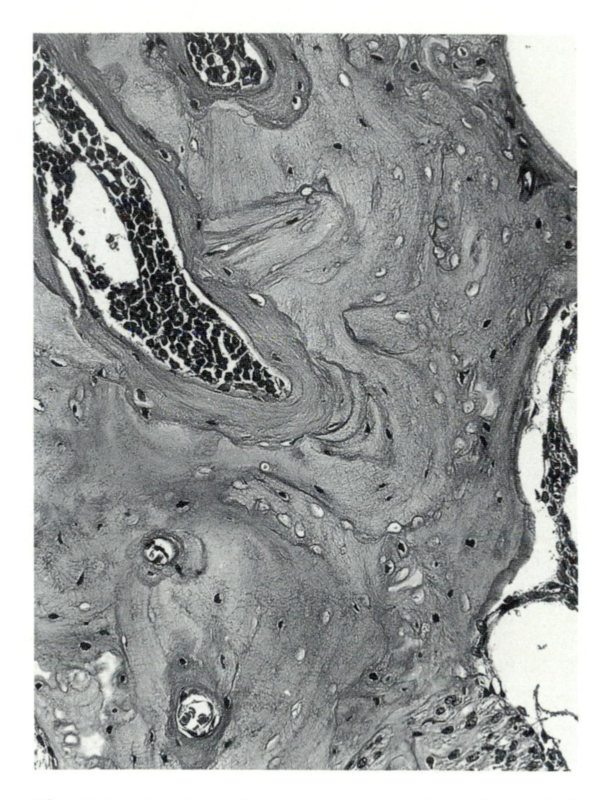

Fig. 1.10. Section of petrous temporal bone from adult mouse, illustrating absence of haversian systems.

Lymphopoietic System

The thymus does not involute completely in adults. Hassall's corpuscles are indistinct. Islands of ectopic parathyroid tissue can be encountered in the septal or surface connective tissue of the thymus, and conversely, thymic tissue can occur in thyroid and parathyroid glands. Epithelial-lined cysts are also common. The splenic red pulp is an active hematopoeitic site throughout life (Fig. 1.11). During disease states and pregnancy, increased hematopoeisis can result in splenomegaly. Lymphocytes tend to accumulate around renal interlobular arteries, salivary gland ducts, urinary bladder submucosa, and other sites, increasing with age. These sites are often involved in generalized lympho-proliferative disorders. Melanosis of the splenic capsule and trabeculae is common in pigmented mice (Fig. 1.12). This must be differentiated from hemosiderin pigment, which tends to accumulate in the red pulp as mice age. Mast cells can be frequent in the spleen of some mouse strains, such as A strain, but not others.

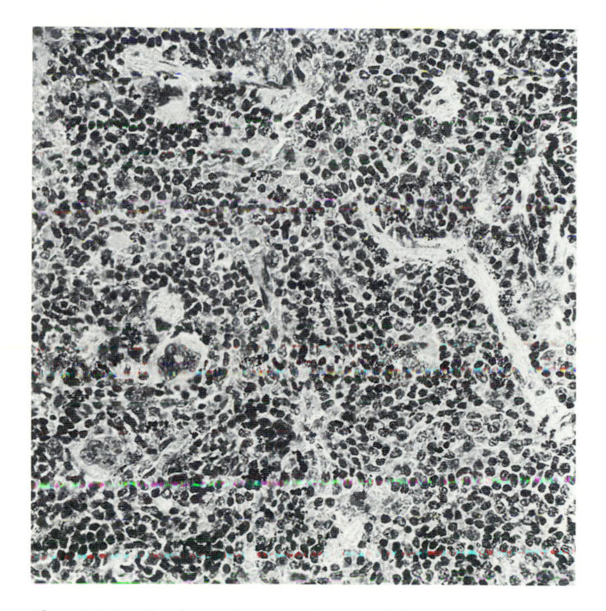

Fig. 1.11. Section of spleen from adult mouse, illustrating the large numbers of hematopoietic cells, including megakaryocytes, in the sinusoids, a common finding throughout life.

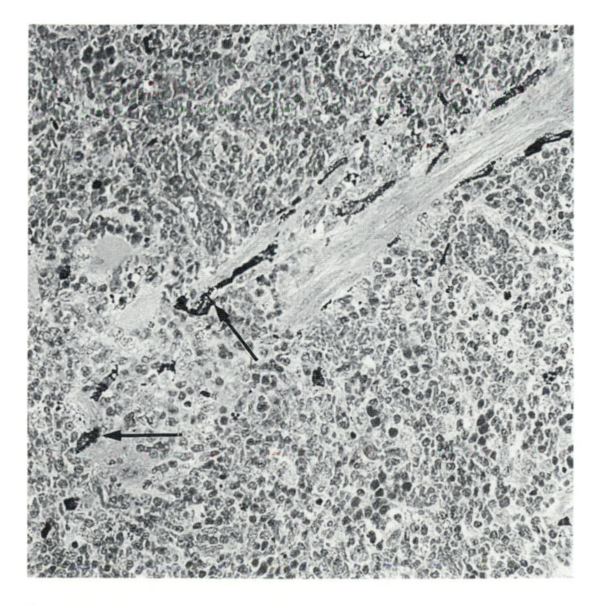

Fig. 1.12. Splenic melanosis in pigmented mouse. Note the pigment-bearing cells (*arrows*) along the splenic trabeculae.

Respiratory System

The microscopic anatomy of the upper and lower respiratory tract of rodents has been thoroughly reviewed (Kuhn 1985; Popp and Monteiro-Riviere 1985). Cross sections of the nose reveal prominent vomeronasal organs, which are important in pheromone sensing and frequent targets of viral attack. Respiratory epithelium can contain eosinophilic secretory inclusions, which may be especially obvious in some strains, such as C57BL. The lungs have a single left lobe and four right lobes. Cartilage envelopes are present only in extrapulmonary airways in mice, rats, and hamsters. Thus, primary bronchi are extrapulmonary. Respiratory bronchioles are short or nonexistent. Cardiac muscle surrounds major branches of pulmonary veins in most rodents (Fig. 1.13) and should not be misconstrued as medial hypertrophy. Bronchus-associated lymphoid tissue is normally present only at the hilus of the lung, except in hamsters. Lymphoid accumulations are often encountered on the visceral pleura of mice, within septal clefts.

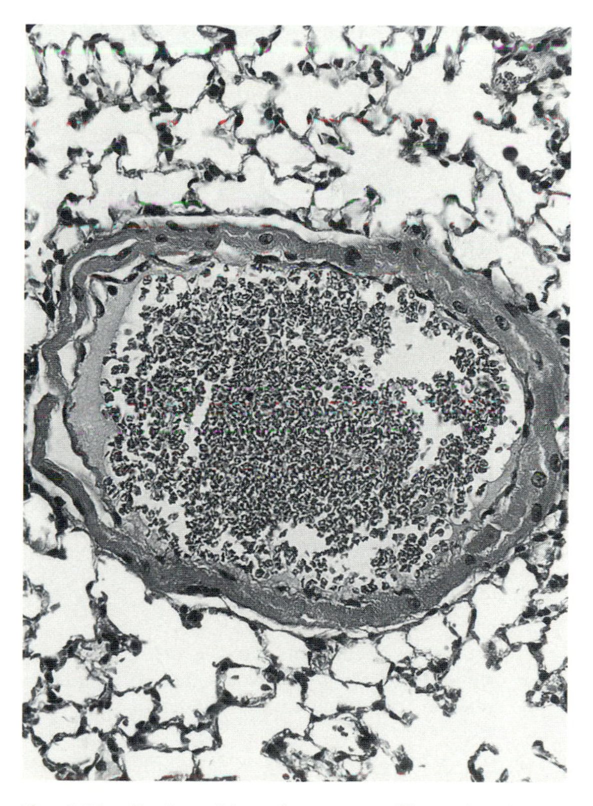

Fig. 1.13. Section of lung from mouse illustrating the extension of cardiac muscle along pulmonary veins, a normal feature in small rodents.

Endocrine System

The mouse adrenal gland has several notable features. Accessory adrenals, either partial or complete, are very common in the adrenal capsule or surrounding connective tissue. The zona reticularis of the adrenal cortex is not discernible from the zona fasciculata. Proliferation of subcortical spindle cells, with displacement of the cortex, is common in mice of all ages. The function of these cells is not known. The most important unique feature of the mouse adrenal is the X zone of the cortex, which surrounds the medulla. The X zone is composed of basophilic cells, and appears in mice around 10 days of age. When males reach sexual maturity and females undergo their first pregnancy, the X zone disappears. The zone disappears gradually in virgin females. During involution, the X zone undergoes marked vacuolation (Fig. 1.14) in females but not in males (Dunn 1970). Residual cells accumulate ceroid. Epithelial cysts are common in the thyroid and pituitary. Thyroid cysts are often lined by ciliated cells.

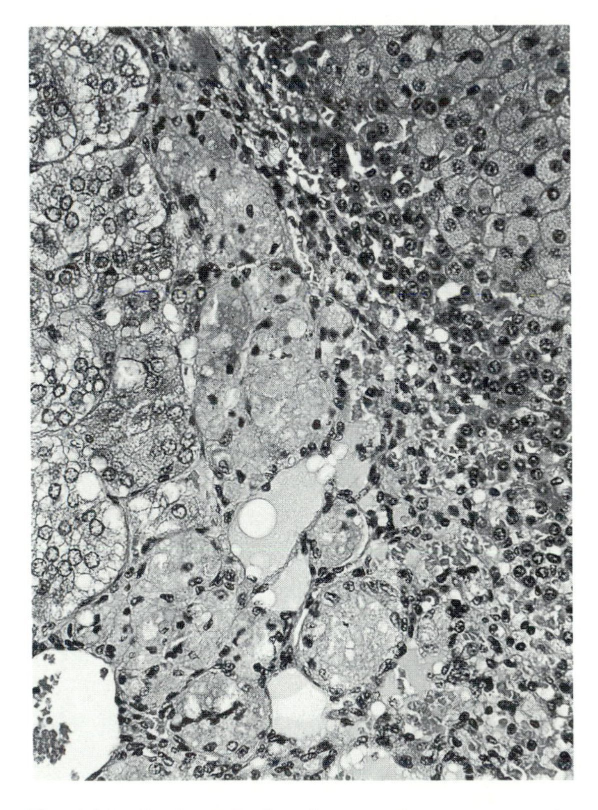

Fig. 1.14. Section of adrenal cortex from young male mouse, illustrating the distinct line of basophilic cells (X zone) at the corticomedullary junction. The medulla is at the left.

Other Anatomic Features

Melanosis occurs in several organs, including the anteroventral meninges of the olfactory bulbs (Fig. 1.15), optic nerves, parathyroid glands, heart valves, and spleens of pigmented mouse strains, such as C57BL mice. Mice have three pectoral and two inguinal pairs of mammary glands, with mammary tissue enveloping much of the subcutis, including the neck. Mammary tissue can be found immediately adjacent to salivary glands. Brown fat is prominent as a subcutaneous fat pad over the shoulders and is also present in the neck, axillae, and peritoneal tissue.

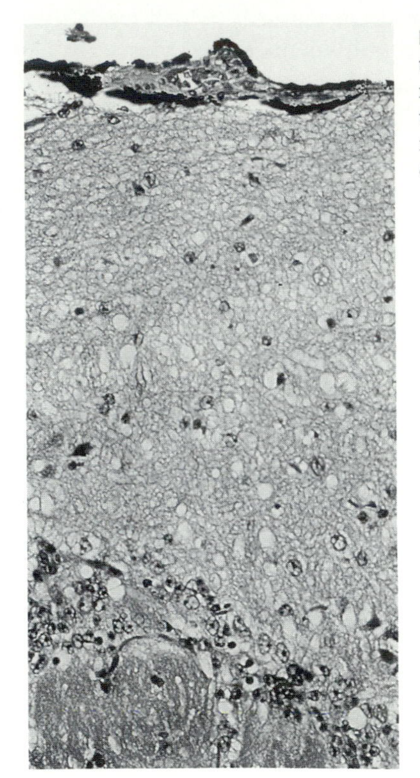

Fig. 1.15. Olfactory bulb from pigmented mouse, illustrating meningeal melanosis, an incidental finding.

Bibliography for Anatomic Features

Bannerman, R.M. 1983. Hematology. In *The Mouse in Biomedical Research. III. Normative Biology, Immunology, and Husbandry*, ed. H.F. Foster et al., pp. 293–312. New York: Academic.

Blumershine, R.V. 1978. Filamentous microbes indigenous to the murine small bowel: A scanning electron microscopic study of their morphology and attachment to the epithelium. Microbial Ecology 4:95–103.

Cook, M.J. 1983. Anatomy. In *The Mouse in Biomedical Research. III. Normative Biology, Immunology, and Husbandry*, ed. H.L. Foster et al., pp. 102–20. New York: Academic.

Crabtree, C.E. 1941. The structure of Bowman's capsule as an index of age and sex variation in normal mice. Anat. Rec. 79:395–413.

Danse, L.H.J.C., and Crichton, D.N. 1990. Pigment deposition, rat, mouse. In *Monographs on Pathology of Laboratory Animals: Hematopoietic System*, ed. T.C. Jones et al., pp. 226–32. New York: Springer-Verlag.

Dunn, T.B. 1970. Normal and pathologic anatomy of the adrenal gland of the mouse, including neoplasms. J. Natl. Cancer Inst. 44:1323–89.

Frith, C.H., and Townsend, J.W. 1985. Histology and ultrastructure, salivary glands, mouse. In *Monographs on Pathology of Laboratory Animals: Digestive System*, ed. T.C. Jones et al., pp. 177–84. New York: Springer-Verlag.

Hummel, K.P. et al. 1975. Anatomy. In *Biology of the Laboratory Mouse*, ed. E.L. Green, pp. 247–307. New York: Dover.

Jones, T.C. 1967. Pathology of the liver of rats and mice. In *Pathology of Laboratory Rats and Mice*, ed. E. Cotchin and F.J.C. Roe, pp. 1–17. Oxford: Blackwell.

Kaplan, H.M.et al. 1983. Physiology. In *The Mouse in Biomedical Research. III. Normative Biology, Immunology, and Husbandry*, ed. H.L. Foster et al., pp. 247–92. New York: Academic.

Kuhn, C., III. 1985. Structure and function of the lung. In *Monographs on Pathology of Laboratory Animals: Respiratory System*, ed. T.C. Jones et al., pp. 89–98. New York: Springer-Verlag.

Liebelt, A.G. 1986. Unique features of anatomy and ultrastructure, kidney, mouse. In *Monographs on Pathology of Laboratory Animals: Urinary System*, ed. T.C. Jones et al., pp. 24–44. New York. Springer-Verlag.

Popp, J.A., and Monteiro-Riviere, N.A. 1985. Macroscopic, microscopic and ultrastructural anatomy of the nasal cavity, rat. In *Monographs on Pathology of Laboratory Animals: Respiratory System*, ed. T.C. Jones et al., pp. 3–10. New York: Springer-Verlag.

Silberberg, M., and Silberberg, R. 1962. Osteoarthritis and osteoporosis in senile mice. Gerontologia 6:91–101.

Staley, M.W., and Trier, J.S. 1965. Morphologic heterogeneity of mouse Paneth cell granules before and after secretory stimulation. Am. J. Anat. 117:365–83.

Wicks, L.F. 1941. Sex and proteinuria in mice. Proc. Soc. Exp. Biol. Med. 48:395–400.

VIRAL INFECTIONS

Laboratory mice are host to a large spectrum of viral agents (Table 1.1). For practical purposes, this list can be abbreviated considerably for a number of reasons. Several agents have been essentially eliminated from contemporary mouse colonies by modern husbandry practices, including mouse adenovirus, K virus, polyomavirus, and transmissible murine mammary tumor virus (MuMTV). They may, however, reappear, since most have been retained or are still being used experimentally. The actual prevalence of the herpesviruses and lactate dehydrogenase elevating virus (LDHE-V) is not known, since these agents are not routinely monitored by serological methods, but they are probably rare. Ectromelia virus and lymphocytic choriomeningitis virus, although significant agents, are also relatively rare. Furthermore, some of the common viruses seldom produce discernible pathology, including minute virus of mice (MVM), and reovirus. Thus, the list of agents likely to cause lesions that will be encountered by the diagnostic pathologist is relatively short. This text will emphasize those agents, but all are discussed.

TABLE 1.1. Viruses of Laboratory Mice

DNA viruses	
Adenovirus	MAd-1(FL), MAd-2(K87)
Herpesvirus	Mouse cytomegalovirus (MCMV), mouse thymic virus (MTV)
Papovavirus	K virus, polyoma virus
Parvovirus	Minute virus of mice (MVM)
Poxvirus	Ectromelia virus
RNA viruses	
Arenavirus	Lymphocytic choriomeningitis (LCMV)
Coronavirus	Mouse hepatitis virus (MHV)
Paramyxovirus	Pneumonia virus of mice (PVM), Sendai virus
Picornavirus	Mouse encephalomyelitis virus (MEV)
Reovirus	Epizootic diarrhea of infant mice (EDIM) virus, reovirus 3
Retrovirus	Murine leukemia virus (MuLV), murine mammary tumor virus (MuMTV)
Togavirus	Lactate dehydrogenase elevating virus (LDHE-V)

Disease expression is significantly influenced by age, genotype, and immune status. Under most circumstances, even the most pathogenic murine viral agents cause minimal clinical disease. However, under select circumstances, the same agents can have devastating consequences. Infant mice less than 2 wk of age are highly susceptible to viral disease and lesions, but are often protected by maternal antibody during this period of vulnerability. Genotype, as already discussed, is also an important factor in host susceptibility. Immune status can be influenced by maternally derived passive immunity, which usually wanes around 4–6 wk of age; actively acquired immunity, which can be virus strain–specific or short-lived; and genetic or induced immunologic aberrations caused by athymia, x-irradiation, corticosteroids, etc. Investigation of host-virus epizootiology by the astute clinician must encompass these three major factors as well as population dynamics and husbandry practices. This is an important component of murine diagnostics.

Most viral infections in mice are acute, or short-term, and lesions are often subtle. Animals submitted for necropsy should be carefully selected to provide maximal opportunity for diagnosis. Live, clinically ill animals should be selected, since they would be most likely to have active lesions. Diagnosis of viral infections in a rodent colony should not be solely dependent upon gross and microscopic pathology. A very useful adjunct is viral serology, but this should not be used alone for diagnosis either. Actively infected mice are likely to be seronegative. Conversely, mice can be seropositive, yet actively infected with a second strain of the same agent, as is the case with MHV. Young mice can be seropositive with passively derived maternal antibody but not actually exposed to the virus in question. These examples underscore that seroconversion to an agent does not imply a cause and effect relationship with disease, unless epizootiology, pathology, and serology are considered collectively.

DNA Viral Infections

Adenoviral Infections

Mice are host to two distinct adenoviruses, MAd-1(FL) and MAd-2(K87), which can be differentiated from one another by both serology and pathology. These agents partially cross-react serologically, depending on method, in a one-sided relationship with antiserum to K87 cross-reacting with FL, but not conversely. Adenoviruses are nonenveloped DNA viruses that replicate in the nucleus and produce characteristic intranuclear inclusions.

EPIZOOTIOLOGY AND PATHOGENESIS. MAd(FL) was first discovered as a contaminating cytopathic agent while attempting to establish Friend leukemia virus in tissue culture (Hartley and Rowe 1960). Naturally occurring clinical disease or lesions due to FL strain have not been described, but experimental inoculation of suckling mice with FL by a variety of routes results in viremia and a fatal, multisystemic infection within 10 days. Inoculation of weanling or adult mice also results in multisystemic infection with prolonged viruria, but with only occasional deaths. Infection is transmitted by direct contact through urine and nasal secretions. Serological surveys have indicated that the FL strain was at one time found in 11% of mouse colonies tested (Parker et al. 1966), but it is now virtually nonexistent in North America and Europe. MAd-2(K87) was initially isolated from feces of an otherwise healthy mouse. MAd(K87) infects only intestinal epithelium, regardless of route of inoculation and is excreted in feces. Following oral inoculation of 4-wk-old or younger mice, virus is excreted in feces for 3 or more wk with peak infection between 7 and 14 days. Mice apparently recover. Clinical signs are usually absent in experimentally or naturally infected mice, but runting can occur. Clinically normal athymic nude mice have been found to be infected. The prevalence of MAd(K87) infection is unknown, since serological monitoring with MAd(K87) antigen has not been routinely done. It has been reported in Japan, Europe, and North America. Seroconversion of rats to MAd-2(K87) has been noted, but they are not susceptible to experimental inoculation, suggesting that they are host to a related, but different adenovirus (see Chap. 2).

PATHOLOGY. Athymic nude mice have been shown to develop hemorrhagic lesions in the duodenum and wasting disease when inoculated with FL strain (Winters and Brown 1980). Mice experimentally inoculated with MAd(FL) develop foci of necrosis in liver and other organs. Gross lesions of MAd(K87) infection are not evident, except that juvenile mice may be runted (Heck et al. 1972). Mice experimentally inoculated with MAd(FL) develop type A intranuclear inclusions in foci of necrosis that occur in multiple organs, including brown fat, myocardium, adrenal gland (Fig. 1.16), spleen, brain, pancreas, liver, intestine, salivary glands, and kidney. Adult athymic nude mice inoculated with MAd(FL) develop inclusions in endothelial and mucosal cells of the duodenum. Lesions have not been described in naturally infected mice.

Mice naturally or experimentally infected with MAd(K87) develop intranuclear inclusions in mucosal epithelial cells of the small intestine, especially in the distal segments, and the cecum. Inclusions are most plentiful in infant mice but can be found in smaller numbers in the mucosa of adult mice as well. Similar

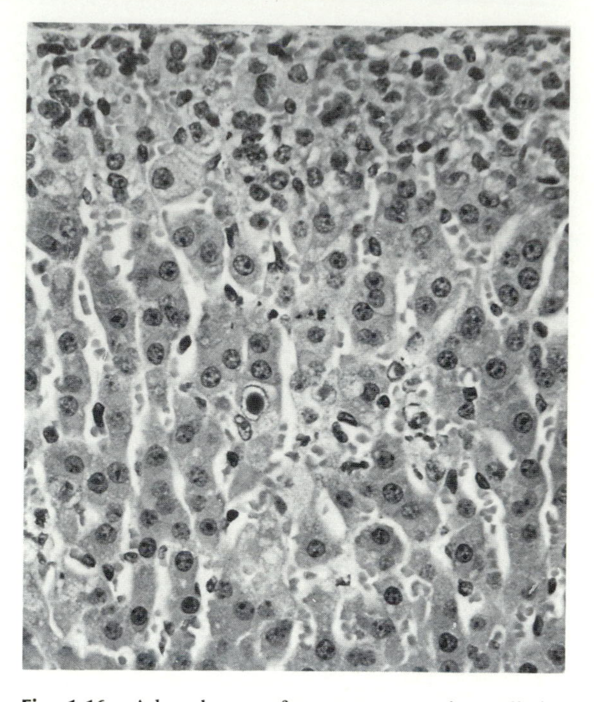

Fig. 1.16. Adrenal cortex from mouse experimentally infected with mouse adenovirus, MAd(FL) strain. There is necrosis with mononuclear cell infiltration in the superficial cortex. A single intranuclear inclusion is present in the zona reticularis region.

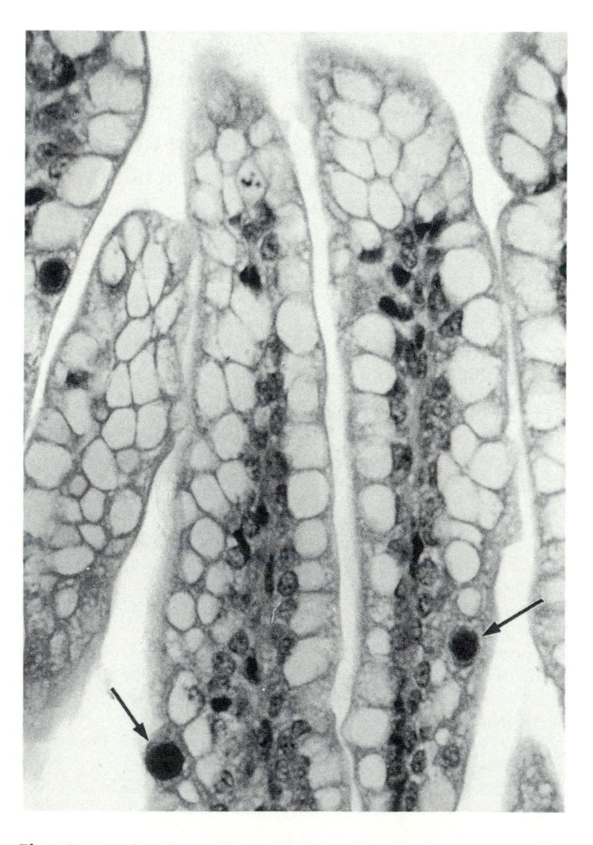

Fig. 1.17. Section of small intestine from mouse with subclinical adenovirus infection, an incidental finding. Note the distinct intranuclear inclusion bodies (*arrows*) in a few enterocytes lining the villi.

inclusions have been noted in athymic nude mice without other detectable lesions. Typically, inclusion-bearing nuclei are often located in the apical portions of cells, rather than in their normal basal location (Fig. 1.17). Such nuclei must be differentiated from mitotic cells and intraepithelial lymphocytes.

DIAGNOSIS. Adenovirus inclusions are often quite obvious and pathognomonic, especially when found in intestinal epithelium. Success at finding inclusions is maximized in infant mice. Both MAd(FL) and MAd(K87) viruses are amenable to in vitro culture. Serological testing is probably the most effective means of screening mouse populations for MAd, using the indirect immunofluorescence assay. Since MAd(FL) is virtually nonexistent and serological cross-reactivity between MAd(FL) and (K87) strains is one-way, K87 antigen should be used. Nevertheless, murine serological testing continues to frequently employ FL antigen, so that K87 infection will continue to be overlooked. *Differential diagnoses*: Multisystemic infections that produce intranuclear inclusions, such as polyomavirus and cytomegalovirus, both of which are also rare, must be differentiated for MAd(FL). Intestinal mucosal MAd(K87) inclusions are pathognomonic and not induced by any other known agent.

SIGNIFICANCE. Enhanced susceptibility to experimental coliform pyelonephritis in adenovirus-infected mice has been reported previously (Ginder 1964). However, MAd(FL) can now be relegated to historical interest. MAd(K87) is innocuous, even in infant and athymic mice.

Mouse Thymic Viral (MTV) Infection

MTV is presumably a herpesvirus, based on morphological criteria. It is biologically and antigenically distinct from another herpesvirus of mice, murine cytomegalovirus. Detailed information about this virus is generally lacking since in vitro methods of propagation have not been identified and little experimental work has been performed. Synonyms include thymic necrosis virus and thymic agent.

EPIZOOTIOLOGY AND PATHOGENESIS. MTV was first discovered during early studies on the etiology of mammary tumor viruses. Inoculation of newborn mice less than 10 days of age resulted in thymic necrosis. This feature of the virus has been emphasized in subsequent studies, but MTV appears to infect salivary glands as its primary target. Outcome of experimental infection is strikingly age-dependent and also influenced by mouse genotype. Intraperitoneal inoculation of newborn mice results in acute thymic necrosis, which is visible grossly as diminished thymic mass, within 14 days. Although largely an experimental phenomenon, thymic necrosis has been encountered in infant mice from naturally infected mouse colonies. Helper T cells are the specific cellular target for MTV. Older mice do not develop thymic necrosis. Mice of all ages develop infection of salivary glands, with persistent virus shedding in saliva for several months or more. The mode of MTV transmission is therefore presumed to be via the saliva. MTV has also been isolated from mammary tissue of an infected mother (Morse 1987) and from mammary tumor extracts, suggesting another probable route of transmission. Vertical (in utero) transmission has not been documented. The current prevalence of MTV in mouse colonies is unknown, but earlier serological surveys indicated that 4 of 15 commercial mouse colonies tested were positive. It is very common among wild mouse populations.

PATHOLOGY. MTV infection of infant mice results in the formation of intranuclear inclusion bodies and necrosis of thymocytes, and to a lesser extent, cells in lymph nodes and spleens. During recovery, there is granuloma formation. Lesions in salivary glands have not been noted.

DIAGNOSIS. Diagnosis of active infections in infant mice can be made by confirmation of thymic necrosis with characteristic intranuclear inclusions. Definitive diagnosis can be made by immunofluorescence of affected thymic tissue or infant mouse bioassay. *Differential diagnoses* include agents that cause thymic necrosis in infant mice, such as coronavirus or stress. Confirmation of infection in mice without thymic lesions is more problematic. Saliva or salivary glands from suspect mice can be inoculated into infant mice as a bioassay, but in vitro propagation is currently not feasible. Serologic tests include complement fixation, serum neutralization (mouse bioassay), and immunofluorescence, but mice infected as neonates may not seroconvert. MTV does not share antigenic cross-reactivity with murine cytomegalovirus.

SIGNIFICANCE. The major concern of MTV infection is the variety of immunosuppressive effects that can be of prolonged duration in infected mice. MTV is a frequent contaminant of murine cytomegalovirus stocks, which are prepared from salivary glands.

Murine Cytomegaloviral (MCMV) Infection

MCMV is a mouse-specific cytomegalovirus. Cytomegaloviruses (CMV) belong to the betaherpesvirus group of the herpesvirus family. Cytomegaloviruses cause cytomegalic inclusion disease, characterized by enlarged cells bearing both intranuclear and intracytoplasmic inclusions, particularly in salivary glands. They have also been termed salivary gland viruses for this reason. MCMV has been studied extensively as an animal model of human CMV infection, but significant biological differences exist. The MCMV literature has been well reviewed (Lussier 1975; Osborne 1982).

EPIZOOTIOLOGY AND PATHOGENESIS. MCMV lesions are common in the salivary glands, and virus can be isolated from salivary glands or saliva in a majority of wild mice but not in laboratory mice. The true prevalence of MCMV infection among laboratory mice is unclear, since serological surveillance is not generally practiced. It is of interest to note that in situ DNA hybridization techniques have recently revealed MCMV DNA in tissues of many specific-pathogen-free laboratory mice, in the absence of virus detectable by other means. A great deal of emphasis has been placed on disease resulting from experimental inoculation, with clear-cut effects of virus strain, dose, route of inoculation, and host factors (age, genotype). Virus is transmitted oronasally by direct contact and is excreted in saliva, tears, and urine. Following experimental inoculation of infant mice, viremia and multisystemic dissemination occur within 1 wk. Maternally derived antibody is protective in newborn mice and probably plays a significant role in protecting young mice from overt disease under natural conditions. In utero transmission does not appear to take place in naturally infected mice. Intranasal or oral inoculation of young adult mice causes subclinical pulmonary infection, viremia, and dissemination, with virus replicating in alveolar macrophages and monocytes. Mice develop alveolar septal thickening and edema. Salivary glands are preferentially infected regardless of host age and other factors and natural infections are localized to salivary glands. Salivary gland infection occurs later than infection of other tissues and persists for many months in this site after virus is cleared from other organs. Persistent salivary gland infections can result in chronic replication and excretion as well as latent infections without virus replication. MCMV latency can also occur in macrophages, B lymphocytes, and reproductive tissues, but not brain, thymus, liver, or kidney. Experimental immunosuppression of chronically or latently infected mice can reactivate virus, with dissemination. Spontaneous disseminated cytomegalic inclusion disease has been reported in an aging laboratory mouse.

PATHOLOGY. Overt disease and disseminated lesions do not usually occur. The most frequently encountered lesions occur in the submaxillary salivary glands and, rarely, in the parotid glands. Eosinophilic intranuclear and intracytoplasmic inclusions are present in acinar epithelial cells with cytomegaly and lymphoplasmacytic infiltration of interstitium (Fig. 1.18 A & B). During the acute disseminated phase in experimentally inoculated infant mice, focal necrosis, cytomegaly, inclusions, and inflammation occur in many tissues, including salivary glands, lacrimal glands, brain, liver, spleen, thymus, lymph nodes, peritoneum, lung, skin, kidney, bowel, pancreas, adrenal, skeletal and cardiac muscle, cartilage, and brown fat. Only a single case of spontaneous disseminated

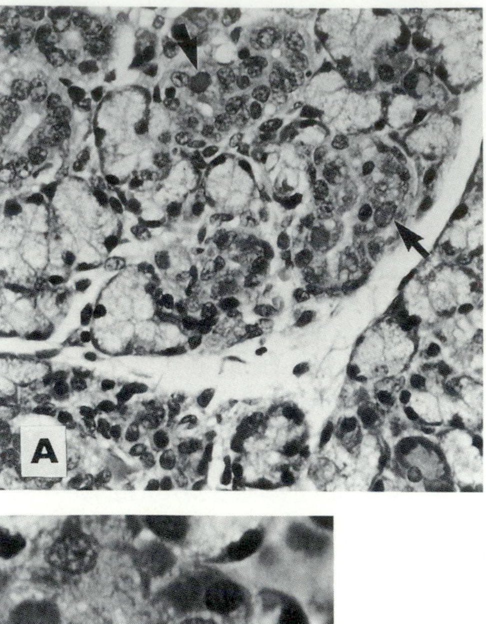

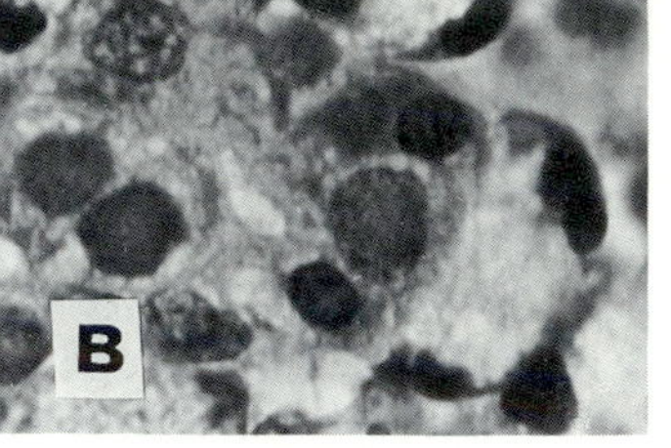

Fig. 1.18. Section of submaxillary (submandibular) salivary gland from animal infected with mouse cytomegalovirus. Intranuclear inclusion (*arrows*) are present in some ductal epithelial cells at a low magnification (*A*) and a higher magnification (*B*).

infection has been reported, involving an aged laboratory mouse.

DIAGNOSIS. Lesions in salivary glands are typical of cytomegalovirus but are not always present in infected animals. Virus can be isolated from saliva or salivary gland and grown in cell culture. In situ DNA hybridization has also been used to detect nonproductive, latent infection of tissues. A variety of serological methods have been developed, including enzyme-linked immunosorbent assay (ELISA) (Lussier et al. 1988), but have not been generally applied as a surveillance method. *Differential diagnosis* for sialoadenitis with inclusion bodies must include polyomavirus. Other viruses that infect salivary glands include reovirus 3, mouse thymic virus, and mammary tumor virus.

SIGNIFICANCE. MCMV seldom causes overt disease in naturally infected mice. MCMV has a number of immunosuppressive effects due to B-cell, T-cell, macrophage, and interferon aberrations. Persistent infections can lead to immune complex glomerulitis and antinuclear antibodies. MCMV has also been shown to have a synergistic effect with *Pseudomonas aeruginosa*. Immunosuppression of naturally infected mice can precipitate disseminated disease.

K Virus Infection

Mouse K virus was initially discovered by Kilham (Kilham and Murphy 1953) following intracerebral inoculation of infant mice with tissue extracts from an adult mouse during experiments on the mammary tumor virus. K virus is a papovavirus of the polyomavirus subgroup, but is distinctly different from polyomavirus of mice, which is the namesake of the genus. It should also not be confused with rat parvovirus, which is often called Kilham rat virus. For all practical purposes, K virus is of historical interest and occurs rarely, if at all, in contemporary laboratory mouse colonies.

EPIZOOTIOLOGY AND PATHOGENESIS. K virus appears to be spread by the orofecal route. When orally inoculated into neonatal mice, virus initially replicates in intestinal capillary endothelium, then disseminates hematogenously to other organs, including liver, lung, spleen, and adrenals, where it replicates in vascular endothelium of these tissues. At 6–15 days after inoculation, there is a sudden onset of dyspnea, due to pulmonary vascular edema and hemorrhage, resulting in rapid death. Pulmonary disease does not occur when older mice are inoculated, with complete resistance evolving between 12 and 18 days of age. Older mice apparently mount an early and effective immune response that prevents the viremic phase of infection. Regardless of age, mice remain persistently infected. Infection of athymic nude mice results in disease similar to that seen in suckling mice. In naturally infected colonies, clinical signs are absent, with dams conferring passive immunity to litters during the susceptible neonatal period.

PATHOLOGY. Gross lesions are restricted to lungs of neonatal mice. Microscopically, intranuclear inclusions are present in vascular endothelium of jejunum, ileum, lung, and liver. Inclusions are poorly discernible and require optimal fixation. Pulmonary lesions consist of congestion, edema, hemorrhage, atelectasis, and septal thickening. Livers of neonatal mice can have sinusoidal leukocytic infiltration and nuclear ballooning of cells lining sinusoids.

DIAGNOSIS. Recognition of diagnostic lesions is difficult at best and is most likely in neonatally infected mice. Immunohistochemistry of infected tissues greatly enhances sensitivity of detecting infected tissues including brain. Serological surveillance can be carried out by a variety of methods, including complement fixation, hemagglutination inhibition, and indirect fluorescent antibody tests. *Differential diagnosis* of multisystemic infection with intranuclear inclusions should include polyomavirus of mice, adenovirus, and MCMV.

SIGNIFICANCE. There is minimal significance. K virus is pathogenic in adult nude mice, but its natural prevalence is so low that infection of nude mice is unlikely.

Polyomavirus Infection

Polyomavirus is a papovavirus of mice, which has been extensively studied as an oncogenic virus that induces many (*poly*) types of tumor (*oma*). Under experimental conditions, it is oncogenic in several different species. It is the type species of the polyomavirus genus of Papovaviridae, to which a number of similar viruses (SV40, BK, JC, etc.) belong. Polyomavirus, originally termed the Stewart-Eddy (SE) polyomavirus and the parotid tumor virus, was initially discovered by Ludwig Gross when newborn mice developed salivary gland tumors following inoculation with filtered extracts of mouse leukemia tissue (Gross 1970). A thorough review of the polyomavirus literature is available elsewhere (Eddy 1982; Gross 1970). The oncogenic activity for which this virus is so well known is a laboratory phenomenon, requiring parenteral inoculation of mice with high titers of virus within the first 24 hr of life.

EPIZOOTIOLOGY AND PATHOGENESIS. Polyomavirus is an environmentally stable virus that is shed primarily in urine, and infection is most efficiently acquired intranasally. Infection of a mouse population requires a continuous source of exposure, which is provided by the repeatedly utilized nesting sites of wild mice. The virus fails to survive under laboratory mouse husbandry conditions and is therefore quite rare in contemporary mouse colonies. Oronasal inoculation of neonatal mice results in virus replication in the nasal mucosa, submaxillary salivary glands, and lungs, followed by viremic dissemination to multiple organs, including kidneys. Mortality can be high at this stage. By day 12, virus is cleared from most sites but persists in lung and especially kidney for months. Infection of older mice is more rapidly cleared, with inefficient virus excretion for short periods. Thus, under natural conditions, maternal antibody from immune dams, coupled with the low level of environmental contamination in a laboratory mouse facility, precludes successful infection of neonatal mice and survival of the virus in the population. If mice are experimentally inoculated parenterally with high doses of virus at less than 24 hr of age, tumors arise in multiple sites, particularly salivary glands. Transplacental transmission does not seem to occur naturally, but virus can be reactivated in the kidneys of adult mice during pregnancy if they were infected as neonates. Since polyomavirus is a widely used experimental virus, contamination of laboratory mice can take place, but consequences are limited, for the reasons just cited. Accidental infection of athymic nude mice has caused multisystemic wasting disease, with paralysis and development of multiple tumors, particularly of uterus and bone.

PATHOLOGY. Under natural conditions, lesions are not likely to be encountered, except in nude mice. Nude mice develop multifocal necrosis and inflammation, followed by tumor formation in multiple tissues reminiscent of experimentally inoculated neonatal mice. Intranuclear inclusions can be observed with difficulty in cytolytic lesions. Nude mice also develop infection of oligodendroglia with demyelination, similar to progressive multifocal leukoencephalopathy in humans. Cytopathic and proliferative changes are especially apparent in bronchial, renal pelvic, and ureteral epithelium. Paralysis is due to vertebral tumors as well as demyelination.

DIAGNOSIS. The presence of polyomavirus in immunocompetent mouse populations is best detected serologically. *Differential diagnoses* of nude mice with wasting disease include primarily mouse hepatitis virus, *Pneumocystis carinii*, Sendai virus, and pneumonia virus of mice. Microscopic lesions containing intranuclear inclusion bodies must be differentiated from lesions caused by K virus, adenovirus, and MCMV.

SIGNIFICANCE. The major significant feature of polyomavirus is its usefulness as a research tool. Its significance in laboratory mouse populations is minimal, except that its polytropism can result in contamination of transplantable tumors and cell lines, which in turn have served as inadvertent sources of contamination of mouse stocks.

Minute Virus of Mice (MVM) Infection

MVM was given its name because of its small size, typical of the parvovirus family to which it belongs. MVM was originally discovered as an inadvertent contaminant of a stock of mouse adenovirus, although it is not an adeno-associated parvovirus. MVM is antigenically distinct from the rat and other parvoviruses. Parvoviruses replicate in rapidly dividing tissues and are dependent upon the S phase of the host cell cycle for replicative function. However, growth is limited to certain differentiated cell types, which bear viral receptors. A thorough review of MVM biology is available (Ward and Tattersall 1982).

EPIZOOTIOLOGY AND PATHOGENESIS. MVM is a common viral infection among laboratory mice but is virtually never associated with either natural disease or lesions. It is a very common contaminant of transplantable tumors and leukemia virus stocks. Many isolates or strains have been obtained from such material, as well as other mouse tissues. Transmission requires close contact between mice. Experimental inoculation of infant mice results in proliferation of virus in multiple tissues with development of cerebellar hypoplasia. Disease in neonatal mice is virus strain– and mouse genotype–dependent. Infection of adult mice results in erythrocyte-associated viremia, and pregnant mice, virus replication may occur in various tissues, including placenta and fetus, without histological evidence of lesions. Neonatal mice can be protected from infection by maternal antibody. Infection appears to be long-term, in spite of seroconversion.

PATHOLOGY. There are no lesions present under natural conditions.

DIAGNOSIS. MVM infection of a mouse is usually diagnosed by seroconversion rather than clinical signs or lesions. HAI has been the traditional serological method of choice, but recently ELISA and IFA have been used. With the advent of IFA, seroconversion to another parvovirus has been detected, called mouse orphan parvovirus (OPV). OPV is a mouse parvovirus that does not share common virus structural antigens with the MVM group but does possess cross-reacting nonstructural antigens. OPV can be detected with IFA, since this assay uses fixed MVM-infected cells that express both structural and non-structural parvoviral antigens.

SIGNIFICANCE. MVM and OPV can be troublesome contaminants of many different mouse tissues, including tumors. MVM can induce oncolysis as well as long-term immunosuppressive effects (McMaster et al. 1981).

Ectromelia Virus Infection

No virus of laboratory mice conjures up an image of ruin like ectromelia virus. Some of this reputation is justified, but most is human in origin. Ectromelia virus is a large DNA virus of the poxvirus family and *Orthopoxvirus* genus, to which vaccinia, variola, monkeypox, cowpox, and others also belong. Orthopoxviruses share antigenic cross-reactivity, but each are distinct species. Unlike many viral infections of laboratory mice, the disease was discovered before the agent. Marchal (1930) reported an epizootic disease with high mortality in adult mice and termed it "infectious ectromelia," because of the frequency of limb amputation (ectromelia) in surviving mice. Frank Fenner has performed the seminal work on pathogenesis of the agent "ectromelia virus," which causes the disease "mousepox," although the terms are often erroneously interchanged. A recent outbreak in the United States (Fenner 1981; New 1981) has stimulated renewed interest in the pathogenesis of mousepox (Bhatt and Jacoby 1987a,b; Wallace and Buller 1985).

EPIZOOTIOLOGY AND PATHOGENESIS. The origin of ectromelia virus remains an enigma, since it has never been found in wild populations of *Mus musculus*. Although laboratory mice have been freely disseminated to all parts of the world, ectromelia virus seems to be enzootic only in Europe and Japan. Outbreaks in the United States have been due to introduction of infected mice or mouse products from Europe. Strains of virus vary in virulence but are serologically indistinguishable from one another. Ectromelia virus does not seem to be highly contagious. It can be experimentally transmitted via a number of routes, but the primary means of natural transmission is probably through cutaneous trauma, which requires direct contact. Epizootiological studies in a large mouse breeding population have indicated that only a few cages of mice were infected at any one time and that transmission was facilitated by handling. Young mice suckling immune dams were protected by maternal antibody from disease but not from infection. Since rate of exposure was low, virus-naive mice remained susceptible when maternal antibody waned within 4 wk (Wallace and Buller 1985). The hypothetical model of infection involves invasion (through skin), local replication, liberation to regional lymph nodes, primary viremia, and then replication in spleen and liver. Between 3 and 4 days after exposure, a secondary viremia ensues, inducing replication of virus in skin, kidney, lung, intestine, and other organs. There is increasing evolution of lesions (disease) days 7–11, including cutaneous rash. This scenario differs markedly between mouse genotypes. Susceptible mouse strains, such as C3H, A, DBA, SWR, and BALB/c, die acutely with minimal opportunity for virus excretion. Many mouse strains develop illness but survive long enough to develop cutaneous lesions with maximal opportunity for virus excretion. Others, such as C57BL/6 mice, are remarkably resistant to disease but allow virus replication and excretion. Thus, a textbook mousepox epizootic requires a select combination of introduction, suitable mouse strains for transmission, and the presence of susceptible strains for disease expression. Immunosuppression will exacerbate disease in mildly or subclinically infected mice. For these reasons, classic outbreaks of high mortality are often not seen. Immunologically competent mice recover completely from infection and do not generally serve as carriers. Therefore, rederivation of virus-free stock can be achieved from immune mice. Athymic nude mice cannot clear virus.

PATHOLOGY. Expression of lesions is dependent upon factors discussed above. Clinical signs range from subclinical infection to sudden death. External lesions during the acute phase of infection in susceptible surviving mice include conjunctivitis, alopecia, cutaneous erythma and erosions, and swelling and dry gangrene of extremities. Internally, livers can be swollen, friable, and mottled with multiple pinpoint white to coalescing hemorrhagic foci. Spleens, lymph nodes, and Peyer's patches are enlarged, with patchy pale or hemorrhagic areas. Intestinal hemorrhage, particularly in the upper small intestine, is common. Microscopic lesions consist of focal coagulative necrosis in liver, spleen, lymph nodes, Peyer's patches, and thymus, as well as other organs (Fig. 1.19). Multiple basophilic intracytoplasmic inclusion bodies (1.5–6 μm) are evident in infected cells, especially hepatocytes at the periphery of necrotic foci (type B pox inclusions or Guarnieri's bodies). These inclusions are poorly discernible with routine staining but can be enhanced by doubling hematoxylin-staining time. Lymphoid tissue can be hyperplastic and/or focally necrotic, with occasional eosinophilic cytoplasmic inclusion bodies (type

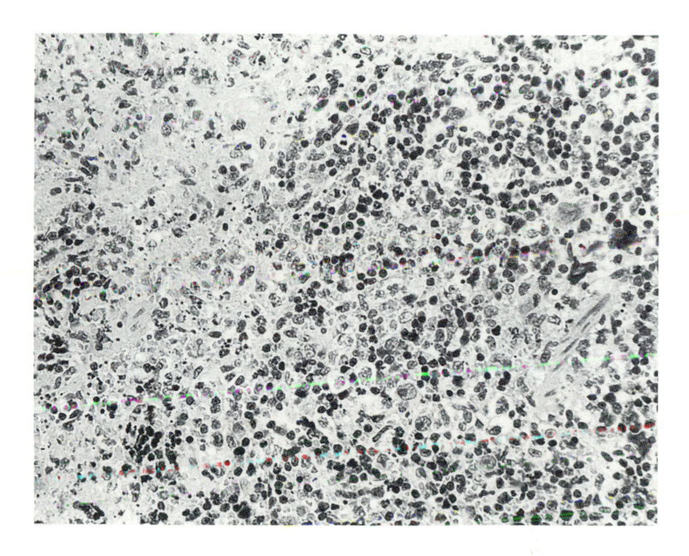

Fig. 1.19. Section of spleen from mouse experimentally infected with ectromelia virus (mousepox). There is extensive splenic necrosis, with karyorrhexis and karyolysis of lymphoid tissue.

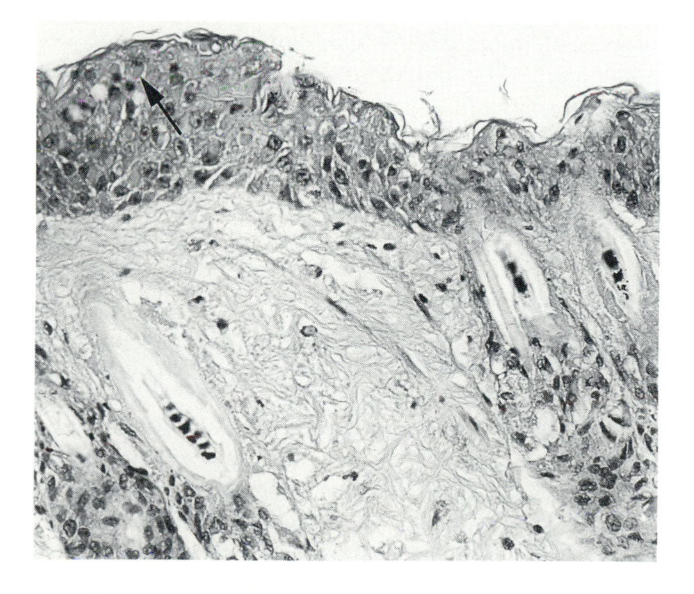

Fig. 1.20. Skin from a DBA mouse post–exposure to ectromelia virus (mousepox). There is epidermal hyperplasia with mononuclear cell infiltration in the underlying dermis. Intracytoplasmic inclusions (*arrows*) are scattered in a few epidermal cells.

A pox inclusions or Marchal bodies). Erosive enteritis, often in association with Peyer's patches, is common, with type A inclusions in enterocytes. Skin lesions consist of focal epidermal hyperplasia, with hypertrophy and ballooning of epithelial cells and formation of prominent large type A inclusions (Fig. 1.20). Later, skin lesions become erosive and inflammatory in character. Inclusions, inflammation, and erosion are also found in conjunctiva, vagina, and nasal mucosa. Recovered mice often have fibrosis of the spleen and can have amputated tails and digits (ectromelia).

DIAGNOSIS. The variable clinical signs and lesions can be problematic, but careful selection of clinically ill mice will enhance an accurate diagnosis. The complex of liver, spleen, and epithelial lesions bearing typical inclusions is pathognomonic. Splenic fibrosis in recovered mice is also unique and pathognomonic. Confirmation can be achieved by electron microscopic identification of the strikingly large poxvirus particles, immunohistochemistry, or virus isolation. Serology is a useful diagnostic adjunct in recovered mice and is an important surveillance tool for monitoring mouse populations. Vaccination is variably practiced and can interfere with interpretation of serology results. It should also be noted that vaccinated mice may be protected from severe disease but still allow active infection. *Differential diagnoses* must include agents that cause hepatitis in adult mice, such as MHV, Tyzzer's disease, salmonellosis, and others. Skin lesions must be differentiated from bite wounds, alopecia, hypersensitivity, and other forms of dermatitis. Gangrene and amputation of digits or tail can also occur due to trauma or "ringtail".

SIGNIFICANCE. Mousepox is an exceptional virus disease of mice, since it can cause high mortality in adult mice. The polytropic nature of the virus and often subclinical but active nature of the infection can allow ready contamination of mouse tissues and biological products and serves as a major mode of spread between laboratories. The drastic measures that have been taken to eliminate ectromelia virus from mouse colonies, particularly in the United States, are clearly disruptive to research.

RNA Viral Infections

Sendai Virus Infection

Sendai virus is a parainfluenza 1 virus that is closely related to other parainfluenza 1 viruses of human origin. It is named after Sendai, Japan, where it was first isolated from laboratory mice inoculated with human lung suspensions and later isolated from naturally infected mice. In some circles, the host of origin is still controversial. Several excellent reviews are available on Sendai virus (Brownstein 1986, 1987; Parker and Richter 1980; Jakob 1981; Ishida and Homma 1978).

EPIZOOTIOLOGY AND PATHOGENESIS. Sendai virus is a labile but highly contagious virus that is aerosol and contact transmitted. In addition to laboratory mice, it is infectious to laboratory rats and hamsters. Guinea pigs seroconvert to Sendai virus, but it is suspected, although not proven, that this is caused by another, related parainfluenza 1 virus. Humans seroconvert to Sendai virus due to infection with related parainfluenza 1 viruses. Sendai virus infects mice throughout the world and is one of the more common infectious disease agents of laboratory mouse populations. It certainly stands out as the most common virus to cause clinical disease in adult, immunocompetent mice.

Mice develop a descending infection of respiratory epithelium, which is abrogated by a cell-mediated immune response that clears the infection but also generates disease. The level to which the infection extends is determined by mouse genotype–related differences in mucociliary clearance and kinetics of immune response. Certain strains of mice, such as DBA/2, and infant mice are exquisitely susceptible to severe disease. These animals have a delayed immune response to the virus, allowing infection to extend deep into the lung but then mount a zealous response that results in severe disease. Other mouse strains often have subclinical infections. Infection is acute, with no persistent carrier state, except in immune-deficient mice. Sendai virus causes little direct damage to target cells, which includes conducting airway epithelium, type II, and to a lesser extent, type I alveolar epithelium. The severe necrotizing and inflammatory lesions typical of this infection are immune-mediated. Peak virus titers occur within 3–6 days, and virus is cleared by 8–12 days. The acute phase of Sendai viral pneumonia occurs in 8–12 days, when immunological attack is under way.

Sendai viral infection is also associated with a number of infectious paraphenomena in mice and rats. It can predispose to the development of bacterial otitis media and interna, as well as precipitate mycoplasma-associated lower respiratory disease in previously subclinically infected mice. Outbreaks of vestibular disease and pneumonia due to *Mycoplasma* or other bacteria can often be associated with recent activity of Sendai virus within the population.

PATHOLOGY. Severely affected mice are dyspneic and have plum-colored consolidation of sharply demarcated foci, anteroventral lung, or entire lung lobes (Fig. 1.21). These consolidated areas may turn gray in surviving mice. Microscopic changes during the acute phase of disease consist of segmental, necrotizing inflammation of nasal and airway epithelium, as well as foci of interstitial pneumonia associated with terminal airways. Infiltrating cells vary with stage of infection, but include neutrophils, lymphocytes, and macrophages (Fig. 1.22). Alveolar spaces may be filled with fibrin, leukocytes, and necrotic cells, with atelectasis. Prior to immune-mediated necrosis, bronchiolar epithelium may be hypertrophic and hyperplastic (Fig. 1.23), contain virus-induced syncytia, and possess poorly organized intracytoplasmic eosinophilic inclusions representing accumulation of viral nucleocapsid material. Intranuclear inclusions have been reported in nude mice. These virus-related changes are most apt to be seen in immature or immunologically deficient mice, such as nude mice, since they are rapidly obscured by immune-mediated necrosis. During resolution,sheets of sloughed airway epithelium are replaced by proliferating hyperplastic epithelium, which may undergo transient but marked nonkeratinizing squamous metaplasia. Alveoli become lined by cuboidal epithelium or filled with metaplastic squamous epithelium. Increased lymphoid cells populate the bronchial tree, adventitia of adjacent blood vessels, and alveolar septa. All of these changes completely resolve by the third or fourth week. Severely affected, but surviving mice can have focal alveolar fibrotic scarring (Fig. 1.24).

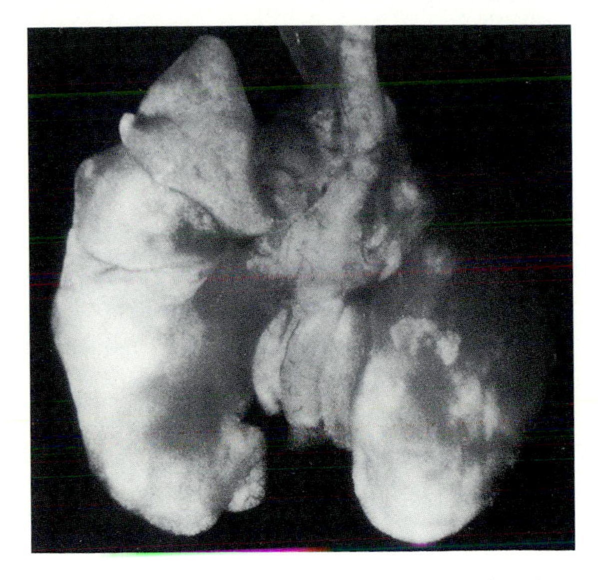

Fig. 1.21. Lungs from DBA mouse killed a few days post–exposure to Sendai virus. Note the dark areas around the hilus of both lungs consistent with congestion and early consolidation.

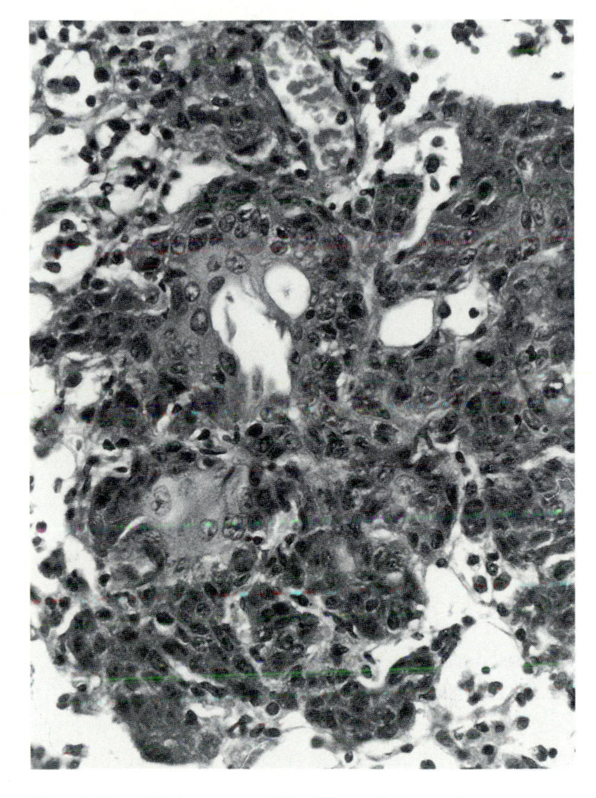

Fig. 1.23. Higher magnification of lung from mouse killed during a naturally occurring outbreak of Sendai virus infection. Note the hypertrophy and hyperplasia, with squamous metaplasia of bronchial epithelial cells.

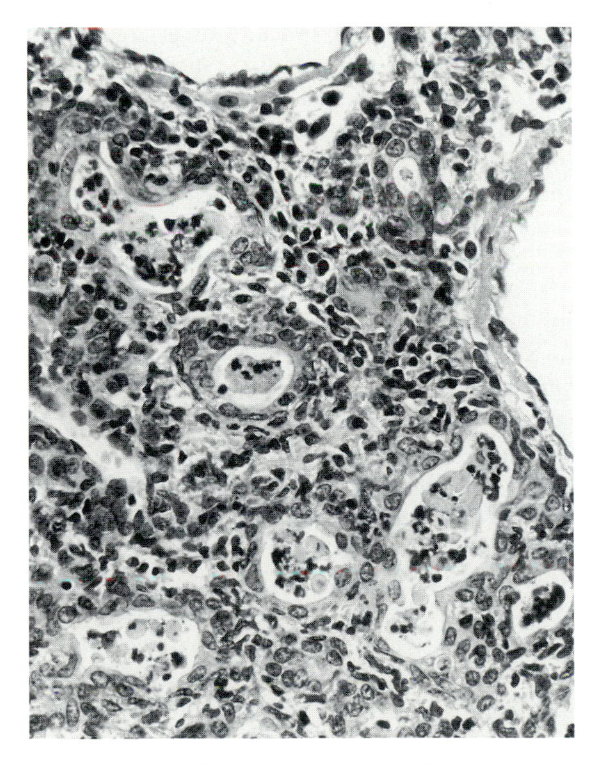

Fig. 1.22. Lung from DBA mouse with Sendai virus infection, illustrating a marked proliferative bronchitis and alveolitis.

Athymic nude and severe combined immunodeficient (SCID) mice develop progressive pulmonary consolidation with wasting. Because they cannot mount an effective immune response, necrotizing changes are minimal and airway epithelium is typically markedly hypertrophic and hyperplastic. They develop progressively severe, diffuse alveolitis.

DIAGNOSIS. The clinical and microscopic features of Sendai viral pneumonia in immunocompetent adult mice are diagnostic and can be confirmed by serocon-

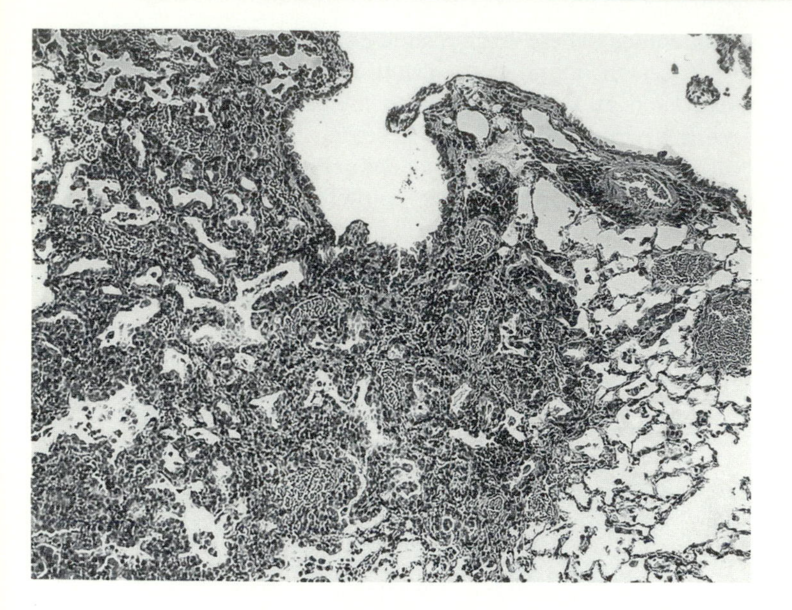

Fig. 1.24. Lung from DBA mouse examined several days after an outbreak of Sendai virus infection. There is marked thickening of alveolar septa. Note the prominent mononuclear cell infiltrates within alveolar septa. Residual lesions of this type may persist for weeks.

version, which generally occurs coincidentally with clinical disease. *Differential diagnoses* include other causes of respiratory disease, such as *Mycoplasma* and *Corynebacterium kutscheri*. Mild respiratory tract lesions can also occur with PVM or MHV. Athymic nude and SCID mice develop wasting disease with progressive pneumonia, which must be differentiated from pneumonia caused by PVM or *Pneumocystis carinii*. Sendai virus and PVM lesions in nude and SCID mice are similar, although in SCID mice, bronchial and bronchiolar lesions are more extensive with Sendai virus infection (see Pneumonia Virus of Mice Infection).

SIGNIFICANCE. Sendai virus can cause overt disease and mortality in genetically susceptible and immune-deficient mice, predispose to bacterial respiratory infections, cause transient infertility, affect the immune response (Kay 1978), and delay wound healing (Kenyon 1983). Sendai infections may also alter the incidence of pulmonary neoplasms in experimental carcinogenesis studies. Its frequency in mouse facilities and its direct and indirect effects render this agent a major concern.

Pneumonia Virus of Mice (PVM) Infection

PVM was originally discovered following serial blind lung passages in mice, which ultimately produced an agent capable of causing pneumonia. The name PVM does not therefore reflect the natural, avirulent behavior of the virus in mice. PVM is a member of the paramyxovirus family, genus *Pneumovirus*, to which respiratory syncytial viruses also belong. PVM is not antigenically related to these viruses.

EPIZOOTIOLOGY AND PATHOGENESIS. PVM is a highly labile virus with a low degree of contagion, requiring close contact between mice. In spite of its name, PVM naturally infects a variety of rodents, including mice, rats, cotton rats, gerbils, guinea pigs, and rabbits. PVM infection occurs in laboratory rodents throughout the world. Clinical disease caused by natural infections with PVM has never been reported, except in athymic nude and SCID mice. Intranasal inoculation of mice results in bronchiolar desquamation and inflammation, followed by nonsuppurative perivascular and interstitial inflammation, which peaks within 2 wk and undergoes resolution within 3 wk. Athymic nude and SCID mice develop progressively severe interstitial pneumonia with wasting syndrome, with PVM antigen confined to alveolar type II cells and occasionally bronchiolar epithelial cells.

PATHOLOGY. Clinical signs of disease and gross lesions are absent in natural infections. Microscopic lesions in intranasally inoculated experimental mice consist of mild necrotizing rhinitis, necrotizing bronchiolitis, and nonsuppurative interstitial pneumonia. Infiltrating leukocytes include neutrophils, but lymphocytes and macrophages predominate. Residual perivascular infiltrates of lymphocytes and plasma cells can persist for several weeks after virus is cleared. Immunocompromised mice manifest chronic wasting with cyanosis and dyspnea. Lungs are dark, fleshy, and firm and do not collapse. Microscopically, alveolar septa are thickened with edema and infiltrating macrophages and leukocytes, and alveolar spaces are collapsed and filled with fibrin, blood, macrophages, and large polygonal mononuclear cells, representing detached alveolar type II cells (Fig. 1.25).

DIAGNOSIS. Most PVM infections are detected retrospectively by seroconversion. Because PVM is not highly contagious, infection within a colony of mice can be focal and the number of seropositive mice can be small. Seropositive mice often have mild perivascular lymphoplasmacytic infiltrates. *Differential diagnoses* for pulmonary disease and wasting syndrome in nude and SCID mice include Sendai virus and *Pneumocystis carinii* infections, which also cause progressive pulmonary disease. PVM lesions are similar to Sendai viral lesions microscopically, but PVM tends not to induce bronchiolar hypertrophy like Sendai virus and can be differentiated by immunohistochemistry, virus isolation, or mouse antibody production (MAP) testing of suspect tissues. Athymic mice do not seroconvert to PVM.

SIGNIFICANCE. PVM appears to have no adverse effect on mice under natural conditions, except in nude and SCID mice. Interspecies transmission may

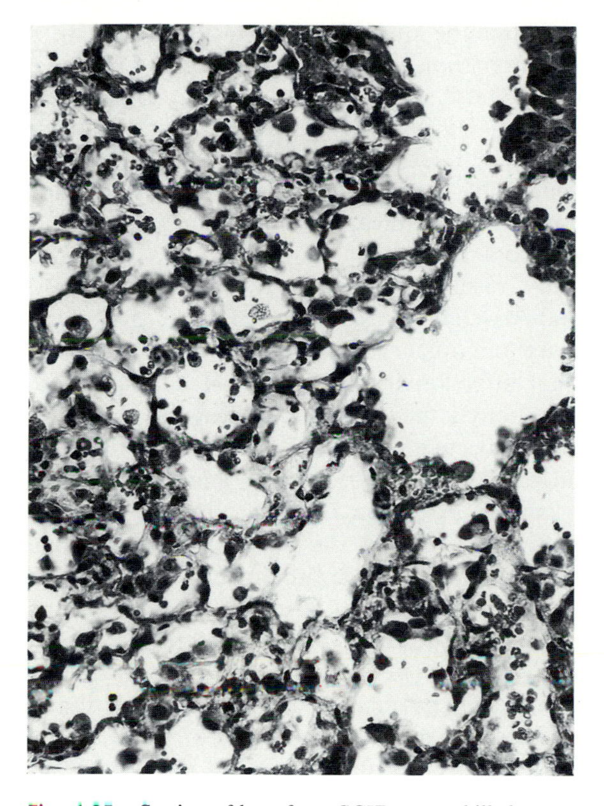

Fig. 1.25. Section of lung from SCID mouse killed post–exposure to PVM. There is marked alveolitis, characterized by hypercellularity of alveolar septa and mobilization of alveolar macrophages.

occur. Rats may develop lesions in the respiratory tract post–exposure to PVM.

Lymphocytic Choriomeningitis

Lymphocytic choriomeningitis virus (LCMV) was named for its ability to produce lymphocytic choriomeningitis upon intracerebral inoculation of mice and monkeys. This is not a major feature of natural infection. LCMV belongs to the family Arenaviridae, named because of the granular-appearing (from the Latin *arenosus* "sandy") ribosomes within virions. Virions are highly pleomorphic, ranging in diameter from 50 to 300 nm, and bud from the cell membrane without cytolysis. LCMV has been studied extensively as a model system of immune-mediated disease, virus persistence, and immune tolerance, resulting in emphasis on aspects of infection that are not necessarily relevant to natural infection.

EPIZOOTIOLOGY AND PATHOGENESIS. There are several laboratory-modified strains of LCMV that vary in pathogenicity but are serologically similar. Some laboratory-adapted strains have been selected to be more virulent ("aggressive") and neurotropic, while wild-type isolates are generally less virulent ("docile") and viscerotropic. The natural host for LCMV is the wild mouse, but it can naturally infect a variety of mammals, including hamsters, guinea pigs, cotton rats, chinchillas, canids, and primates, including humans. Newborn rats can be infected experimentally, but this species seems to be refractory to natural infection. Among mice, the highly labile virus can be transmitted by direct contact with nasal secretions as well as urine and saliva. However, in utero infection is the major means of transmission within infected mouse populations. LCMV is rare among contemporary laboratory mice and occurs sporadically among wild mouse populations.

The course of LCMV infection is dependent upon virus strain, mouse age, and genotype, as well as upon route of inoculation. Since the virus is generally not directly cytolytic, disease is the result of host immune response to infected cells. Infection of fetal or neonatal mice results in a state of immune tolerance to LCMV, with multisystemic, persistent, subclinical infection. Eventually, tolerance breaks down, resulting in chronic illness, with lymphocytic infiltrates occurring in multiple tissues and immune complex glomerulonephritis ("late disease"). Runting occurs in mice infected in utero or as neonates due to the direct, nonlytic effect of virus upon endocrine tissue function. Prior to onset of illness, mice transmit virus, especially by the in utero route. Infection of immunocompetent, older mice is inconsequential, since they mount an efficient immune response that clears the infection without a carrier state. Effective immunity is through cell-mediated immunity. Experimental inoculation of LCMV into neonatal mice results in minimal acute disease and persistence, while inoculation of immunocompetent, older mice causes severe immune-mediated disease several days after inoculation, including lymphocytic choriomeningitis if inoculated intracerebrally.

PATHOLOGY. Clinical signs of natural LCMV infection are minimal but can include runting in infant mice and chronic wasting in older mice if infection is occurring in utero within the colony. Microscopic lesions are likewise nonspecific and most likely to be found in persistently infected aged mice, which can have vasculitis and lymphocytic infiltration in multiple tissues, including brain, liver, adrenal, kidney, and lung. Some mice can develop immune complex glomerulonephritis. Acute disease is largely an experimental phenomenon but can include necrotizing hepatitis and a generalized lymphoid depletion. Natural infection of nude mice has not been described but their T-cell-deficient nature allows virus persistence experimentally.

DIAGNOSIS. Definitive diagnosis of LCMV infection cannot be based on pathology. Serology is also problematic, since horizontal infection among adult mice is inefficient and likely to cause seroconversion among a very few mice within a population. Mice infected in utero or as neonates are immune-tolerant to LCMV and do not seroconvert, or circulating antibody is complexed with antigen in late infections. Thus, serological testing must be applied to a large sample size and can be enhanced by cohousing adult sentinel mice with mice suspected to be persistently infected. As adults, these mice will seroconvert to LCMV when exposed. LCMV can be confirmed in suspect tissues with a variety of bioassay approaches, such as MAP testing, that is, inoculation of immunized and nonimmune adult mice, etc. LCMV is the only virus of mice currently recognized that will kill adults but not neonates when inoculated intracerebrally, a commonly used diagnostic bioassay. *Differential diagnoses* for runting in infant mice include a number of other viral infections. Chronic illness in older mice must be differentiated from generalized lymphoproliferative disorders, amyloidosis, glomerulonephritis, and chronic renal disease of aging mice.

SIGNIFICANCE. The overwhelming significant feature of LCMV is its zoonotic potential. This is enhanced when infections involve laboratory hamsters (see Chap. 3), and recently infection of nude mice has been associated with transmission to humans (Dykewicz et al. 1992). For this reason alone, LCMV is an unacceptable agent in laboratory animal facilities and its eradication should be aggressively effected. The polytropic nature of LCMV and its wide host range allow this virus to readily infect transplantable tumors and cell lines, which can serve as a source of contamination for mouse colonies. The effects of LCMV on the immune system are well documented (Thomsen et al. 1982).

Mouse Hepatitis Viral (MHV) Infection

MHV is a coronavirus with numerous antigenically and genetically related strains that vary considerably in their virulence and organotropism. The name "MHV" was given to a group of virus isolates with similar biological behavior, although it is now known that many MHV strains are not hepatotropic. The misleading name is probably here to stay. MHV shares antigenic cross-reactivity with human coronavirus OC43, bovine coronavirus, rat coronavirus, and hemagglutinating encephalomyelitis virus of swine, but not with other coronaviruses. The significance of this relatedness is unknown; MHV, however, is biologically distinct from these other agents.

EPIZOOTIOLOGY AND PATHOGENESIS. MHV is highly contagious by the oronasal route and prevalent in laboratory and wild mouse populations throughout the world. Under select circumstances, it can cause significant disease. Despite its prevalence and potential pathogenicity, clinical MHV disease is not common. This is due to the interaction of different MHV strains on host variables, which include age, genotype, and immune status. MHV strains appear to possess primary tropism for upper respiratory or enteric mucosa. Those strains with respiratory tropism replicate in nasal mucosa. After infection with virulent MHV strains or infections in mice less than 2 wk of age, in genetically susceptible strains of mice, or in immunocompromised mice, virus rapidly disseminates from the nose by viremia and lymphatics throughout the body. Virus then secondarily replicates in endothelium and parenchyma, causing disease of brain, liver, lymphoid organs, bone marrow, and other sites. Intestinal infection is largely restricted to gut associated lymphoid tissue. Infection of brain by viremic dissemination occurs primarily in neonatal, but not older, mice. However, infection of adult mouse brain can occur by extension of virus along olfactory neural pathways, even in the absence of dissemination to other organs. After approximately 5–7 days, immune-mediated clearance of virus begins, with no persistence or carrier state beyond 3–4 wk after oronasal inoculation. Infection of post-weaning-age mice is usually subclinical, particularly in natural infections caused by generally nonvirulent strains of virus. The obvious exception is athymic or SCID mice, which cannot clear the virus and develop progressively severe disease. They can die acutely or develop chronic wasting disease if infected with avirulent strains of MHV.

In contrast, other MHV strains are enterotropic and selectively infect intestinal mucosal epithelium. All ages and strains of mice are susceptible to infection, but disease is age-related. Infection of neonatal mice results in severe necrotizing enterocolitis with high mortality within 48 hr after inoculation. Before MHV was known to be the causative agent, the unclassified agent of high mortality and enteritis in infant mice was known as lethal intestinal virus of infant mice (LIVIM). Mortality and lesion severity diminish rapidly with advancing age at inoculation. Adult mice develop minimal lesions, but replication of equal or higher titers of virus occurs, compared with neonates. Enterotropic MHV strains tend not to disseminate to other organs; this varies with virus strain. The severity of intestinal disease appears to be associated with age-related intestinal mucosal proliferative kinetics. To underscore this point, disease is minimal in athymic or SCID mice infected as adults. Recovery from enterotropic MHV is immune-mediated and requires functional T cells. Again, no persistent carrier state seems to occur in recovered, immunocompetent mice, but chronic infections occur in nude and SCID mice. Enterotropic MHV infections are often complicated by other opportunistic pathogens, including *Escherichia coli* and *Spironucleus muris*.

A number of generalizations can be made about the MHV complex that provide insight into its epizootiology. Mice are susceptible to infection at any age and allow virus replication and excretion, but disease generally occurs in mice less than 2 wk of age or in immunocompromised mice. Disease severity is mouse genotype-dependent, but this effect is specific to the infecting MHV strains. In other words, mouse genotypes susceptible to one MHV strain may be resistant to another. Passive immunity from immune dams readily protects infant mice from MHV infection, allowing them to acquire active infections later when they are resistant to disease. Infection is acute, with no carrier state in immunocompetent mice. Recovered immune mice are resistant to reinfection with the same MHV strain but are susceptible to repeated infections with different strains of MHV. Finally, MHV is highly mutable and new strains of virus are constantly evolving, particularly when subjected to immune selective pressure.

PATHOLOGY. The majority of MHV infections are subclinical, with mild or no discernible lesions. In susceptible strains, including athymic mice, there may be wasting and prominent multifocal necrosis may be evident at necropsy (Fig. 1.26). On microscopic examination, susceptible mice actively infected with respiratory MHV strains have acute necrosis and syncytia formation in liver (Fig.1.27), splenic red pulp, and lymphoid tissue in spleen, lymph nodes (Fig. 1.28), and gut associated lymphoid tissue and bone marrow (Fig. 1.29). Neonatally infected mice can have a vascular-oriented necrotizing encephalitis with spongiosis

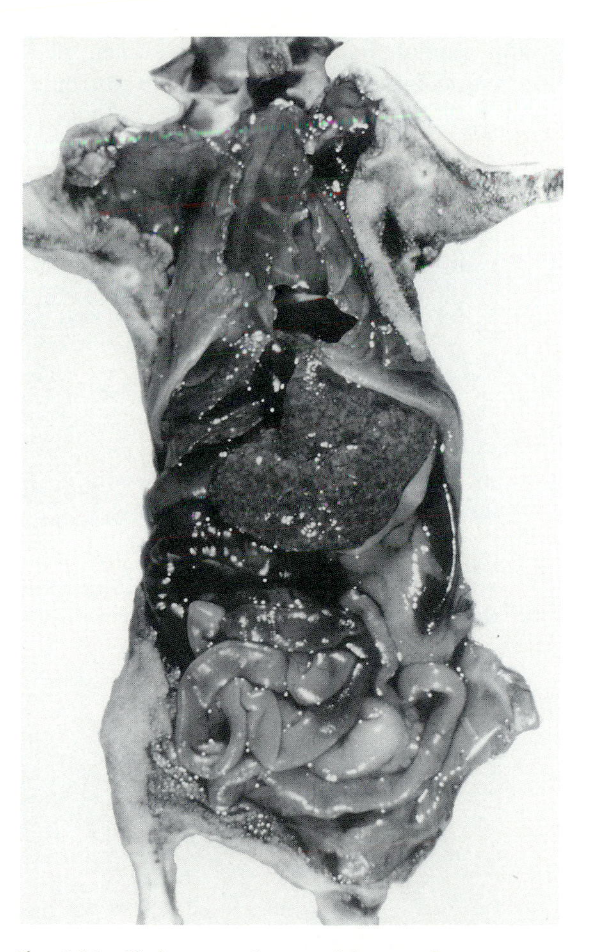

Fig. 1.26. Nude mouse that was doing poorly post–exposure to mouse hepatitis virus (MHV) during a naturally occurring outbreak of the disease and was euthanized in extremis. There are multifocal areas of necrosis and hemorrhage throughout the parenchyma of the liver.

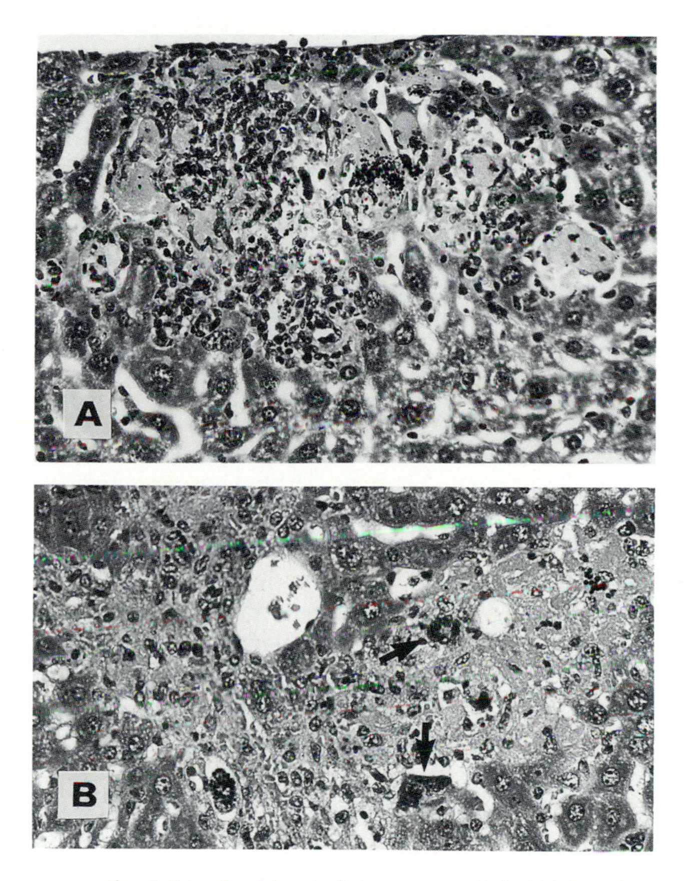

Fig. 1.27. Focal hepatitis in mouse with MHV infection (*A*). Note the granules and debris, and the prominent syncytial giant cells (*arrows*) present within the lesions (*B*).

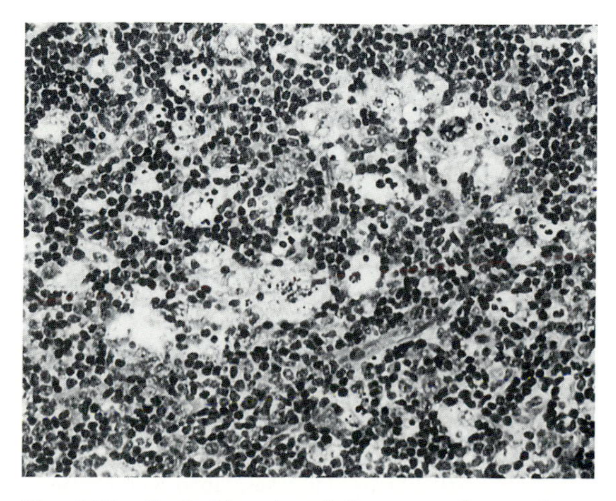

Fig. 1.28. Cervical lymph node from case of mouse hepatitis. Note the pyknosis and karyorrhexis of lymphoid tissue.

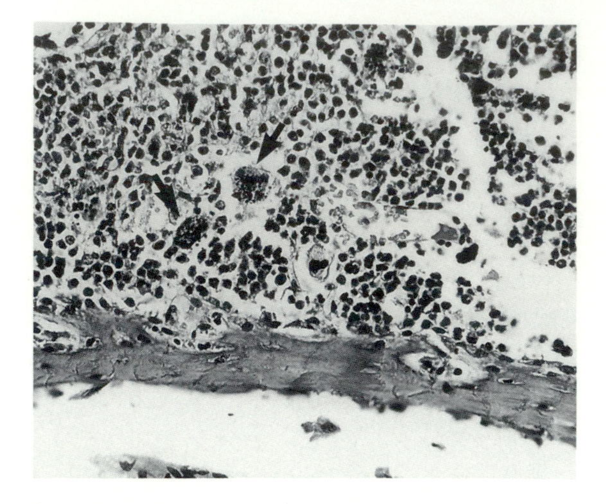

Fig. 1.29. Bone marrow from mouse with MHV infection. There is destruction of hematopoietic cells, and syncytial giant cells (*arrows*) are scattered in the area.

and demyelination in the brain stem. Lesions in peritoneum, bone marrow, thymus, and other tissues can be variably present. Mice can develop nasoencephalitis due to localized infection of olfactory nerves, olfactory bulbs, and olfactory tracts of the brain, with meningoencephalitis and demyelination. This pattern of infection occurs regularly after intranasal inoculation of many MHV strains but is a relatively rare event by natural exposure. Hallmark lesions include virus-induced syncytia arising from endothelium, parenchyma, or leukocytes in target organs, including central nervous system (Fig. 1.30). These cells often display nuclear pyknosis, with dense basophilic bodies. Lesions are transient and seldom fully developed in adult immunocompetent mice, but they are fully manifest in immunocompromised mice. In such animals, vascular endothelial syncytia and hematopoietic involvement are particularly apparent. Athymic nude mice can develop chronic, nodular hepatitis and splenomegaly due to compensatory hematopoiesis.

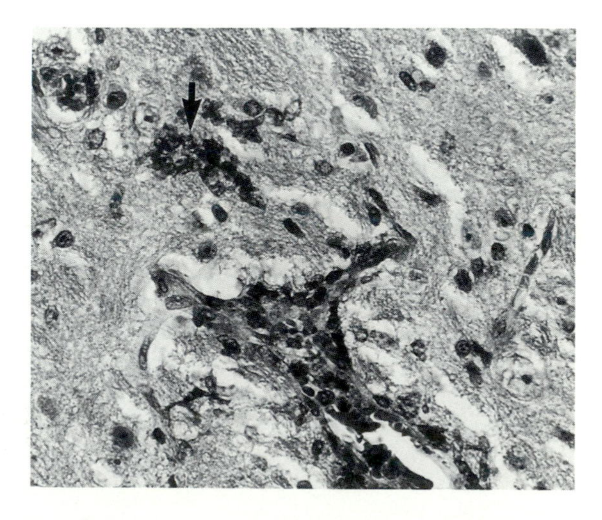

Fig. 1.30. Section of brain, hippocampal region, from a case of MHV-associated encephalitis. There are foci of hypercellularity, with multinucleate giant cell formation (*arrows*).

Lesions due to enterotropic MHV depend primarily upon age of the host. Neonatal mice in naive mouse colonies can experience massive outbreaks of high mortality. These mice have segmentally distributed areas of villus attenuation and atrophy, enterocytic syncytia (balloon cells), and mucosal necrosis (Fig. 1.31). Eosinophilic intracytoplasmic inclusions are present but are not as diagnostic as syncytia. Mesenteric lymph nodes usually contain lymphocytic syncytia, and mesenteric vessels may contain endothelial syncytia. Surviving mice develop compensatory intestinal mucosal hyperplasia. Lesions are most likely to be found in the terminal small intestine, cecum, and ascending colon. Lesions are progressively milder with increasing age at the time of exposure. Adult mice usually have minimal lesions, except for enterocytic syncytia in surface mucosa, particularly in cecum and ascending colon (Fig. 1.32). Athymic and SCID mice develop similar but progressive lesions, depending upon age at inoculation. Natural enterotropic MHV

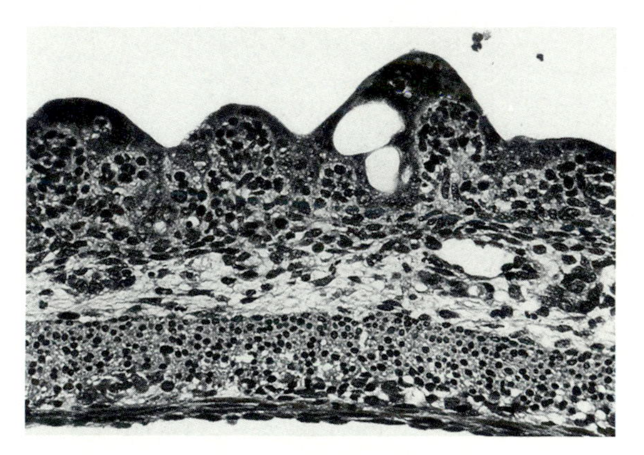

Fig. 1.31. Section of ileum from infant mouse with MHV-associated infantile diarrhea. There is marked villous atrophy with effacement of the normal architecture.

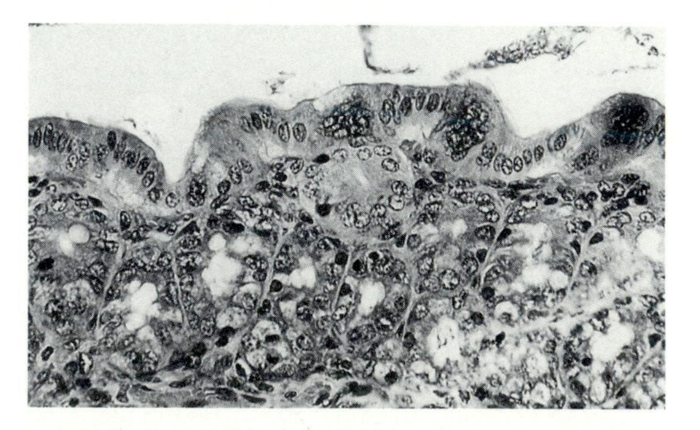

Fig. 1.32. Section of colon from nude mouse with chronic MHV infection. Syncytial cells are present on the mucosal surface.

infection in adult nude mice may cause chronic hyperplastic typhlocolitis with syncytia and mesenteric lymphadenopathy, but intestinal lesions can also be remarkably mild with minimal hyperplasia. Enterotropic MHV strains do not generally disseminate, but hepatitis and encephalitis can occur with some virus strains in certain mouse genotypes.

DIAGNOSIS. Diagnosis during the acute stage of infection can be made by visualization of characteristic lesions with syncytia in target tissues. This can be confirmed by immunohistochemistry or by virus isolation. Recovered mice may have perivascular lymphocytic infiltrates in the lung and microgranulomas in the liver. In general, respiratory strains of MHV are polytropic and grow in a number of established cell lines in vitro, but enterotropic MHV strains are far more restrictive and fastidious in vitro. Virus in suspect tissue can be confirmed by a variety of bioassay methods, such as MAP testing or infant or nude mouse inoculation. Amplification by passage in nude mice will increase the likelihood of in vitro isolation from infected tissue. Serology is a useful means of surveillance for retrospective infection in a colony. Seropositive mice are poor candidates for pathology workup, since they are likely to be recovered, but on occasion they can be actively infected with a second strain of virus. Nude mice can develop antibody that can be detected by IFA or ELISA, although their antibody response is unpredictable. There is little merit in attempting to identify MHV strains serologically, since all strains are broadly cross-reactive and antigenic relatedness does not predict virulence or organotropism. *Differential diagnoses* include salmonellosis, Tyzzer's disease, and mousepox in adult mice, as well as reovirus, cytomegalovirus, and adenovirus in infant mice. Mice with enteritis must be differentiated from EDIM, salmonellosis, Tyzzer's disease, and reovirus infection. Demyelinating lesions must be differentiated from those caused by mouse encephalomyelitis virus, LDHE-V in immunosuppressed mice, or polyomavirus in nude mice.

SIGNIFICANCE. The polytropic potential, ubiquity, and contagiousness of MHV make it the most probable virus to interfere with biological responses of mice. A number of research effects have been documented, including immunomodulation (Boorman et al. 1982; Carrano et al. 1984) and contamination of transplantable tumors and cell lines. Infection of nude and SCID mice or the tumors generated in them, including hybridomas, can be troublesome.

Mouse Encephalomyelitis Viral (MEV) Infection

MEV is also commonly referred to as mouse poliovirus or Theiler's virus, after its initial discoverer, Max Theiler. There are numerous strains of MEV that infect mice, including TO (Theiler's original), GDVII, FA, and DA, among others. Some of these strains have been studied extensively as a model of viral encephalitis and demyelination, resulting in emphasis of the neurological disease, which is only rarely a component of natural infection. MEV strains are closely related antigenically, but they can be differentiated by serum neutralization and differ in pathogenicity. They belong to the family Picornaviridae. In the past, they have been considered to be members of the genus *Enterovirus*, but it is now apparent that MEVs are a *Cardiovirus*. They are serologically related to encephalomyocarditis virus (EMC), which has a less-selective host range but does not infect laboratory mice. MEV and EMC can be distinctly differentiated by serum neutralization.

EPIZOOTIOLOGY AND PATHOGENESIS. MEV is largely regarded as a mouse virus, but serum antibody reactivity to MEV has also been detected in sera from rats and guinea pigs. At least one rat agent (MHG) has been isolated from naturally infected rats that is experimentally pathogenic in both rats and mice and is closely related antigenically to the MEV group. Hamsters are susceptible to experimental infection with MEV. MEVs are enteric viruses that induce virtually no adverse intestinal effects. Virus excretion from the intestine is highly variable among mice, but it is often prolonged and intermittent. Transmission is apparently inefficient, often with only a small percentage of seropositive mice within a population. In utero infection does not occur. Infected mice develop a transient viremia that is limited by host immune response. Occasionally, virus gains access to the central nervous system. Vascular endothelial cells appear to serve as a conduit for entry into the brain. Virulent strains of virus, such as GDVII or FA, induce severe fatal encephalitis, regardless of route of inoculation. Most other strains are less virulent and can cause biphasic disease, consisting initially of acute poliomyelitis, followed later by late-onset demyelinating disease. Virus can persist in the central nervous system for over a year, but virus titers decline markedly and residual virus is restricted to white matter, where it replicates in macrophages, leukocytes, astrocytes, and oligodendrocytes. Immune attack on infected white matter results in demyelination and motor dysfunction, with gait disorders, tremors, ataxia, extensor spasm, urinary incontinence, and other signs. The neurological manifestations of MEV are grossly overemphasized because of their experimental value. Under natural conditions, only 1 in 1000–10,000 infected mice develop clinical signs of the nervous system, and this is invariably flaccid paralysis associated with the acute, poliomyelitis phase.

PATHOLOGY. Lesions are not present in the intestine. During the acute central nervous system (poliomyelitis) phase, the virus attacks neurons and glia of the hippocampus, thalamus, brain stem, and spinal cord. Neuronolysis, neuronophagia, microgliosis, nonsuppurative meningitis, and perivasculitis are typical changes seen microscopically (Fig. 1.33). These changes are most prevalent in the brain stem and ventral horns of the spinal cord. During the demyelinating phase, foci of demyelination are present in the white matter of the spinal cord, brain stem, and cerebellum. Demyelinating lesions are not a likely component of natural infection.

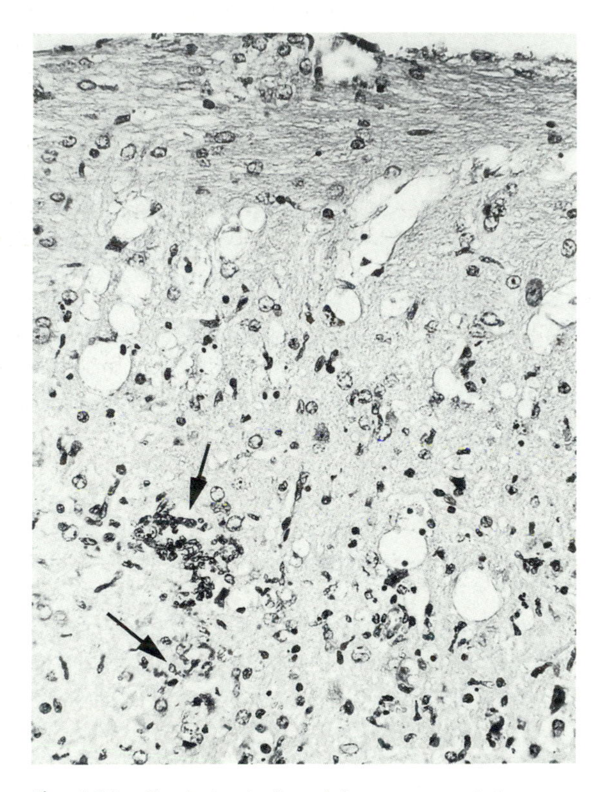

Fig. 1.33. Cervical spinal cord from young adult mouse with acute murine poliomyelitis virus infection. There is a nonsuppurative poliomyeltiis, with focal microgliosis (*arrows*).

DIAGNOSIS. MEV infection is usually diagnosed serologically. Seropositive mice should be considered to be actively infected with virus. Diagnosis can also be achieved by neurological signs and lesions. *Differential diagnoses* for neurological disease include trauma, neoplasia, otitis, MHV, and in nude mice, polyomavirus.

SIGNIFICANCE. MEV has been an inadvertent contaminant of clinical specimens inoculated into mice. Most strains of MEV display low virulence and are thus nearly innocuous.

Epizootic Diarrhea of Infant Mice (EDIM)

EDIM virus is a group A rotavirus that shares a common inner capsid antigen with other group A rotaviruses of humans, nonhuman primates, cattle, sheep, horses, pigs, dogs, cats, turkeys, chickens, and rabbits. Each of these viruses is relatively host-specific, but interspecies infection can be shown experimentally with high doses of virus.

EPIZOOTIOLOGY AND PATHOGENESIS. EDIM virus is highly contagious and prevalent among both laboratory and wild mice, but disease manifestations are relatively rare. Rotaviruses are shed copiously in feces and transmission is by the orofecal route. Clinical disease ranges from inapparent to severe, depending primarily upon age and immune status. All ages of mice are susceptible to infection; however, disease is limited to mice less than 2 wk of age. Virus selectively infects terminally differentiated enterocytes of villi and surface mucosa of the small and large intestine, respectively. These cells are most plentiful and widespread in the neonatal bowel and diminish in number, distribution, and degree of terminal differentiation as mucosal proliferative kinetics accelerate with acquisition of intestinal microflora. Infection does take place in older mice, but the target cell population is limited in number and to small intestine. Thus, functional disturbances tend not to be noted in older mice. Regardless of age at infection, recovery from diarrhea occurs at 14–17 days of age. Active immunity plays a role in resolution of infection but not susceptibility. Infection of SCID mice follows the same age-related pattern of disease as immunocompetent mice. Pups born to immune dams are protected against EDIM during their period of disease susceptibility. Typically, clinical signs of EDIM occur in naive breeding populations, but once infection is enzootic within the colony, EDIM disease is no longer apparent, although EDIM virus remains. The duration of EDIM virus infection in an individual mouse has not been definitively determined.

PATHOLOGY. Clinically affected mice are runted and pot-bellied, with loose, mustard-colored feces staining the perineum. Steatorrhea and oily hair may also be apparent. The bowel is flaccid and distended with fluid and gas, but mice continue to suckle. Some deaths can occur due to obstipation caused by fecal caking around the anus. In infant mice, EDIM virus causes hydropic change and vacuolation of enterocytes at the tips of villi and large intestinal surface mucosa. Some nuclei may be pyknotic. Acidophilic intracytoplasmic inclusions have been described but are not diagnostic. In addition, the lamina propria is edematous and lymphatics are dilated, although inflammation is minimal. These changes are difficult to discern under the best of circumstances and are not apparent in mice older than 14 days of age. Remarkably, mice can manifest significant diarrhea with minimal microscopic lesions. Malabsorption and osmotic diarrhea, with overgrowth of *E. coli*, appear to be major components of the disease process.

DIAGNOSIS. EDIM can be diagnosed presumptively on the basis of age, clinical signs, and lesions. *Differential diagnoses* include enterotropic mouse hepatitis virus, mouse adenovirus, reovirus, salmonellosis, and Tyzzer's disease. Vacuolation and intracytoplasmic inclusions must be differentiated from so-called absorption vacuoles of the neonatal enterocytic apical tubular system that occur in the distal small intestine and contain solitary eosinophilic globules. Definitive diagnosis can be achieved by electron microscopy of intestinal mucosa or feces. Rotavirus antigen can be detected in feces by ELISA. This can be accomplished with commercially available rotavirus diagnostic kits, but false-positive reactions can occur with certain mouse diets. Careful controls are therefore advised. Serology (IFA) for EDIM virus antibody is useful for surveillance and retrospective confirmation of infection.

SIGNIFICANCE. EDIM can be clinically significant, but its effects are transient. Infected infant mice often have severe transient thymic necrosis of unknown pathogenesis that can potentially cause immunologic aberrations. EDIM has been studied as an animal model for rotaviral infections in other species.

Reoviral Infection

The name reovirus (respiratory enteric orphan virus) was proposed for a group of viruses associated with respiratory and enteric infections in humans. Reoviruses have since been isolated from a wide variety of mammals, birds, reptiles, insects, and other species. Mammalian reoviruses are divided into three serotypes, based upon hemagglutination inhibition and neutralization tests. They are commonly referred to as reovirus 1, reovirus 2, and reovirus 3, although they actually represent a spectrum of viruses that are antigenically cross-reactive by other serological methods. Reovirus 3 has been associated with natural infections in laboratory rodents and as a pathogen in mice. A number of reovirus 3 strains have been studied experimentally.

EPIZOOTIOLOGY AND PATHOGENESIS. Reovirus seroconversion is a frequent finding in mouse colonies. Reoviruses are transmitted by the orofecal and aerosol routes, as well as through arthropod vectors. Transmission among mice seems to occur primarily by direct contact among young mice. Contact transmission between adult mice is inefficient. Mice of all ages are susceptible to experimental infection by a variety of routes, but only neonatal mice develop disease. Following oral inoculation of neonatal mice, virus enters through the upper intestine and disseminates to multiple organs. Experimental studies have shown that reovirus 1 replicates in intestinal epithelium, but reovirus 3 does not. Reovirus 3 replication occurs in multiple organs of neonatal mice in the absence of significant lesions, until around day 10–12, when mice become clinically ill and develop lesions in multiple tissues, followed by recovery. This suggests an immune-mediated mechanism for development of lesions during the process of recovery, but these mechanisms have not been defined. Pups can develop steatorrhea secondary to liver and pancreatic lesions. Pups born to immune dams do not develop disease.

PATHOLOGY. Disease and lesions occur only in infant mice from colonies previously unexposed to reovirus. At around 2 wk of age, mice can be runted, jaundiced, and uncoordinated and have only matted hair (caused by steatorrhea). Surviving mice can remain runted and have transient dorsal alopecia. The most significant microscopic lesion is acute diffuse encephalitis that has a vascular distribution. Mice also develop focal necrotizing myocarditis, variable necrosis of lymphoid tissue, focal necrotizing hepatitis, portal hepatitis, acinar pancreatitis, and sialodacryoadenitis.

DIAGNOSIS. Reovirus infection is usually documented in a mouse population serologically, in the absence of obvious disease, although outbreaks of disease can occur among neonates. Virus can be isolated from infected tissues or antigen can be visualized by immunohistochemistry. Lesions are nonspecific. *Differential diagnoses* of neonatal disease with steatorrhea include mouse hepatitis virus, EDIM virus, and salmonella infections.

SIGNIFICANCE. Reoviruses are not generally significant pathogens in laboratory mice. Many of the lesions attributed to reovirus in the historic literature are likely due to other murine viruses, such as MHV. Reoviruses frequently contaminate transplantable tumors.

Murine Retroviral Infection

Although retroviruses play a significant role in disease pathogenesis in mice, they are seldom considered in the same way as other viruses. This is because they are incorporated in the mouse genome and transmitted genetically as inherited Mendelian characteristics. In many respects, they represent specific characteristics of different mouse strains. They are known as the murine leukemia viruses (MuLVs) and the murine mammary tumor viruses (MuMTVs). This section provides a brief summary of retroviruses in mice; the major lesions they induce are described in the Neoplasms section of this chapter.

EPIZOOTIOLOGY AND PATHOGENESIS. MuLVs possess two replication strategies. Endogenous MuLVs are integrated into the host genome (provirus) and depend upon their chromosomally integrated state for transmission. Exogenous MuLVs behave like conventional, transmissible viruses. Both endogenous and exogenous MuLVs are genetically, morphologically, and antigenically homologous. All wild and laboratory mice harbor multiple copies of endogenous MuLV in their genomes. Today exogenous MuLVs occur only in wild mice, but several laboratory MuLV isolates (Friend, Moloney, and Rausker LVs) obtained in the early years of tumor biology are more closely related to exogenous MuLV. Endogenous MuLVs have integrated into the mouse genome randomly, and once integrated, they have remained stable. No two endogenous genomes are known to be located in the same chromosomal site among different mouse strains. Although endogenous MuLVs are transmitted genetically, they are expressed in somatic cells, and this expression is mouse strain–, age-, and tissue-specific. When expressed, they have a tendency for recombination or transduction with host genes to produce new variants that produce disease. Endogenous MuLVs have three categories of host-range polymorphism. Ecotropic viruses can infect mouse cells but not cells of other species. Xenotropic viruses can infect cells of other species but not those of mice. Amphotropic (or polytropic) viruses can infect both mouse and xenogeneic cells. Tropism is also restricted to tissue types and mouse genotype. For example, B(BALB)-tropic viruses can only infect mice that are homozygous at the Fv-1^b locus and N(NIH)-tropic viruses at the Fv-1^n locus.

The significance of these characteristics is best provided by example. AKR mice develop a high prevalence of thymic lymphomas at 6–12 mo of age. They carry an endogenous nononcogenic ectotropic MuLV that is expressed in multiple tissues at birth. They also carry an endogenous nononcogenic xenotropic MuLV that is expressed in thymus at 6 mo of age. At 6 mo, the concomitant expression of these viruses allows recombinant viruses to be expressed that are polytropic and oncogenic, capable of infecting cells and causing neoplasia. BALB/c mice develop a late-onset lymphoma. They carry nononcogenic, N-ecotropic, and xenotropic viruses that form recombinants that are oncogenic and B-ecotropic, resulting in disease. Each mouse genotype possesses its own set of circumstances, with varying degrees of MuLV expression and disease that are mouse strain–, age-, tissue-, and cell type–specific. Not all outcomes are neoplastic. MuLV reintegrations can occur randomly in both germ cells and somatic cells, with varied results, including most often no effect, or altered coat color and consistency, central nervous system disease, premature graying, etc. Another group of viruses, called murine sarcoma viruses (MuSV), have undergone similar recombination and/or transduction, incorporating portions of the host genome. These viruses are missing components of their own genome and are unable to replicate on their own (defective) without the assistance of another, nondefective MuLV (helper) virus.

Likewise, MuMTVs are both endogenous and exogenous in nature. Exogenous MuMTV is transmitted in the milk, and to a lesser extent saliva, as a conventional virus, inducing mammary neoplasia. This virus, which is known as the MuMTV-S (standard), the "milk factor," or the Bittner agent, has been eliminated from modern mouse populations by cesarean rederivation or foster nursing, unless intentionally maintained for experimental purposes. All strains of laboratory mice possess endogenous MuMTV provirus in their genome and expression of disease (mammary tumors) is mouse strain–, age-, and cell type–specific.

PATHOLOGY. See the Neoplasms section of this chapter.

DIAGNOSIS. Diagnosis of MuLV or MuMTV is not necessary, since all mice are infected. Electron microscopic examination of normal and neoplastic mouse tissues frequently reveals C-type (MuLV) and B-type (MuMTV) particles as an incidental finding. Certain specific disease entities are known to be caused by these viruses, but the relationship of retroviruses to disease expression in mice is complex and largely undefined.

SIGNIFICANCE. MuLvs and MuMTVs are part of mouse biology and inseparable features of the laboratory mouse. They become significant when they induce life-limiting disease, as well as valuable models of retrovirus pathogenesis.

Lactate Dehydrogenase Elevating Virus (LDH-EV) Infection

LDH-EV, currently classified as togavirus, was originally discovered as a contaminant of a transplantable tumor that caused significant elevation of LDH enzyme activity in serum of inoculated mice.

EPIZOOTIOLOGY AND PATHOGENESIS. LDH-EV is probably not common among contemporary mouse colonies, but its true prevalence is unknown since serological surveillance seldom includes this agent. It is apparently a virus of mice only. The primary means of natural transmission is through bite wounds among fighting mice. In the absence of fighting, LDH-EV is inefficiently transmitted by direct contact, even though virus is excreted in feces, urine, milk, and saliva. The most important means of transmission is through experimental inoculation of mice with biological products derived from infected mice. Viremia peaks within 12 to 24 hr after infection, and then it drops but persists for the lifetime of the mouse, complexed with antibody. LDH-EV infects macrophages and monocytes without overt lesions. In addition to elevations in LDH enzyme, several other enzymes are also significantly elevated. Enzyme elevations are believed to be due to diminished enzyme clearance function of the infected reticuloendothelial system rather than to direct effects of the virus, but these mechanisms are unclear. The antibody generated against LDH-EV is nonneutralizing and does not fix complement, and it is complexed with the virus. Immune complex disease does not occur in chronically infected mice. Immunosuppressed C58 and AKR mice experimentally inoculated with LDH-EV develop poliomyelitis during the acute phase of LDH-EV infection. Inoculation of nonimmunosuppressed C57BR and related C57BL strains with LDH-EV causes subclinical leptomeningitis and myelitis.

PATHOLOGY. No lesions have been seen in naturally infected mice. Experimentally inoculated mice can develop transient splenomegaly, necrosis of T-cell areas of lymphoid tissues, pyknosis of reticuloendothelial cells, and leukopenia within 72 hr after inoculation. Central nervous system disease in experimentally infected, immunosuppressed C58 and AKR mice consist of mononuclear leukocytic infiltrates in the ventral, and to a lesser extent dorsal, horns of the spinal cord, scattered neuronolysis, and perivasculitis. C57 mice infected with LDH-EV develop a mild to moderate nonsuppurative leptomeningitis, myelitis, and occasionally radiculitis without clinical signs.

DIAGNOSIS. Serological methods for LDH-EV have not been generally used because of difficulties with antigen-antibody complexes and low avidity of antibody for antigen. LDH-EV replicates in vitro but causes no cytopathic effect. The standard method of LDH-EV diagnosis is by measuring plasma LDH in mice given serial dilutions of test material (Rowson and Mahy 1975). *Differential diagnoses* include any agent that causes enzyme elevations, but other enzyme elevations are not as high or persistent.

SIGNIFICANCE. LHD-EV causes no clinical disease or lesions but significantly alters macrophage function and immune response, rendering infected mice useless for immunological research. LDH-EV can be a contaminant of murine biological products, including transplantable tumors.

Bibliography for Viral Infections

Adenoviral Infection

Barthold, S.W. 1985. Adenovirus infection, intestine, mouse. In *Monographs on Pathology of Laboratory Animals: Digestive System*, ed. T.C. Jones et al., pp. 325–27. New York: Springer-Verlag.

Cohen, B.J., and de Groot, F.G. 1976. Adenovirus infection in athymic (nude) mice. Lab. Anim. Sci. 26:955–56.

Ginder, D.R. 1964. Increased susceptibility of mice infected with mouse adenovirus to *Escherichia coli*–induced pyelonephritis. J. Exp. Med. 120:1117–28.

Hartley, J.W., and Rowe, W.P. 1960. A new mouse adenovirus apparently related to the adenovirus group. Virology 11:645–47.

Hashimoto, K., et al. 1966. An adenovirus isolated from feces of mice. I. Isolation and identification. Jpn. J. Microbiol. 10:115–25.

Heck, F.C., Jr., et al. 1972. Pathogenesis of experimentally-produced mouse adenovirus infection in mice. Am. J. Vet. Res. 33:841–46.

Leuthans, T.N., and Wagner, J.E. 1983. A naturally occurring intestinal mouse adenovirus infection associated with negative serologic findings. Lab. Anim. Sci. 33:270–72.

Lussier, G., et al. 1987. Serological relationship between mouse adenovirus strains FL and K87. Lab. Anim. Sci. 37:55–57.

Margolis, G., et al. 1974. Experimental adenovirus infection of the mouse adrenal gland. I. Light microscopic observations. Am. J. Pathol. 75:363–72.

Parker, J.C., et al. 1966. Prevalence of viruses in mouse colonies. Natl. Cancer Inst. Monogr. 20:25–36.

Smith, A.L., and Barthold, S.W. 1987. Factors influencing susceptibility of laboratory rodents to infection with mouse adenovirus strains K87 and FL. Arch. Virol. 95:143–48.

Sugiyama, T., et al. 1967. An adenovirus isolated from the feces of mice. II. Experimental infection. Jpn. J. Microbiol. 11:33–42.

Takeuchi, A., and Hashimoto, K. 1976. Electron microscope study of experimental enteric adenovirus infection in mice. Infect. Immunol. 13:569–80.

van der Veen, J., and Mes, A. 1974. Serological classification of two mouse adenoviruses. Arch. Virol. 45:386–87.

______. 1973. Experimental infection with mouse adenovirus in adult mice. Arch. Virol. 42:235–41.

Wigand, R., et al. 1977. Biological and biophysical characteristics of mouse adenovirus, strain FL. Arch. Virol. 54:131–42.

Winters, A.L., and Brown, H.K. 1980. Duodenal lesions associated with adenovirus infection in athymic "nude" mice. Proc. Soc. Exp. Biol. Med. 164:280–86.

Mouse Thymic Viral (MTV) Infection

Cohen, P.L., et al. 1975. Immunologic effects of neonatal infection with mouse thymic virus. J. Immunol. 115:706–10.

Cross, S.S., et al. 1979. Biology of mouse thymic virus, a herpesvirus of mice, and the antigenic relationship to mouse cytomegalovirus. Infect. Immunol. 26:1186–95.

Cross, S.S., et al. 1976. Neonatal infection with mouse thymic virus. Differential effects on T cells mediating the graft-versus-host reaction. J. Immunol. 117:635-38.

Morse, H.C., III, et al. 1976. Neonatal infection with mouse thymic virus: Effects on cells regulating the antibody response to type III pneumococcal polysaccharide. J. Immunol. 116:1613–17.

Morse, S.S. 1987. Mouse thymic necrosis virus: A novel murine lymophotropic agent. Lab. Anim. Sci. 37:717–25.

Morse, S.S., and Valinsky, J.E. 1989. Mouse thymic virus (mTLV): A mammalian herpesvirus cytolytic for CD4+ (L3T4+) T lymphocytes. J. Exp. Med. 169:591–96.

Parker, J.C., et al. 1973. Classification of mouse thymic virus as a herpesvirus. Infect. Immunol. 7:305–8.

Rowe, W.P., and Capps, W.I. 1961. A new mouse virus causing necrosis of the thymus in newborn mice. J. Exp. Med. 113:831–44.

Wood, B.A., et al. 1981. Neonatal infection with mouse thymic virus: Spleen and lymph node necrosis. J. Gen. Virol. 57:139–47.

Murine Cytomegaloviral (MCMV) Infection

Brautigam, A.R., et al. 1979. Pathogenesis of murine cytomegalovirus infection: The macrophage as a permissive cell for cytomegalovirus infection, replication and latency. J. Gen. Virol. 44:349–59.

Brodsky, I., and Rowe, W.P. 1958. Chronic subclinical infection with mouse salivary gland virus. Proc. Soc. Exp. Biol. Med. 99:654–55.

Brody, A.R., and Craighead, J.E. 1974. Pathogenesis of pulmonary cytomegalovirus infection in immunosuppressed mice. J. Infect. Dis. 129:677–89.

Chen, H.C., and Cover, C.E. 1988. Spontaneous disseminated cytomegalic inclusion disease in an ageing laboratory mouse. J. Comp. Pathol. 98:489–93.

Cheung, K.-S., et al. 1980. Murine cytomegalovirus: Detection of latent infection by nucleic acid hybridization technique. Infect. Immunol. 27:851–54.

Gardner, M.B., et al. 1974. Induction of disseminated virulent cytomegalovirus infection by immunosuppression of naturally chronically infected wild mice. Infect. Immunol. 10:966-69.

Hamilton, J.R., and Overall, J.C., Jr. 1978. Synergistic infection with murine cytomegalovirus and *Pseudomonas aeruginosa* in mice. J. Infect. Dis. 137:775–82.

Hudson, J.B. 1979. The murine cytomegalovirus as a model for the study of viral pathogenesis and persistent infections. Arch. Virol. 62:1–29.

Jordan, M.C. 1978. Interstitial pneumonia and subclinical infection after intranasal inoculation of murine cytomegalovirus. Infect. Immunol. 21:275–80.

Klotman, M.E., et al. 1990. Detection of mouse cytomegalovirus nucleic acid in latently infected mice by in vitro enzymatic amplification. J. Infect. Dis. 161:220–25.

Lussier, G. 1975. Murine cytomegalovirus (MCMV). Adv. Vet. Comp. Med. 10:223–47.

Lussier, G., et al. 1988. Detection of antibodies to mouse thymic virus by enzyme-immunosorbent assay. Can. J. Vet. Res. 52:236–38.

McCordock, H.A., and Smith, M.G. 1963. The visceral lesions produced in mice by the salivary gland virus of mice. J. Exp. Med. 63:303–10.

Mannini, A., and Medearis, D.N., Jr. 1961. Mouse salivary gland virus infections. Am. J. Hyg. 73:329–43.

Medearia, D.N., Jr. 1964. Mouse cytomegalovirus infection: Attempts to produce intrauterine infections. Am. J. Hyg. 80:113–20.

Mims, C.A., and Gould, J. 1979. Infection of salivary glands, kidneys, adrenals, ovaries and epithelia by murine cytomegalovirus. J. Med. Microbiol. 12:113–22.

Olding, L.B., et al. 1976. Pathogenesis of cytomegalovirus infection: Distribution of viral products, immune complexes and autoimmunity during latent murine infection. J. Gen. Virol. 33:267–80.

Osborne, J.E. 1982. Cytomegalovirus and other herpesviruses. In *The Mouse in Biomedical Research. II. Diseases*, ed. H.L. Foster, et al., pp. 267–92. New York: Academic.

K Virus Infection

Fisher, E.R., and Kilham, L. 1953. Pathology of a pneumotropic virus recovered from C3H mice carrying the Bittner milk agent. Arch. Pathol. 55:14–19.

Greenlee, J.E. 1981. Effect of host age on experimental K virus infection in mice. Infect. Immunol. 33:297–303.

———. 1979. Pathogenesis of K virus infection in newborn mice. Infect. Immunol. 26:705–13.

Kilham, L., and Murphy, H.W. 1953. A pneumotropic virus isolated from C3H mice carrying the Bittner milk agent. Proc. Soc. Exp. Biol. Med. 82:133–37.

Kraus, G.E., et al. 1968. Uber Vorkommen, Diagnostik und Pathologie latenter Infektionen mit Kilham-virus in Laboratoriums Mausen. Arch. Exp. Veterinaermed. 22:1203–10.

Margolis, G., et al. 1976. Oxygen tension and the selective tropism of K virus for mouse pulmonary endothelium. Am. Rev. Respir. Dis. 114:45–51.

Mokhtarian, F., and Shah, K.V. 1983. Pathogenesis of K papovavirus infection in athymic nude mice. Infect. Immunol. 41:434–36.

———. 1980. Role of antibody response in recovery from K papovavirus infection in mice. Infect. Immunol. 29:1169–79.

Polyomavirus Infection

Allison, A.C. 1980. Immune responses to polyoma virus and polyoma virus–induced tumors. In *Viral Oncology*, ed. G. Klein, pp. 481–87. New York: Raven.

Buffet, R.F., and Levinthal, J.D. 1962. Polyoma virus infection in mice. Arch. Pathol. 74:513–26.

Dawe, C.J. 1979. Tumours of the salivary and lachrymal glands, nasal fossa and maxillary sinuses. In *Pathology of Tumours in Laboratory Animals. II. Tumours of the Mouse*, ed. V.S. Turusov. Lyon: IARC Scientific Publications.

Dubensky, T.W., et al. 1984. Detection of DNA and RNA virus genomes in organ systems of whole mice: Patterns of mouse organ infection by polyomavirus. J. Virol. 50:779–83.

Eddy, B.E. 1982. Polyomavirus. In *The Mouse in Biomedical Research. II. Diseases*, Ed. H.L. Foster et al., pp. 293–311. New York: Academic.

Gross, L. 1970. The parotid tumor (polyoma) virus. In *Oncogenic Viruses*, by L. Gross, ed., 2d ed., pp. 651–750. London: Pergamon.

McCance, D.J., and Mims, C.A. 1979. Reactivation of polyomavirus in kidneys of persistently infected mice during pregnancy. Infect. Immunol. 25:998–1002.

McCance, D.J., et al. 1983. A paralytic disease in nude mice associated with polyoma virus infection. J. Gen. Virol. 64:57–67.

Rowe, W.P. 1961. The epidemiology of mouse polyoma virus infection. Bacteriol. Rev. 25:18–31.

Sebesteny, A., et al. 1980. Demyelination and wasting associated with polyomavirus infection in nude (nu/nu) mice. Lab. Anim. Sci. 14:337–45.

Stanton, M.F., et al. 1959. Oncogenic effect of tissue culture preparations of polyomavirus on fetal mice. J. Natl. Cancer Inst. 23:1441–75.

Stewart, S.E. 1960. The polyoma virus. Adv. Virus Res. 7:61–90.

Vandeputte, M., et al. 1974. Induction of polyoma tumors in athymic nude mice. Int. J. Cancer. 14:445–50.

Minute Virus of Mice (MVM) Infection

Bonnard, G.D., et al. 1976. Immunosuppressive activity of a subline of the mouse EL-4 lymphoma. Evidence for minute virus of mice causing the inhibition. J. Exp. Med. 143:187–205.

Collins, M.J., Jr., and Parker, J.C. 1972. Murine virus contaminants of leukemia viruses and transplantable tumors. J. Natl. Cancer Inst. 49:1139–43.

Crawford, L.V. 1966. A minute virus of mice. Virology 29:605–12.

Harris, R.E., et al. 1974. Erythrocyte association and interferon production of minute virus of mice. Proc. Soc. Exp. Biol. Med. 145:1288–92.

Kilham, L., and Margolis, G. 1971. Fetal infections of hamsters, rats, and mice induced with the minute virus of mice (MVM). Teratology 4:43–62.

———. 1970. Pathogenicity of minute virus of mice (MVM) for rats, mice and hamsters. Proc. Soc. Exp. Biol. Med. 133:1447–52.

Kimsey, P.B., et al. 1988. Immunological characteristics of a rodent parvovirus. Abstract. Lab. Anim. Sci. 38:488.

Kimsey, P.B., et al. 1986. Pathogenicity of fibroblast and lymphocyte-specific variants of minute virus of mice. J. Virol. 59:8–13.

McMaster, G.K., et al. 1981. Characterization of an immunosuppressive parvovirus related to minute virus of mice. J. Virol. 38:317–26.

Parker, J.C., et al. 1970a. Minute virus of mice. II. Prevalence, epidemiology, and occurrence as a contaminant of transplanted tumors. J. Natl. Cancer Inst. 45:305–10.

———. 1970b. Minute virus of mice. I. Procedures for quantitation and detection. J. Natl. Cancer Inst. 45:297–303.

Smith, A.L., et al. 1988. Acute and chronic effects of minute virus of mice (MVM) infection of neonatal inbred mice. Abstract. Lab. Anim. Sci. 38:488.

Ward, D.C., and Tattersall, P.J. 1982. Minute virus of mice. In *The Mouse in Biomedical Research. II. Diseases*, ed. H.L. Foster et al., pp. 313–34. New York: Academic.

Ectromelia Virus Infection

Allen, A.M., et al. 1981. Pathology and diagnosis of mousepox. Lab. Anim. Sci. 31:599–608.
Bhatt, P.N., and Jacoby, R.O. 1987a. Effect of vaccination on the clinical response, pathogenesis and transmission of mousepox. Lab. Anim. Sci. 37:610–14.
______. 1987b. Mousepox in inbred mice innately resistant or susceptible to lethal infection with ectromelia virus. 1. Clinical responses. Lab. Anim. Sci. 37:11–15.
Fenner, F. 1981. Mousepox (infectious ectromelia): Past, present, and future. Lab. Anim. Sci. 31:553–59.
Marchal, J. 1930. Infectious ectromelia: A hitherto undescribed virus disease of mice. J. Pathol. Bacteriol. 33:713–18.
New, A.E. 1981. Ectromelia (mousepox) in the United States. Proceedings of seminar, 31st annual meeting, American Association for Laboratory Animal Science. Lab. Anim. Sci. 31:549–622.
Wallace, G.W., and Buller, R.M.L. 1985. Kinetics of ectromelia virus (mousepox) transmission and clinical response in C57BL/6J, BALB/cByJ and AKR/J inbred mice. Lab. Anim. Sci. 35:41–46.

Sendai Virus Infection

Brownstein, D.G. 1987. Resistance/susceptibility to lethal Sendai virus infection genetically linked to a mucociliary transport polymorphism. J. Virol. 61:1670–71.
______. 1986. Sendai virus. In *Viral and Mycoplasmal Infections of Laboratory Rodents: Effects on Biomedical Research*, ed. P.N. Bhatt et al., pp. 37–61. New York: Academic.
______. 1985. Sendai virus infection, lung, mouse and rat. In *Monographs on Pathology of Laboratory Animals: Respiratory System*, ed. T.C. Jones et al., pp. 195–203. New York: Springer-Verlag.
Brownstein, D.G., and Weir, E.C. 1987. Immunostimulation in mice infected with Sendai virus. Am. J. Vet. Res. 48:1692–98.
Brownstein, D.G., and Winkler, S. 1986. Genetic resistance to lethal Sendai virus pneumonia: Virus replication and interferon production in C57/6J and DBA/2J mice. Lab. Anim. Sci. 36:126–29.
Brownstein, D.G., et al. 1981. Sendai virus infection in genetically resistant and susceptible mice. Am. J. Pathol. 105:156–63.
Ishida, N., and Homma, M. 1978. Sendai virus. Adv. Virus Res. 23:349–83.
Jakob, G. 1981. Interactions between Sendai virus and bacterial pathogens in the murine lung: A review. Lab. Anim. Sci. 31:170–77.
Kay, M.M.B. 1978. Long term subclinical effects of parainfluenza (Sendai) infection on immune cells of aging mice. Proc. Soc. Exp. Biol. Med. 158:326–31.
Kenyon, A.J. 1983. Delayed wound healing in mice associated with viral alteration of macrophages. Am. J. Vet. Res. 44:652–56.
Parker, J.C., and Richter, C.B. 1982. Viral diseases of the respiratory system. In *The Mouse in Biomedical Research. II. Diseases*, ed. H.L. Foster et al., pp. 109–34. New York: Academic.
Peck, R.M., et al. 1983. Influence of Sendai virus on carcinogenesis in strain A mice. Lab. Anim. Sci. 33:154–56.
Roberts, N.J. 1982. Different effects of influenza virus, respiratory syncytial virus, and Sendai virus on human lymphocytes and macrophages. Infect. Immunol. 35:1142–46.

Pneumonia Virus of Mice (PVM) Infection

Berthiaume, L., et al. 1974. Comparative structure, morphogenesis and biological characteristics of the respiratory syncytial (RS) virus and the pneumonia virus of mice (PVM). Arch. Virusforsch. 45:39–51.
Carthew, P., and Sparrow, S. 1980. A comparison in germ-free mice of the pathogenesis of Sendai virus and mouse pneumonia virus infections. J. Pathol. 130:153–58.
______. 1980. Persistence of pneumonia virus of mice and Sendai virus in germ-free (nu/nu) mice. Br. J. Pathol. 61:172–75.
Horsfall, F.L., and Curnen, E.C. 1946. Studies on pneumonia virus of mice (PVM). II. Immunological evidence of latent infection with the virus in numerous mammalian species. J. Exp. Med. 83:43–64.
Horsfall, F.L., and Hahn, R.G. 1940. A latent virus in normal mice capable of producing pneumonia in its natural host. J. Exp. Med. 71:391–408.
Richter, C.B., et al. 1988. Fatal pneumonia with terminal emaciation in nude mice caused by pneumonia virus of mice. Lab. Anim. Sci. 38:255–61.
Weir, E.C., et al. 1988. Respiratory disease and wasting in athymic mice infected with pneumonia virus of mice. Lab. Anim. Sci. 38:133–37.

Lymphocytic Choriomeningitis

Buchmeier, M.J. 1980. The virology and immunobiology of lymphocytic choriomeningitis virus infection. Adv. Immunol. 30:275–331.
Christofferson, P.J., et al. 1976. Immunological unresponsiveness of nude mice to LCM virus infection. Acta Pathol. Microbiol. Scand. [C] 84:520–23.
Dalton, A.J., et al. 1968. Morphological and cytochemical studies on lymphocytic choriomeningitis virus. J. Virol. 2:1465–78.
Dykewicz, C.A., et al. 1992. Lymphocytic choriomeningitis outbreak associated with nude mice in a research institute. J. Am. Med. Assoc. 267:1349–53.
Findlay, G.M., and Stern, R.O. 1936. Pathological changes due to infection with the virus of lymphocytic choriomeningitis. J. Pathol. Bacteriol. 43:327–38.
Lehman-Grube, F. 1971. Lymphocytic choriomeningitis virus. Virol. Monogr. 10:3–173.
Lilly, R.D., and Armstrong, C. 1945. Pathology of lymphocytic choriomeningitis in mice. Arch. Pathol. 40:141–52.
Oldstone, M.B.A., and Dixon, F.J. 1970. Pathogenesis with persistent lymphocytic choriomeningitis viral infection. II. Relationship of tissue injury in chronic lymphocytic choriomeningitis disease. J. Exp. Med. 131:1–19.
______. 1969. Pathogenesis of chronic disease associated with persistent lymphocytic choriomeningitis viral infection. I. Relationship of antibody production to disease in neonatally infected mice. J. Exp. Med. 129:483–505.
Oldstone, M.B.A., et al. 1982. Virus-induced alterations in homeostasis: Alterations in differentiated functions of infected cells in vivo. Science 218:1125–27.
Thomsen, A.R., et al. 1982. Lymphocytic choriomeningitis virus-induced immunosuppression: Evidence for viral interference with T-cell maturation. Infect. Immunol. 37:981–86.
Traub, E. 1936. The epidemiology of lymphocytic choriomeningitis in white mice. J. Exp. Med. 64:183–200.
van der Zeijst, B.A.M., et al. 1983. Persistent infection of some standard cell lines by lymphocytic choriomeningitis virus: Transmission of infection by an intracellular agent. J. Virol. 48:249–61.
Volkert, M., and Lundstedt, C. 1971. Tolerance and immunity to the lymphocytic choriomeningitis virus. Ann. N.Y. Acad. Sci. 181:183–95.

Mouse Hepatitis Viral (MHV) Infection

Bailey, O.T., et al. 1949. A murine virus (JHM) causing disseminated encephalomyelitis with extensive destruction of myelin.

II. Pathology. J. Exp. Med. 90:195–221.

Barthold, S.W. 1988. Olfactory neural pathway in mouse hepatitis virus nasoencephalitis. Acta Neuropathol. 76:502–506.

______. 1986. Mouse hepatitis virus biology and epizootiology. In *Viral and Mycoplasmal Infections of Laboratory Rodents: Effects on Biomedical Research*, ed. P.N. Bhatt et al., pp. 571–601. New York: Academic.

______. 1985a. Mouse hepatitis virus infection, liver, mouse. In *Monographs on Pathology of Laboratory Animals. III. Digestive System*, ed. T.C. Jones et al., pp. 134–39, New York: Springer-Verlag.

______. 1985b. Research complications and state of knowledge of rodent coronaviruses. In *Complications of Viral and Mycoplasmal Infections in Rodents to Toxicology Research and Testing*, ed. T.E. Hamm, pp. 53–89. Washington, D.C.: Hemisphere.

Barthold, S.W., and Smith, A.L. 1989. Virus strain specificity of challenge immunity to coronavirus. Arch. Virol. 104:187–96.

______. 1987. Response of genetically susceptible and resistant mice to intranasal inoculation with mouse hepatitis virus. Virus Res. 7:225–39.

Barthold, S.W., et al. 1985. Enterotropic mouse hepatitis virus infection in nude mice. Lab. Anim. Sci. 35:613–18.

Barthold, S.W., et al. 1982. Epizootic coronaviral typhlocolitis in suckling mice. Lab. Anim. Sci. 32:376–83.

Biggers, D.C., et al. 1964. Lethal intestinal virus in mice (LIVIM): An important new model for study of the response of the intestinal mucosa to injury. Am. J. Pathol. 45:413–27.

Boorman, G.A., et al. 1982. Peritoneal macrophage alterations caused by naturally occurring mouse hepatitis virus. Am. J. Pathol. 106:11–117.

Carrano, V.A., et al. 1984. Alteration of viral respiratory infections in mice by prior infection with mouse hepatitis virus. Lab. Anim. Sci. 34:573–76.

Ward, J.M., et al. 1977. Naturally occurring mouse hepatitis virus infection in the nude mouse. Lab. Anim. Sci. 27:372–76.

Mouse Encephalomyelitis Viral (MEV) Infection

Abzug, M.J., et al. 1989. Demonstration of a barrier to transplacental passage of murine enteroviruses in late gestation. J. Infect. Dis. 159:761–65.

Brownstein, D., et al. 1989. Duration and patterns of transmission of Theiler's mouse encephalomyelitis virus infection. Lab. Anim. Sci. 39:299–301.

Jacoby, R.O. 1988. Encephalomyelitis, Theiler's virus, mouse. In *Monographs on Pathology of Laboratory Animals: Nervous System*, ed. T.C. Jones, et al., pp. 175–79. New York: Springer-Verlag.

Lipton, H.L., and Rozhon, E.J. 1986. The Theiler's murine encephalomyelitis viruses. In *Viral and Mycoplasmal Infections of Laboratory Animals: Effects on Biomedical Research*, ed. P.N. Bhatt et al., pp. 253–75. New York: Academic.

Pevear, D.C., et al. 1987. Analysis of the complete nucleotide sequence of the picornavirus Theiler's mouse encephalomyelitis virus indicated that it is closely related to cardioviruses. J. Virol. 61:1507–16.

Zurbriggen, A., and Fujinami, R.S. 1988. Theiler's virus infection in nude mice: Viral RNA in vascular endothelial cells. J. Virol. 62: 3589–96.

Epizootic Diarrhea of Infant Mice (EDIM)

Coelho, K.I.R., et al. 1981. Pathology of rotavirus infection in suckling mice: A study by conventional histology, immunofluorescence, and scanning electron microscopy. Ultrastructural Pathol. 2:59–80.

Riepenhoff-Tatty, M., et al. 1987. Rotavirus infection in mice: Pathogenesis and immunity. Adv. Exp. Med. Biol. 216:1015–23.

Sheridan, J.F., and Vonderfecht, S. 1986. Mouse rotavirus. In *Viral and Mycoplasmal Infections of Laboratory Rodents: Effects on Biomedical Research*, ed. P.N. Bhatt et al., pp. 217–43. New York: Academic.

Reoviral Infection

Bennette, J.G., et al. 1967a. Characteristics of a newborn runt disease induced by neonatal infection with an oncolytic strain of reovirus type 3 (REO3MH). I. Pathological investigations in rats and mice. Br. J. Exp. Pathol. 48:251–66.

______. 1967b. Characteristics of a newborn runt disease induced by neonatal infection with an oncolytic strain of reovirus type 3 (REO3MH). II. Immunological aspects of the disease in mice. Br. J. Exp. Pathol. 48:267–84.

Branski, D., et al. 1980. Reovirus type 3 infection in a suckling mouse: The effects on pancreatic structure and enzyme content. Pediatr. Res. 14:8–11.

Cook, I. 1963. Reovirus type 3 infection in laboratory mice. Aust. J. Exp. Biol. 41:651–59.

Joske, R.A., et al. 1966. Murine infection with reovirus. IV. Late chronic disease and the induction of lymphoma after reovirus type 3 infection. Br. J. Exp. Pathol. 47:337–46.

Papadimitriou, J.M. 1968. The biliary tract in acute murine reovirus 3 infection. Light and electron microscopic study. Am. J. Pathol. 52:595–611.

Papadimitriou, J.M., and Walters, M.N.-I. 1967. Studies on the exocrine pancreas. II. Ultrastructural investigation of reovirus pancreatitis. Am. J. Pathol. 51:387–403.

Phillips, P.A., et al. 1969. Chronic obstructive jaundice induced by reovirus type 3 in weanling mice. Pathology 1:193–203.

Stanley, N.F. 1974. The reovirus murine models. Prog. Med. Virol. 18:257–72.

Stanley, N.F., et al. 1966. The association of murine lymphoma with reovirus type 3 infection. Proc. Soc. Exp. Biol. Med. 121:90–93.

Stanley, N.F., et al. 1964. Murine infections with reovirus: II. The chronic disease following reovirus type 3 infection. Br. J. Exp. Pathol. 45:142–49.

Stanley, N.F., et al. 1954. Studies on the hepatoencephalomyelitis virus (HEV). Aust. J. Exp. Biol. 32:543–62.

______. 1953. Studies on the pathogenesis of a hitherto undescribed virus (hepatoencephalomyelitis) producing unusual symptoms in suckling mice. Aust. J. Exp. Biol. 31:147–59.

Tyler, K.L., and Fields, B.N. 1986. Reovirus infection in laboratory rodents. In *Viral and Mycoplasmal Infections of Laboratory Rodents: Effects on Biomedical Research*, ed. P.N. Bhatt et al., pp. 277–303. New York: Academic.

Walters, M.N.-I., et al. 1963. Murine infection with reovirus. I. Pathology of the acute phase. Br. J. Exp. Pathol. 44:427–36.

Murine Retroviral Infection

Bentvelzen, P., and Hilgers, J. 1980. Murine mammary tumor virus. In *Viral Oncology*, ed. G. Klein, pp. 311–55. New York: Raven.

Lilly, F., and Mayer, A. 1980. Genetic aspects of murine type-C viruses and their hosts in oncogenesis. In *Viral Oncology*, ed. G. Klein, pp. 89–108. New York: Raven.

Morse, H.C., III, and Hartley, J.W. 1986. Murine leukemia viruses. In *Viral and Mycoplasmal Infections of Laboratory Rodents: Effects on Biomedical Research*, ed. P.N. Bhatt et al., pp. 349–88. New York: Academic.

Shih, T.Y., and Scolnick, E.M. 1980. Molecular biology of mammalian sarcoma viruses. In *Viral Oncology*, ed. G. Klein, pp. 135–60. New York: Raven.

Lactate Dehydrogenase Elevating Virus (LDH-EV) Infection

Brinton, M. 1986. Lactate dehydrogenase-elevating virus. In *Viral and Mycoplasmal Infections of Laboratory Rodents: Effects on Biomedical Research*, ed. P.N. Bhatt et al., pp. 389–420. New York: Academic.

Martinez, D., et al. 1980. Identification of lactate dehydrogenase-elevating virus as the etiologic agent of the genetically restricted age-dependent polioencephalomyelitis of mice. Infect. Immun. 27:979–87.

Notkins, A.L. 1965. Lactic dehydrogenase virus. Bacteriol. Rev. 29:143–60.

Riley, V. 1974. Persistence and other characteristics of lactic dehydrogenase-elevating virus (LDH-virus). Prog. Med. Virol. 18:198–213.

Rowson, K.E.K., and Mahy, B.W.J. 1975. Lactic dehydrogenase virus. Virol. Monogr. 13:1–121.

Snodgrass, M.J., et al. 1972. Changes induced by lactic dehydrogenase virus in thymus and thymus-dependent areas of lymphatic tissue. J. Immunol. 108:877–92.

Stroop, W.G., and Brinton, M.A. 1983. Mouse strain–specific central nervous system lesions associated with lactate dehydrogenase-elevating virus infection. Lab. Invest. 49:334–45.

General Bibliography

Collins, M.J., and Parker, J.C. 1972. Murine virus contaminants of leukemia viruses and transplantable tumors. J. Natl. Cancer Inst. 49:1139–43.

Lussier, G. 1991. Detection methods for the identification of rodent viral and mycoplasmal infections. Lab. Anim. Sci. 41:199–225.

______. 1988. Potential detrimental effects of rodent viral infections on long term experiments. Veter. Res. Communications 12:199–217.

van Nunen, M.C.J., et al. 1978. Prevalence of viruses in colonies of laboratory rodents. Z. Versuch. 20:201–8.

BACTERIAL INFECTIONS

Mycoplasmosis

Mycoplasma is a bacterium of the family Mycoplasmataceae. These pleomorphic organisms lack a cell wall and are enclosed by a single limiting membrane. Laboratory mice are host to several *Mycoplasma* species including *M. pulmonis*, *M. arthritidis*, *M. neurolyticum*, and *M. collis*. *M. pulmonis*, *M. arthritidis*, and *M. neurolyticum* inhabit the upper respiratory tract and *M. collis* inhabits the genital tract. Only *M. pulmonis* is a significant natural pathogen.

EPIZOOTIOLOGY AND PATHOGENESIS. Prior to and during the 1960s, infections with *M. pulmonis* were widespread in many colonies of mice. With marked improvements in health quality assessment and improved management practices, there has been a marked reduction in the incidence of clinically affected colonies of mice. However, based on some serological surveys, a relatively high percentage of laboratory mice may have detectable antibody to *M. pulmonis*, or to *M. arthritidis*, which cross-reacts with *M. pulmonis* antigenically. Exposure occurs by aerosol transmission. Newborn animals frequently become infected during the first few weeks of life from contact with affected mothers. Compared to the laboratory rat, mice are relatively resistant to the disease. In one study, intranasal inoculation with 10^4 colony-forming units or less of *M. pulmonis* resulted in disease confined to the upper respiratory tract and tympanic bullae. Higher doses resulted in lower respiratory tract disease, with mortality in some animals. Chronic suppurative arthritis has been produced in mice inoculated intravenously with *M. pulmonis*. However, it appears that spontaneous cases of *M. pulmonis*–associated arthritis rarely occur. In natural outbreaks of disease due to *M. pulmonis*, affected animals may exhibit weight loss, dyspnea, and a characteristic "chattering" sound on clinical examination. *M. pulmonis* colonizes the apical cell membranes of respiratory epithelium and interferes with mucociliary clearance. Mycoplasmosis is exacerbated by viral infections, such as Sendai virus and MHV, by other bacteria, and by environmental ammonia levels. These cofactors probably play a significant role in driving subclinical mycoplasmal infections into overt disease. In outbreaks of chronic respiratory disease and/or otitis media, other contributing factors such as concurrent infections with Sendai virus are important considerations. Sendai virus infection has been shown to have a synergistic effect in *M. pulmonis*–infected mice (Saito et al. 1981). The role and importance of *P. pneumotropica* in chronic respiratory disease is still open to speculation. Environmental factors, such as high cage-ammonia levels may play a significant role in the development of the disease and warrant investigation. For additional information on mycoplasmal infections, consult *M. pulmonis* infections in the rat in Chap. 2.

PATHOLOGY. Infection is often subclinical or mild. At necropsy, mucopurulent exudate may be present in the nasal passages, with variable involvement of the trachea and major airways. In advanced cases, cranioventral consolidation, bronchiectasis, and abscessation may be evident on gross examination (Fig. 1.34). Otitis media is another frequent manifestation of murine mycoplasmosis (Fig. 1.35). On microscopic examination, suppurative rhinitis, with polymorphonuclear and lymphocytic infiltration and hyperplasia of submucosal glands are characteristic findings (Fig. 1.36). In the respiratory epithelium of the nasal passages and major airways, there may be loss of cilia and flattening of epithelial lining cells. In association with the chronic suppurative process, syncytia may be present in affected nasal mucosa and larynx (Fig. 1.37). In the lower respiratory tract, lesions vary from discrete peribronchial and perivascular lymphocytic and plasma cell infiltration, to chronic sup-

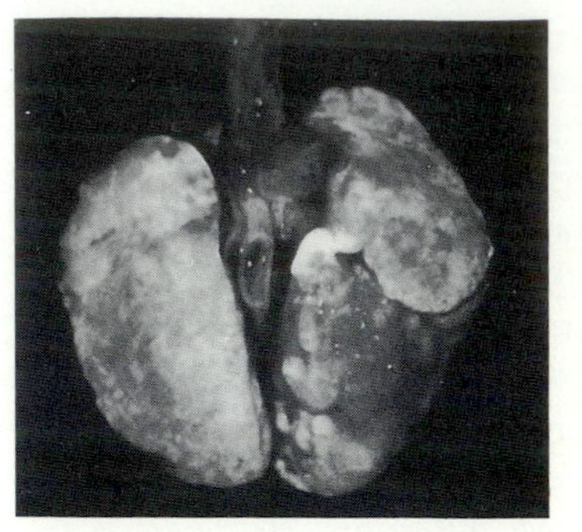

Fig. 1.34. Lungs from mouse with advanced mycoplasmal infection. There is marked mottling and discoloration, particularly in the right lung.

purative bronchitis, bronchiolitis, and alveolitis, with mobilization of alveolar macrophages (Fig. 1.38). In advanced cases, there may be squamous metaplasia of respiratory epithelium, bronchiectasis, and abscessation, with obliteration of the normal architecture. Purulent otitis media is a frequent histological finding in chronic mycoplasmosis. Mice do not develop the intense peribronchial lymphocytic infiltrates and severe bronchiectasis that are common in mycoplasmosis in the rat.

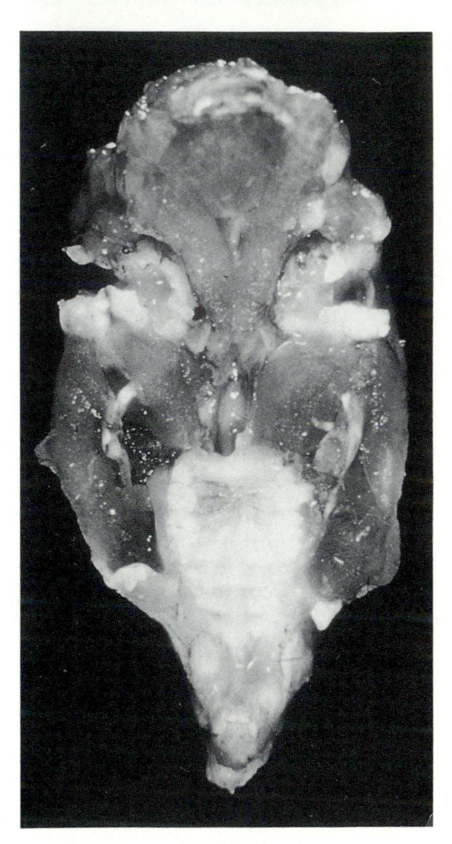

Fig. 1.35. Chronic suppurative otitis media associated with chronic mycoplasmosis. Note the marked thickening of the tympanic membranes and the presence of exudate in the opened tympanic bullae.

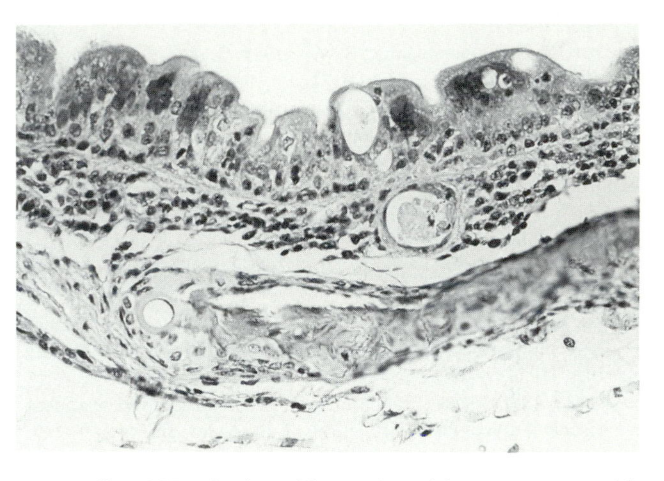

Fig. 1.37. Section of larynx from laboratory mouse with chronic mycoplasmal infection. Multinucleated giant cells are present in the respiratory epithelium. This is a common feature of the disease in this species.

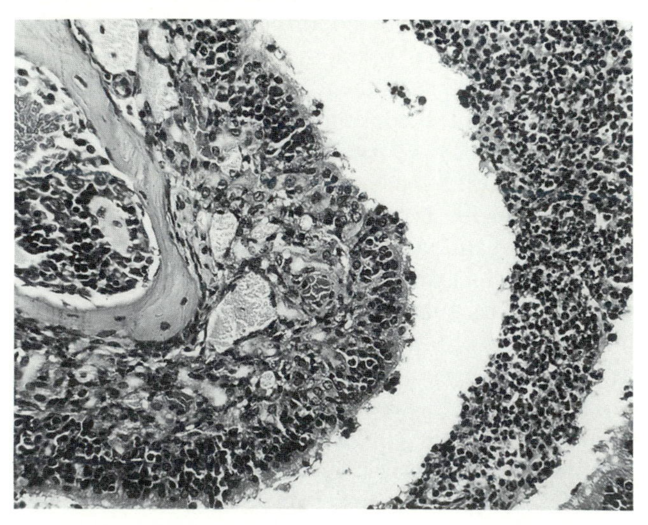

Fig. 1.36. Turbinates from adult mouse with chronic mycoplasmosis, illustrating a suppurative rhinitis with hyperplasia of respiratory epithelium.

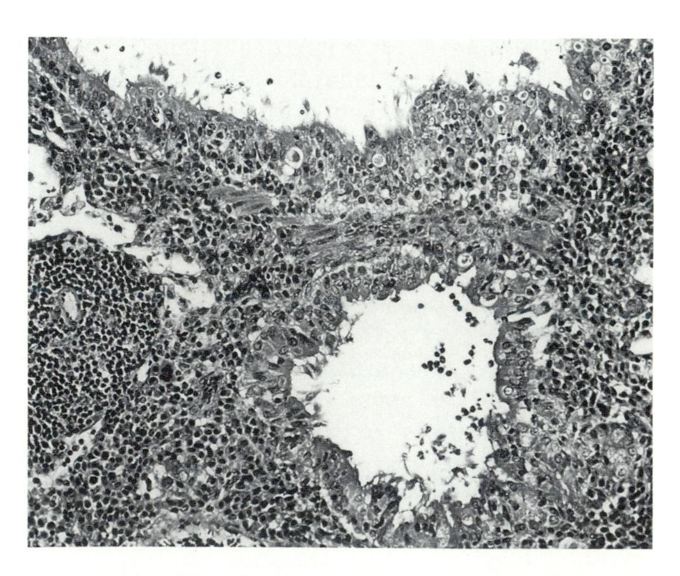

Fig. 1.38. Chronic respiratory mycoplasmosis in adult mouse: section of lower respiratory tract, illustrating an advanced case with marked peribronchial lymphocytic infiltration, hyperplasia of bronchial mucosa, and mobilization of alveolar macrophages.

DIAGNOSIS. Confirmation of mycoplasmal infection by recovery of the organism or immunofluorescence microscopy and/or serological testing are essential steps in order to confirm the diagnosis. For culture, nasopharyngeal flushing and tracheobronchial lavages with *Mycoplasma* broth or phosphate-buffered saline are recommended procedures. Cultures are often negative for *Mycoplasma* in affected animals. Thus additional tests are required. In serological testing, infected mice may have relatively low antibody titers to *M. pulmonis*; therefore procedures such as the ELISA are required in order to consistently detect seropositive animals. In the interpretation of seropositive cases, it is important to be aware of the serological cross-reactivity of *M. pulmonis* and *M. arthritidis*. Cell-mediated immunity to *M. pulmonis* may be suppressed in mice with advanced pulmonary lesions (Kishima et al. 1989). The immunofluorescence or the immunoperoxidase techniques are sensitive methods used for the detection of mycoplasmal antigens in tissue fluids or infected tissues. In view of the recognized role of secondary bacterial invaders in the disease, the respiratory tract should also be cultured for bacteria such as *P. pneumotropica*. Histological assessment should include a search for syncytia in the upper respiratory tract, a characteristic of the disease. Staining procedures such as the Warthin-Starry method should be performed on tissue sections of major airways in order to screen for possible infections with the cilia-associated respiratory (CAR) bacillus, an organism associated with some outbreaks of murine respiratory disease in mice (Griffith et al. 1988). *Differential diagnoses* include bronchopneumonia associated with infections with the CAR bacillus, and primary infections with Sendai virus with secondary bacterial invaders.

SIGNIFICANCE. Although mice are relatively resistant to clinical disease following infection with *M. pulmonis*, lesions of chronic respiratory disease occasionally occur, particularly in older mice. There is evidence that *M. pulmonis* infections may complicate certain types of research, such as tumor metastases studies (Lai et al. 1986), and infections may depress humoral and cell-mediated responses (Lai et al. 1989).

SIGNIFICANCE OF OTHER MURINE MYCOPLASMAL INFECTIONS. *M. neurolyticum* is the causative agent of "rolling disease," a term used to denote the neurologic signs associated with the exotoxin that follows experimental inoculation of the organism in mice. Spontaneous outbreaks of conjunctivitis have been associated with *M. neurolyticum* infection in young mice (Harkness and Ferguson 1982), but the organism appears to be relatively nonpathogenic under most conditions and is exceedingly rare or nonexistent anymore. *M. arthritidis* is antigenically related to *M. pulmonis* and causes arthritis when inoculated intravenously into mice but is nonpathogenic under natural conditions. *M. collis* has been isolated from the genital tract, without known adverse effect in mice.

Cilia-associated Respiratory (CAR) Bacillus Infection

CAR bacillus is a widespread and significant respiratory pathogen in the rat (see Chap. 2) and probably infects mice at a higher rate than is currently recognized. The organism has been associated with chronic respiratory disease in conventional and obese mice dying with the disease. Chronic suppurative anteroventral bronchopneumonia and marked peribronchial infiltration with lymphocytes and plasma cells are evident on microscopic examination. Filamentous bacteria were demonstrated in association with the ciliated respiratory epithelium of bronchi and bronchioles. In one study, representative animals were seropositive for Sendai and pneumonia virus of mice, and negative for *M. pulmonis* (Griffith et al. 1988). Thus it is possible that more than one organism was involved in the evolution of lesions in the respiratory tract. Chronic respiratory disease and seroconversion have been produced in BALB/c mice inoculated intranasally with the CAR bacillus. The infection may be transmitted by intracage contact, but airborne transmission to adjacent cages does not appear to occur in mice (Matsushita et al. 1989). Mice appear to be one of the most susceptible species (Shoji-Darkye et al. 1992).

Salmonellosis

Salmonella, gram-negative member of the Enterobacteriaceae, is currently classified into three species: *S. cholerasuis*, *S. typhi*, and *S. enteritidis*. Mice are infected with *S. enteritidis*. They are non–lactose fermenters and consist of over 1600 recognized serotypes. During the first half of this century, sporadic outbreaks of salmonellosis were a relatively common occurrence in conventional colonies of mice. With improved quality control and health assurance programs and good housing, husbandry, and feeding practices, recognized outbreaks now rarely occur.

EPIZOOTIOLOGY AND PATHOGENESIS. Exposure is considered to occur primarily by ingestion of contaminated feed or bedding, although conjunctival inoculation requires fewer organisms to establish an infection. Contaminated feed and bedding are the usual sources of the infection. In one study, *Salmonella* spp. were isolated from around 20% of feed samples tested. Pelleting resulted in a 1000-fold reduction in Enterobacteriaceae (Stott et al. 1975). Susceptibility/resistance depends on a variety of factors, including age (weanlings are more susceptible than adult mice), gut microflora, strain of mice, virulence and dose of organism, route of inoculation, intercurrent infections, and manipulations that impair the immune

differences include nonmotility and lack of indole production. *Differential diagnoses* include other agents that cause enteritis in the mouse, including rotavirus, coronavirus, adenovirus, and reovirus in young mice and *Salmonella* and *Bacillus piliformis* in older mice. Rectal prolapse is frequently associated with TMCH but can also occur spontaneously or in association with enteritis of other causes. Hyperplastic colitis has been observed in athymic (nude) mice chronically infected with enterotropic coronavirus (see Mouse Hepatitis Virus Infection). A hyperplastic lesion of unknown etiology in the distal colon of mice with severe combined immunodeficiency has also been observed.

SIGNIFICANCE. *C. freundii* infections are rare, and infection is transient but can cause low mortality, permanent rectal prolapse, and runting. TMCH represents a possible complication in certain types of research, such as carcinogenesis studies.

Escherichia coli Typhlocolitis in Combined Immunodeficient Mice

E. coli is a common gut organism in mice that has rarely been considered to be a primary pathogen in this species. A syndrome resembling *Citrobacter freundii* colonic hyperplasia in immunodeficient mice has been associated with an atypical, lactose-negative *E. coli* (Waggie et al. 1988).

EPIZOOTIOLOGY AND PATHOGENESIS. In published reports, hyperplastic lesions were observed primarily in young adult triple-deficient N:NIH(s) (homozygous for nu, xid, bg) and to a lesser extent double-deficient mice. Other immunocompetent and partially deficient mice were infected without significant hyperplastic lesions. Bacteria were located in the gut lumen and intracellularly within enterocytes. This syndrome has also been observed in severe combined immunodeficient (SCID) mice.

PATHOLOGY. Mice are depressed, with perianal fecal staining. Gross necropsy findings are limited to mild to moderate thickening of segments of colon or cecum and occasional blood-tinged feces. Microscopic findings consist of mucosal hyperplasia in one or all segments of colon, with variable inflammation and erosion (Fig. 1.47). *E. coli* are present in the gut lumen, attached to the surface and within enterocytes of superficial mucosa of both small and large intestine.

DIAGNOSIS. Segmental hyperplastic lesions in the colon and cecum of immunodeficient mice and isolation of atypical *E. coli* are required in order to confirm the diagnosis. The causative agent is non-lactose-fermenting, an unusual feature of *E. coli*. *Differential diagnoses* must include hyperplastic typhlocolitis caused by *C. freundii* (which ferments lactose and is pathogenic in immunocompetent mice) and enterotropic mouse hepatitis virus in immunodeficient mice.

SIGNIFICANCE. This organism can potentially cause significant illness in immunodeficient mice but is nonpathogenic in other mice.

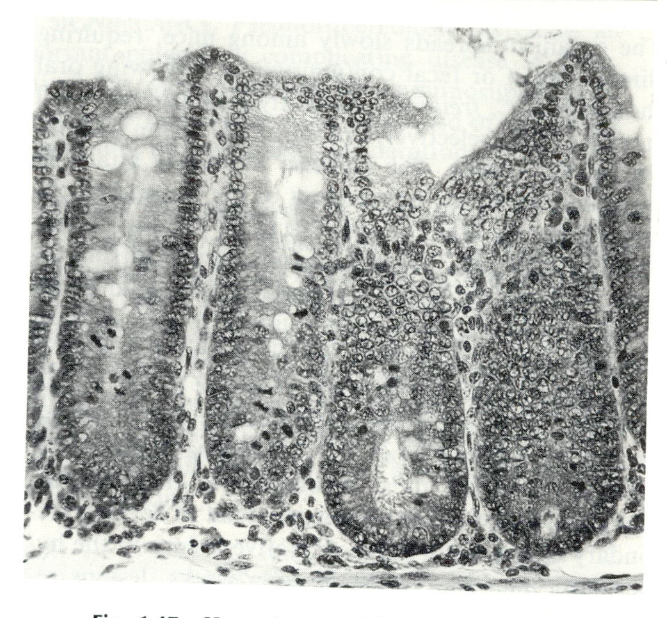

Fig. 1.47. Hyperplastic typhlitis due to *Escherichia coli* infection in SCID mouse. There is marked hyperplasia of enterocytes lining crypts. Mucous and bacterial colonies are present in the lumen.

Proteus mirabilis Infection

Proteus infection has been observed in both immunocompetent and immunodeficient laboratory mice. In immunocompetent animals, C3H/HeJ mice, the disease was observed most frequently in female mice (Jones et al. 1972). *P. mirabilis* has been isolated from the intestinal tract of both clinically affected and asymptomatic animals, but the nasopharynx may be another important portal of entry (Wensinck 1961). In spontaneous cases in severe combined immunodeficient (SCID) mice, clinical signs were characterized by weight loss, hunched posture, and dehydration (Scott et al. 1991; Scott 1989).

PATHOLOGY. Suppurative pyelonephritis and septicemia may occur, and there is some evidence that the renal lesions are hematogenous in origin. In immunodeficient mice, splenomegaly and multifocal hepatic lesions are typical macroscopic findings. In some cases, fibrinopurulent exudate is present in the peritoneal cavity. On microscopic examination, there are multifocal areas of coagulation necrosis present in the subcapsular regions of the liver and around central veins and minimal to moderate infiltration with neutrophils (Fig. 1.48). Septic thrombi may be present in vessels of tissues such as liver, intestinal serosa, and pancreas. Pulmonary lesions, when present, are characterized by alveolar flooding and mobilization of alveolar macrophages (Fig. 1.49).

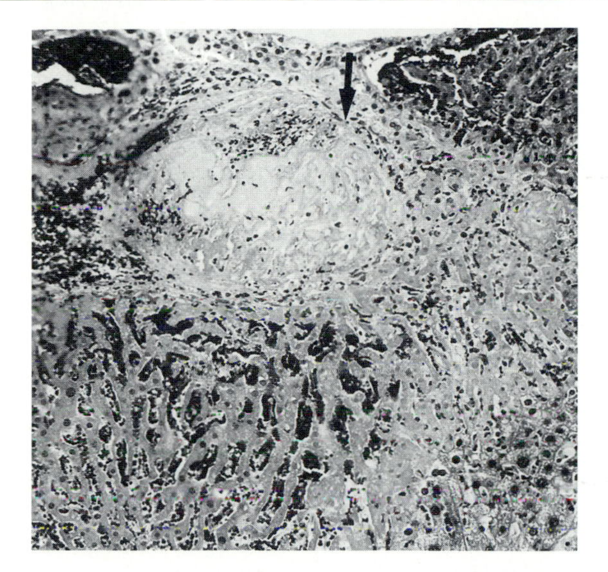

Fig. 1.48. Liver from SCID/beige mouse with systemic *Proteus mirabilis* infection. There is thrombophlebitis with coagulation necrosis of hepatic parenchyma and thrombosis of a hepatic vein (*arrow*).

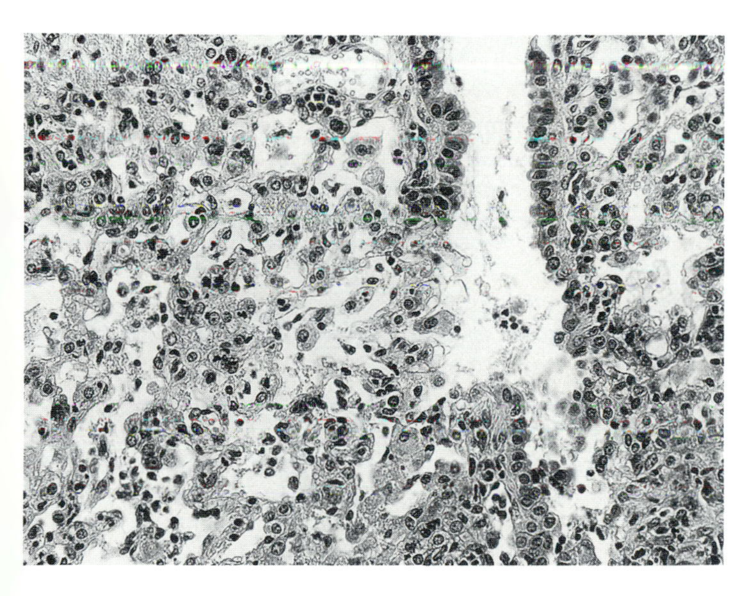

Fig. 1.49. Interstitial pneumonitis in SCID mouse with systemic *Proteus mirabilis* infection. Alveolar septa are hypercellular, with marked mononuclear and polymorphonuclear cell infiltration. Bacterial colonies are present in many alveolar macrophages.

DIAGNOSIS. In addition to the presence of histological lesions consistent with bacterial sepsis, the recovery of large numbers of *P. mirabilis* from sites such as liver, peritoneal cavity, and intestine will serve to confirm the diagnosis.

SIGNIFICANCE. Infections with *P. mirabilis* may cause significant mortality in colonies of SCID mice, and infections have also been observed in immunocompetent mice. Meticulous sanitation practices and reduced population densities should alleviate the problem.

Leptospirosis

EPIZOOTIOLOGY AND PATHOGENESIS. Leptospira are represented by a large number of species or serotypes with a wide host range. Individual serotypes tend to prefer a single primary host but can infect many different host species. Mice can be infected with a number of leptospira serotypes, but *L. balllum* is common (Friedman et al. 1973; Stoenner and Maclean 1958). Infection of laboratory mice now appears to be quite rare. Rodents do not become clinically ill when infected and can shed organisms in their urine throughout life.

PATHOLOGY. Lesions are absent.

DIAGNOSIS. The most accurate means of diagnosis is kidney culture, which should be performed on serial 10-fold dilutions of tissue homogenates because growth inhibition can occur in undiluted samples. Serology is also possible; however, mice infected as neonates may become persistently infected but never seroconvert. Under natural conditions, this phenomenon is common.

SIGNIFICANCE. Leptospirosis is a zoonotic disease. Humans can become infected and develop clinical illness when handling asymptomatic mice.

Mycobacterium Infection

Although laboratory mice are susceptible to experimental infections with *Mycobacterium*, naturally occurring infections are apparently exceedingly rare. A single outbreak of infection in laboratory mice with *M. avium-intracellulare* has been documented (Waggie et al. 1983a).

EPIZOOTIOLOGY AND PATHOGENESIS. *M. avium-intracellulare* complex organisms can be isolated from soil, water, and sawdust, which was the presumed origin of the infection in mice. Infection was observed in C57BL/6N mice, but not in C3H/HeN or B6C3F1 mice or F344 rats housed in the same room.

PATHOLOGY. Infected mice displayed no clinical signs, but necropsy findings consisted of few subpleural 1- to 5-mm-diameter tan-colored masses in lung. Microscopic findings consisted of focal accumulations of epithelioid cells, foamy macrophages and lymphocytes in alveolar spaces and septa, with variable amounts of necrosis and neutrophilic leukocyte infiltration. Langhans' giant cells were present in lesions of some mice. Many of the mice also had microgranulomas with occasional giant cells in liver parenchyma and mesenteric lymph nodes. Small numbers of acid-fast bacilli were visualized in some, but not all, lesions.

DIAGNOSIS. Definitive diagnosis can be made by demonstrating acid-fast organisms in granulomas and isolation of *Mycobacterium*. Organisms can be grown from tracheal washes on *Mycoplasma* agar incubated for 1 wk at 37°C in humidified atmosphere containing 10% CO_2 or cultured from a suspension of tissue homogenates in sterile saline on blood agar. *Differential diagnoses* for pulmonary granulomas should include *M. pulmonis* and *Corynebacterium kutscheri* infections.

SIGNIFICANCE. Mycobacteria are ubiquitous in the environment, yet infections in rodents are rare, minimizing the significance of this disease. When infection is present, careful attention should be paid to husbandry practices.

Streptobacillosis

EPIZOOTIOLOGY AND PATHOGENESIS. Streptobacillosis is a septicemic disease of mice caused by *Streptobacillus moniliformis*, which is carried in the nasopharynx of wild rats and some populations of laboratory rats. This organism is no longer common in laboratory rats, and the disease in mice has likewise become rare. In documented outbreaks of streptobacillosis in mice, carrier rats were maintained in the same room, but mouse-to-mouse transmission occurs readily (Wullenweber et al. 1990). Following oral inoculation, *S. moniliformis* can be isolated from submaxillary and cervical lymph nodes within 48 hr, with subsequent septicemia. A large number of organisms can be isolated from the blood during this stage.

PATHOLOGY. Clinical signs include diarrhea, hemoglobinuria, and conjunctivitis. There are disseminated foci of necrosis and inflammation in liver, spleen, and lymph nodes, with petechial and ecchymotic hemorrhages on serosal surfaces. Some mice, particularly those that survive the acute phase, develop suppurative polyarthritis (Freundt 1959).

DIAGNOSIS. Definitive diagnosis of streptobacillosis is made by isolation from infected tissues, using blood agar. Organisms are nonmotile, gram-negative rods that are highly pleomorphic, with long filamentous forms under ideal growth conditions. *Differential diagnoses* must include other forms of septicemic disease that cause disseminated lesions, such as salmonellosis, pseudomoniasis, and Tyzzer's disease.

SIGNIFICANCE. Streptobacillosis is an unlikely pathogen in contemporary laboratory mouse colonies but is a major reason for not cohabitating mice and rats in the same room. *S. moniliformis* causes rat-bite fever in humans (Anderson et al. 1983).

Klebsiella oxytoca-associated Utero-Ovarian Infection

EPIZOOTIOLOGY AND PATHOGENESIS. An unusually high prevalence of suppurative female reproductive tract lesions has been reported in a large population of aging B6C3F_1 mice (Rao et al. 1987). *K. oxytoca* was the most frequently isolated organism from these lesions, but disease could not be experimentally reproduced, suggesting that other factors may be involved. Other organisms isolated from affected mice included *K. pneumoniae*, *Escherichia coli*, *Enterobacter*, and others.

PATHOLOGY. Aged female mice had suppurative endometritis, salpingitis, and perioophoritis and/or peritonitis, often resulting in the formation of abscesses and adhesions. Lesions were severe enough to be life-limiting in some mice.

DIAGNOSIS. Diagnosis is based on recognition of lesions and isolation of the agent and is not necessarily specific for *K. oxytoca*.

SIGNIFICANCE. Utero-ovarian infection can be life-limiting in aging studies.

Necrotizing Enteritis Associated with Clostridial Infection

Sporadic cases of necrotizing enteritis have been observed in SPF mice that died during the postweaning period. At necropsy, the small intestines in affected animals were dilated, and contained blood-stained fluid contents. In affected regions of the small and large intestine, there were fibrinous exudation and effacement of the normal architecture. Large numbers of gram-positive bacilli were present in the exudate and *Clostridium perfringens* were isolated consistently from the intestinal contents (Matsushita and Matsumoto 1986).

Rickettsial (*Eperythrozoon coccoides*) Infection

Eperythrozoon is naturally transmitted by the louse *Polypax serrata*; both the infection and the carrier are now nonexistent in laboratory mice. With Giemsa and Romanowsky stains, the organism can be found attached to erythrocytes as well as free in the plasma of peripheral blood. In the early infection, a high level of parasitemia occurs within a few days, with clinical signs ranging from inapparent to severe anemia and death. Splenomegaly is a prominent feature of this infection, and this organ plays a central role in clearance of the parasite from the blood.

BIBLIOGRAPHY FOR BACTERIAL INFECTIONS

Mycoplasmosis

Griffith, J.W., et al. 1988. Cilia-associated respiratory (CAR) bacillus infection in obese mice. Vet. Pathol. 25:72–76.

Harkness, J.E., and Ferguson, F.G. 1982. Bacterial, mycoplasmal, and mycotic diseases of the lymphoreticular, musculoskeletal, cardiovascular, and endocrine systems. In *The Mouse in Biomedical Research. II. Diseases*, ed. H.L. Foster et al., pp. 83–97. New York: Academic.

Kishima, M., et al. 1989. Cell-mediated and humoral immune responses in mice during experimental infection with *Mycoplasma pulmonis*. Lab. Anim. 23:138–42.

Lai, W.C., et al. 1989. *Mycoplasma pulmonis* depresses humoral and cell-mediated responses in mice. Lab. Anim. Sci. 39:11–15.

Lai, W.C., et al. 1986. *Mycoplasma pulmonis* infection of mice influences tumor metastasis research. Lab. Anim. Sci. 36:568.

Saito, M., et al. 1981. Synergistic effect of Sendai virus on *Mycoplasma pulmonis* infection in mice. Jpn. J. Vet. Res. 43:43–50.

Taylor, G., and Taylor-Robinson, D. 1976. Effects of active and passive immunization on *Mycoplasma pulmonis*–induced arthritis in mice. Ann. Rheum. Dis. 36:232–38.

Cilia-associated Respiratory (CAR) Bacillus Infection

Griffith, J.W., et al. 1988. Cilia-associated respiratory (CAR) bacillus infection in obese mice. Vet. Pathol. 25:72–76.

Matsushita, S., et al. 1989. Transmission experiments of cilia-associated respiratory bacillus in mice, rabbits and guinea pigs. Lab. Anim. 23:96–102.

Shoji-Darkye, Y., et al. 1992. Pathogenesis of CAR bacillus in rabbits, guinea pigs, Syrian hamsters, and mice. Lab. Anim. Sci. 41:567–71.

Salmonellosis

Caseboldt, D.B., and Schoeb, T.R. 1988. An outbreak in mice of salmonellosis caused by *Salmonella enteritidis* serotype *enteritidis*. Lab. Anim. Sci. 38:190–92.

Lin, F., et al. 1987. Electron microscopic studies on the location of bacterial proliferation in the liver in murine salmonellosis. Br. J. Exp. Pathol. 68:539–50.

Margard, W.L., et al. 1963. Salmonellosis in mice-diagnostic procedures. Lab. Anim. Care 13:144–65.

Stott, J.A., et al. 1975. Incidence of salmonellae in animal feed and the effect of pelleting on content of Enterobacteriaceae. J. Appl. Bacteriol. 39:41–46.

Tannock, G.W., and Smith, J.M.B. 1971. A *Salmonella* carrier state involving the upper respiratory tract of mice. J. Infect. Dis. 123:502–6.

Pasteurella pneumotropica Infection

Brennan, P.C., et al. 1969. Role of *Pasteurella pneumotropica* and *Mycoplasma pulmonis* in murine pneumonia. J. Bacteriol. 97:337–49.

Davis, J.K., et al. 1987. The role of *Klebsiella oxytoca* in utero-ovarian infection of B6C3F1 mice. Lab. Anim. Sci. 37:159–66.

Needham, J.R., and Cooper, J.E. 1975. An eye infection in laboratory mice associated with *Pasteurella pneumotropica*. Lab. Anim. 9:197–200.

Wagner, J.E., et al. 1969. Spontaneous conjunctivitis and dacryoadenitis of mice. J. Am. Vet. Med. Assoc. 155:1211–17.

Ward, G.E.R., et al. 1978. Abortion in mice associated with *Pasteurella pneumotropica*. J. Clin. Microbiol. 8:177–80.

Weisbroth, S.H., et al. 1969. *Pasteurella pneumotropica* abscess syndrome in a mouse colony. J. Am. Vet. Med. Assoc. 155:1206–10.

Staphylococcal Infections

Cooper, J.E. 1977. Furunculosis in the mouse. Vet. Rec. 101:433.

Streptococcal Dermatitis

Stewart, D.D., et al. 1975. An epizootic of necrotic dermatitis in laboratory mice caused by Lancefield group G streptococci. Lab. Anim. Sci. 25:296–302.

Systemic Streptococcal Infections

Duignan, P.J., and Percy, D.H. 1992. Diagnostic exercise: Unexplained deaths in recently acquired C3H3 mice. Lab. Anim. Sci. 42:610–11.

Percy, D.H., and Barta, J.R. 1993. Spontaneous and experimental infections in SCID and SCID/beige mice. Lab. Anim. Sci. 43: 127–32.

Tyzzer's Disease

Fries, A.S. 1977. Studies on Tyzzer's disease: Application of immunofluorescence for detection of *Bacillus piliformis* and for the demonstration and determination of antibodies to it in sera from mice and rabbits. Lab. Anim. 11:69–73.

Ganaway, J.R., et al. 1971. Tyzzer's disease. Am. J. Pathol. 64:717–32.

Motzel, S.L., et al. 1991. Detection of serum antibodies to *Bacillus piliformis* in mice and rats using an enzyme-linked immunoabsorbent assay. Lab. Anim. Sci. 41:26–30.

Tyzzer, E.E. 1917. A fatal disease of the Japanese waltzing mouse caused by a spore-bearing bacillus (*Bacillus piliformis* N.sp.). J. Med. Res. 37:307–38.

Waggie, K.S.,et al. 1981. A study of mouse strain susceptibility to *Bacillus piliformis* (Tyzzer's disease): The association of B-cell function and resistance. Lab. Anim. Sci. 31:139–42.

Pseudomoniasis

Brownstein, D.G. 1978. Pathogenesis of bacteremia due to *Pseudomonas aeruginosa* in cyclophosphamide-treated mice and potentiation of virulence of endogenous streptococci. J. Infect. Dis. 137:795–801.

Lindsey, J.R. 1986. Prevalence of viral and mycoplasmal infections in laboratory rodents. In *Viral and Mycoplasmal Infections of Laboratory Rodents: Effects on Biomedical Research*, ed. P.N. Bhatt et al., pp. 801–8. New York: Academic.

Pseudotuberculosis (*Corynebacterium kutscheri Infection*)

Weisbroth, S.H. 1979. Bacterial and mycotic diseases. In *The Laboratory Rat. I. Biology and Diseases*, ed. H.J. Baker et al., pp. 214–19. New York: Academic.

Coryneform Hyperkeratosis in Nude Mice

Richter, C.B., et al. 1991. D2 coryneforms as a cause of severe hyperkeratotic dermatitis in athymic nude mice. Lab. Anim. Sci. 40:545.

Colonic Hyperplasia (*Citrobacter freundii* Infection)

Barthold, S.W. 1980. The microbiology of transmissible murine colonic hyperplasia. Lab. Anim. Sci. 30:167–73.

Barthold, S.W., and Adams, R.L. 1986. *Citrobacter freundii*. In *Manual of Microbiologic Monitoring of Laboratory Animals*, ed. A.M. Allen and T. Nomura. Natl. Inst. Health Pub. No. 86-2498. Washington, D.C.: Government Printing Office.

Barthold, S.W., et al. 1978. Transmissible murine colonic hyperplasia. Vet. Pathol. 15:223–36.

Barthold, S.W., et al. 1977. Dietary, bacterial, and host genetic interactions in the pathogenesis of transmissible murine colonic hyperplasia. Lab. Anim. Sci. 27:938–45.

Barthold, S.W., et al. 1976. The etiology of transmissible murine colonic hyperplasia. Lab. Anim. Sci. 26:889–94.

Bienick, H., and Tober-Meyer, B. 1976. Zur Atiologie der Colitis und des Prolapses recti bei der Maus. Z. Versuch. 18:337–48.

Brennan, P.C., et al. 1965. *Citrobacter freundii* associated with diarrhea in laboratory mice. Lab. Anim. Care 15:266–75.

Ediger, R.D., et al. 1974. Colitis in mice with a high incidence of rectal prolapse. Lab. Anim. Sci. 24:488–94.

Silverman, J., et al. 1979. A natural outbreak of transmissible murine colonic hyperplasia in A/J mice. Lab. Anim. Sci. 29:209–13.

Escherichia coli Typhlocolitis in Combined Immunodeficient Mice

Waggie, K.S., et al. 1988. Cecocolitis in immunodeficient mice associated with an enteroinvasive lactose negative *E. coli*. Lab. Anim. Sci. 38:389–93.

Proteus mirabilis Infection

Jones, J.B., et al. 1972. *Proteus mirabilis* infection in a mouse colony. J. Am. Vet. Med. Assoc. 161:661–64.

Scott, R.A.W. 1989. Fatal *Proteus mirabilis* infection in a colony of SCID/bg immunodeficient mice. Lab. Anim. Sci. 39:470–71.

Scott, R.A.W., et al. 1991. Diagnostic exercise: Hepatitis in SCID-beige mice. Lab. Anim. Sci. 41:166–68.

Wensinck, F. 1961. The origin of endogenous *Proteus mirabilis* bacteremia in irradiated mice. J. Pathol. Bacteriol. 81:395–401.

Leptospirosis

Birnbaum, S., et al. 1972. The influence of maternal antibodies on the epidemiology of leptospiral carrier state in mice. Am. J. Epidemiol. 96:313–17.

Friedman, C.T.H., et al. 1973. *Leptospirosis ballum* contracted from pet mice. Calif. Med. 118:51–52.

Stoenner, H.G. 1957. The laboratory diagnosis of leptospirosis. Vet. Med. 52:540–42.

Stoenner, H.G., and Maclean, D. 1958. *Leptospirosis (ballum)* contracted from Swiss albino mice. Arch. Intern. Med. 101:706–10.

Torten, M. 1979. Leptospirosis. In *CRC Handbook Series in Zoonoses*, ed. J.H. Steel, pp. 363–421. Cleveland: CRC.

Mycobacterium Infection

Waggie, K.S., et al. 1983a. A naturally occurring outbreak of *Mycobacterium avium-intracellulare* infections in C57BL/6N mice. Lab. Anim. Sci. 33:249–53.

Waggie, K.S., et al. 1983b. Experimental murine infections with a *Mycobacterium avium-intracellulare* complex organism isolated from mice. Lab. Anim. Sci. 33:254–57.

Streptobacillosis

Anderson, L.C., et al. 1983. Rat-bite fever in animal research laboratory personnel. Lab. Anim. Sci. 33:292–94.

Freundt, E.A. 1959. Arthritis caused by *Streptobacillus moniliformis* and pleuropneumonia-like organisms in small rodents. Lab. Invest. 8:1358–75.

Wullenweber, M., et al. 1990. *Streptobacillus moniliformis* epizootic in barrier-maintained C57BL/6J mice and susceptibility to infection of different strains of mice. Lab. Anim. Sci. 90: 608–12.

Klebsiella oxytoca–associated Utero-Ovarian Infection

Davis, J.K., et al. 1987. The role of *Klebsiella oxytoca* in utero-ovarian infection of B6C3F1 mice. Lab. Anim. Sci. 37:159–66.

Rao, G.N., et al. 1987. Utero-ovarian infection in aged B6C3F1 mice. Lab. Anim. Sci. 37:153–58.

Necrotizing Enteritis Associated with Clostridial Infection

Matsushita, S., and Matsumoto, T. 1986. Spontaneous necrotic enteritis in young RFM/Ms mice. Lab. Anim. 20:114–17.

Rickettsial Infection

Baker, H.J., et al. 1971. Research complications due to *Hemobartonella* and *Eperythrozoon* infections in experimental animals. Am. J. Pathol. 64:625–56.

General Bibliography

Caseboldt, D.B., et al. 1988. Prevalence rates of infectious agents among commercial breeding populations of rats and mice. Lab. Anim. Sci. 38:327–29.

Ganaway, J.R. 1982. Bacterial and mycotic diseases of the digestive system. In *The Mouse in Biomedical Research. II. Diseases*, ed. H.L. Foster, pp. 1–20. New York: Academic.

Shultz, L.D., and Sidman, C.L. 1987. Genetically determined murine models of immunodeficiency. Annu. Rev. Immunol. 5:367–403.

Sparrow, S. 1976. The microbiological and parasitological status of laboratory animals from accredited breeders in the United Kingdom. Lab. Anim. 10:365–73.

Williford, C.B., and Wagner, J.E. 1982. Bacterial and mycotic diseases of the integumentary system. In *The Mouse in Biomedical Research. II. Diseases*, ed. H.L. Foster, pp. 55–75. New York: Academic.

MYCOTIC INFECTIONS

Dermatomycosis

EPIZOOTIOLOGY AND PATHOGENESIS. *Trichophyton mentagrophytes* is the predominant dermatophyte among mice, although other organisms have been isolated. Two varieties of *T. mentagrophytes* have been recovered from mice: *T. mentagrophytes* var. *quinckeanum* and *T. mentagrophytes* var. *mentagrophytes*. *Trichophyton mentagrophytes* occurs worldwide. Its true prevalence in laboratory mouse colonies is unclear, since the great majority of infections are subclinical, especially among adult mice. The most severe manifestation, favus, is usually associated with *T. mentagrophytes* var. *quinckeanum*. Other predisposing factors probably play a role in this disease manifestation.

PATHOLOGY. Favus is characterized by yellowish, cuplike crusts or scutula on the muzzle, head, ears, face, tail, and extremities. These crusts are composed of epithelial debris, exudate, mycelia, and masses of arthrospores, with underlying dermatitis. Hair invasion has not been observed in mouse favus. Other lesions attributed to *T. mentagrophytes* include alopecia and focal crusts, particularly on the head, but the majority of infections are subclinical.

DIAGNOSIS. Lesions of favus are characteristic. Arthrospores and mycelia can be visualized with fungal strains and are Schiff-positive. *Differential diagnoses* must include other forms of dermatitis and alopecia. *Trichophyton* can be readily grown on Sabaroud's agar.

SIGNIFICANCE. *Trichophyton* appears to be nearly nonpathogenic in mice. Subclinical carriers are the norm among mice and have been shown to occur in high prevalence in some mouse populations. *Trichophyton* is nonselective in its host range and can infect other laboratory animals and humans.

BIBLIOGRAPHY FOR MYCOTIC INFECTIONS

Blank, F. Favus in mice. Can. J. Microbiol. 3:885–96.

Dolan, M.M., et al. 1958. Ringworm epizootics in laboratory mice and rats: Experimental and accidental transmission of infection. J. Invest. Dermatol. 30:23–25.

Donald, G.F., et al. *T. mentagrophytes* and *T. mentagrophytes* var. *quinckeanum* infections in South Australian mice. Aust. J. Dermatol. 7:133–40.

Mackenzie, D.W.R. 1961. *Trichophyton mentagrophytes* in mice: Infections of humans and incidence amongst laboratory animals. Sabouradia 1:178–82.

PARASITIC DISEASES

ECTOPARASITIC INFECTIONS

Mite Infections (Acariasis)

Mice are commonly infested with mixed populations of fur mites, including *Myobia musculi*, *Radfordia affinis*, and *Myocoptes musculinis* (Weisbroth 1982). Other mites that may be encountered include *Psorergates simplex* and *Trichoecius romboutsi*, which are either rare or overlooked in laboratory mice. *Myobia musculi* is the most clinically significant mouse mite.

EPIZOOTIOLOGY, LIFE CYCLES, AND PATHOGENESIS. *Myobia* infestations are widespread in mouse populations. Eggs are laid on hair shafts adjacent to the epidermis. Larvae hatch in 7–8 days, and egg-laying adults may evolve as early as 16 days after the eggs are laid. *Myobia* mites feed on skin secretions and interstitial fluid but apparently not on blood. This intimate feeding pattern is unique to *Myobia*, resulting in immunization of the host and a high frequency of immune-mediated paraphenomena. Transmission is by direct transfer of adult mites. Adults may migrate to sucklings from infested mothers at around 1 wk. Infestation corresponds with the appearance of pelage on the young mice. The presence of hair shafts is critical for successful colonization. Newborn mice are not susceptible until pelage erupts, and nude mice are resistant to experimental infestation. In newly infested mice, the mite numbers increase for the first 8–10 wk, but host immunity diminishes the populations to a point of equilibrium. This state of equilibrium persists for months to years, with cyclic variations corresponding to waves of egg hatches (Weisbroth et al. 1974) . Factors recognized to influence parasite load include strain of mouse, age, self-grooming, and mutual grooming. Impairment of grooming function by procedures such as hind toe amputation or Elizabethan collars result in increased parasite load (Weisbroth et al. 1974).

Adverse effects of *Myobia* infestation are highly varied and often difficult to prove with certainty. *Myobia* can sensitize the host, resulting in severe pruritis, with self-inflicted ulcerative lesions that often have secondary bacterial components. Sensitivity is genetically associated (Weisbroth et al. 1976; Friedman and Weisbroth 1975), and strains such as C57BL/6 are highly prone to hypersensitivity dermatitis. Lesion susceptibility is affected by a non-H-2-linked gene or gene combination shared by all C57BL background strains. Manifestations range from ruffled fur and alopecia in the head, neck, or shoulder regions to severe ulcerative dermatitis with marked pruritis, occasionally resulting in traumatic amputation of the ear pinnae. Self-trauma is an important factor in the development of these lesions (Fig. 1.50). Other adverse effects include reduced life span, weight loss, and infertility.

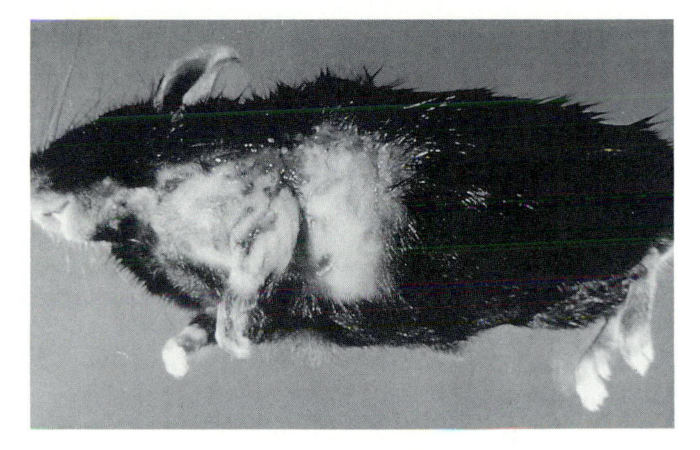

Fig. 1.50. Ulcerative dermatitis with denuding of hair associated with *Myobia musculi* infestation in a young male C57BL mouse. Hypersensitivity with severe pruritis may occur in mice of this strain.

Radfordia affinis is also common among mice, but its life cycle is not well studied. Its close resemblance to *Myobia* has led to confusion to its actual prevalence and effects. It does not induce overt disease like *Myobia* and often exists in mixed infestations.

Myocoptes musculinis is the most common of the mouse fur mites and usually exists as a mixed infestation with *Myobia*. *Myocoptes* is a surface dweller and feeds upon material in the superficial epidermis. Transmission occurs by close contact, and mites can be transferred within 1 wk of birth to newborns. *Myocoptes* tends to be peripatetic, spreading all over the body. In mixed infestations, *Myobia* tends to dominate the head and shoulder pelage, and *Myocoptes* can be found in the inguinal, ventral abdomen, and back. Clinical signs are mild, including patchy hair loss, erythema, and mild pruritis. *Trichoecius romboutsi* closely resembles *Myocoptes*, and its actual prevalence is therefore unknown.

Psorergates simplex was once common among laboratory mice but is now rare. It remains common in wild and pet mice. This small mite inhabits hair follicles, inciting the formation of comedones in the skin of the head, shoulders, and lumbar areas, and (less commonly) elsewhere. The life cycle of this mite is not known, but all of its life stages can be found within a single hair follicle. *Pathology*: Mice infested with fur mites display varying degrees of pruritis, with hyperactive, agitated behavior. Skin lesions include scruffiness, varying degrees of alopecia and dermatitis. Mice highly sensitized to *Myobia* develop self-inflicted ulcerative dermatitis, with secondary pyoderma. Pruritis may be intense in these mice, resulting in self-mutilation. Regional lymph nodes are often enlarged. *Psorergates* infection results in the formation of follicular cysts, which can be seen as white nodules on the subcutaneous side of the dermis. These are most common around the head and neck.

Microscopic examination of fur mite–induced skin lesions will reveal mild epidermal hyperplasia and hyperkeratosis, with variable dermal infiltrates of mononuclear leukocytes and mast cells. In ulcerated lesions, exudation and secondary bacterial colonization are often present, with underlying fibrovascular proliferation, mixed leukocyte infiltration, and hyperplasia of the adjacent intact epidermis. Mites may be present on the surface of the lesions, particularly in early, mild lesions (Fig. 1.51).

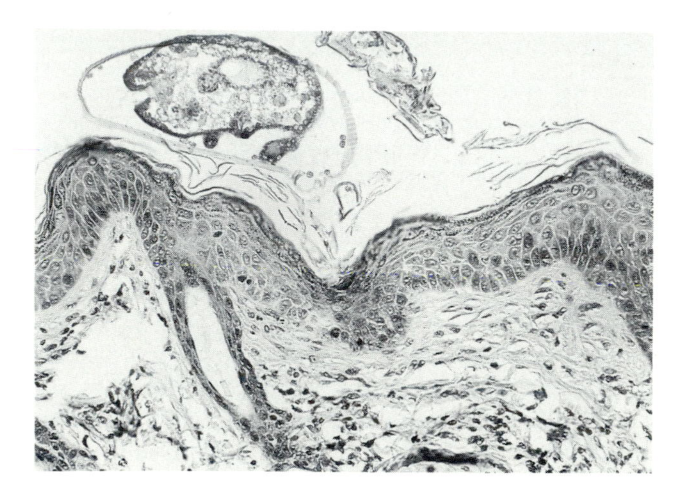

Fig. 1.51. Histological section of skin from a case of acariasis in C57BL mouse. There is epidermal hyperplasia, with mononuclear cell infiltration in the dermis. A mite is present on the surface of the lesion.

DIAGNOSIS. Fur mites can be demonstrated by placing the mouse or a portion of the skinned pelt (head and shoulder regions) in a Petri dish for 1 or more hr. The mites will climb up the hair shafts and can then be visualized under a dissecting microscope (Fig. 1.52), collected, and identified under a light microscope. Cellophane tape can also be applied to the hair and then placed on a glass slide. A number of points are important to consider in the diagnosis of acariasis. The number of mites will be greatest in young mice, before immune-mediated equilibrium has occurred. For this reason, the number of mites on mice with severe hypersensitivity-induced lesions may be exceedingly few. Infestations are usually mixed, so identification of a single mite will not reflect the true population. Finally, *Myobia* is the most clinically significant, but clinical signs are extremely variable, depending on host factors. One does not have to be a sophisticated acarologist to identify mouse fur mites. A few distinguishing features allow simple speciation. *Myobia* and *Radfordia* are remarkably similar in morphology, with slightly elongated bodies possessing bulges between their legs. If the second pair of legs is carefully examined, *Myobia* has a single terminal tarsal claw, while *Radfordia* has two of unequal length. *Myocoptes* is oval, with heavily chitinized, pigmented third and fourth legs and suckers on its tarsi. *Psorergates* infestation can be diagnosed by microscopic examination of cystic hair follicles or their contents. Follicles are filled with keratin squames and mites are present along the epidermis. *Differential diagnoses* for fur mite infestation includes pediculosis, trauma, bacterial dermatitis, dermatophytosis, hair chewing, and mechanically induced muzzle alopecia.

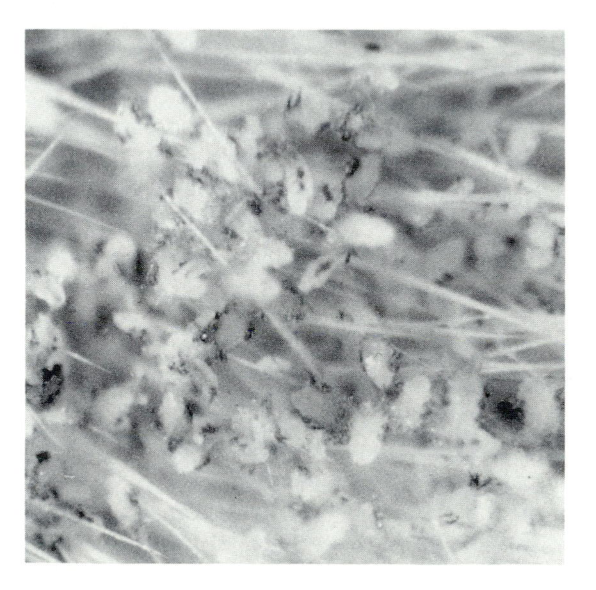

Fig. 1.52. *Mycoptes musculinus* infestation. Large numbers of mites are present on the pellage (dissecting microscope preparation). (Courtesy of J.P. Lautenslager)

SIGNIFICANCE. Complications include reduced life span, infertility, weight loss, modified immune responsiveness, and secondary amyloidosis. The association with amyloidosis is not absolute but is suspected.

Ornithonyssus bacoti **Infection.** *O. bacoti*, or tropical rat mite, is a blood-sucking mesostigmate mite that infests wild rats, as well as other species. It is nonselective in its host range. It inhabits its host only to feed and then hides in nearby niches. It causes intense pruritis, and its presence in a rodent population is often first manifest on human handlers. Its complete life cycle can occur within 2 wk, allowing massive infestation to occur within a short period of time. Because of its nonselective nature, *Ornithonyssus* has been found in laboratory mouse colonies.

Louse Infection (Pediculosis): *Polyplax serrata*

Polyplax serrata is a relatively common louse of wild mice, and at one time of infested laboratory mice throughout the world. It is now essentially nonexistent in laboratory mouse colonies. Eggs are attached to the base of hair shafts and hatch through an operculum at their top. Stage I nymphs can be found over the entire body, but the later 4 stages tend to prefer the anterior dorsum of the body. Eggs hatch within 5–6 days, and nymphs develop into adults within 1 wk. Transmission is by direct contact. Host immunity appears to develop, as parasite numbers diminish with time. As sucking lice, heavy infestations can result in anemia and debilitation. Bites are pruritic, resulting in intense scratching and dermatitis. *Polyplax* once played a significant role as a vector of *Eperythrozoon coccoides*.

ENDOPARASITIC INFECTIONS

Protozoal Infections

Spironucleus muris, *Giardia muris*, and *Cryptosporidium muris* are examples of protozoal organisms found in the intestinal tract that are considered to be relatively nonpathogenic. Under some circumstances, they may represent opportunistic infections associated with overt disease. *Eimeria muris* is more overtly pathogenic, but it is rare in laboratory mice. Another coccidian, *Klossiella muris*, is also rare in laboratory mice.

Spironucleus (Hexamita) muris Infection

EPIZOOTIOLOGY AND PATHOGENESIS. This flagellated protozoan parasite is frequently present in the alimentary tract of clinically normal mice. In surveys of commercial suppliers, over 60% of premises studied were positive for *Spironucleus* (Lindsey 1986; Sparrow 1976). Other species commonly infected include rats and hamsters. Both *Spironucleus* and *Giardia* are transmissible from hamsters to mice. The organism is usually associated with clinical disease only in young mice, and often there are identifiable predisposing factors. The organism colonizes in the small intestine, primarily in the crypts in the duodenum. *Spironucleus* divides by longitudinal fission. Animals become infected by the ingestion of trophozoites or cysts. Clinical manifestations of hexamitiasis are usually associated with immunosuppression or environmental stress, and there may be a concomitant infection with other pathogens (Meshorer 1969). Animals 3–6 wk of age are particularly at risk. Clinical signs include depression, weight loss, dehydration, hunched posture, diarrhea, and mortality rates of up to 50% in young animals (Flatt et al. 1978).

PATHOLOGY. At necropsy, the small intestine is distended with dark red to brown watery contents and gas. In tissue sections of small intestine examined microscopically from animals with the acute form of the disease, there may be edema of the lamina propria, with mild leukocytic infiltration, neutrophils predominating. Crypts and intervillous spaces are distended with elongated, pear-shaped trophozoites (Fig. 1.53). Organisms may also be present between enterocytes and within the lamina propria. In the chronic form of the disease, the cellular infiltrate consists primarily of lymphocytes and plasma cells. Scattered duodenal crypts may be markedly dilated and contain leukocytes and cellular debris. The trophozoites stain well with the PAS staining technique, while the organism is poorly delineated in H & E–stained preparations.

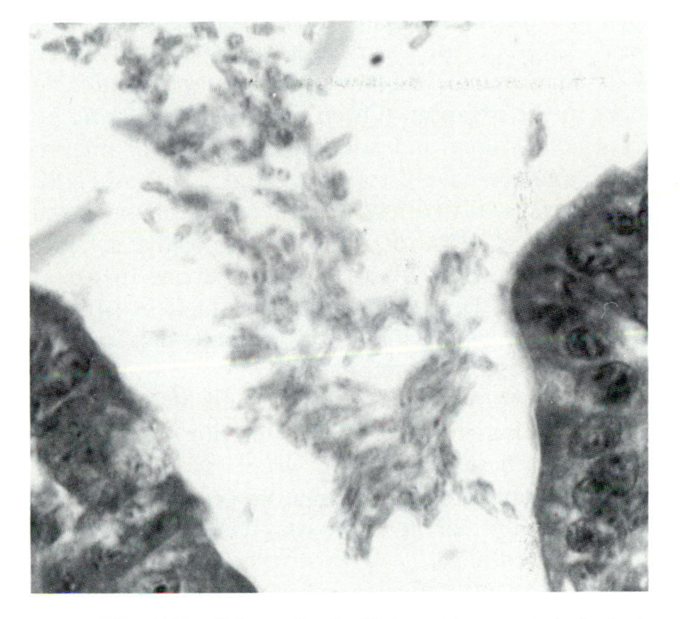

Fig. 1.53. Spironucleosis (*Spironucleus muris* infection) in young laboratory mouse with diarrhea. In this section of duodenum, large numbers of trophozoites are present on the mucosal surface. Note the hyperplasia of enterocytes lining villi.

DIAGNOSIS. Trophozoites with fast straight or zigzag movements can be visualized microscopically on direct wet mount smears prepared from small intestine. Typical banded "Easter egg" cysts may be present in the intestinal contents.

SIGNIFICANCE. It is frequently difficult to determine the significance of *Spironucleus* infections in this species, since infected animals are normally asymptomatic. Investigations should include a search for underlying disease or other predisposing factors. Copathogens may include concurrent viral infections, such as enterotropic murine coronavirus. Complications in research associated with heavy infestations with *Spironucleus* include impaired immune response and macrophage function in euthymic mice (Ruitenberg and Kruyt 1975), chronic doubling of enterocyte turnover in the small intestine (MacDonald and Ferguson 1978), and shortened life span in athymic mice.

Interspecies transmission has been demonstrated between hamsters and mice (Sebesteny 1979), but not to rats (Schagemann et al. 1990).

Giardiasis

EPIZOOTIOLOGY AND PATHOGENESIS. *Giardia muris* is a flagellate that normally resides primarily in the lumen of the duodenum. Mice, hamsters, rats, and other rodents are natural hosts. In the naturally occurring disease, trophozoites proliferate in the small intestine and adhere to the microvilli of enterocytes near the base of villi by means of concave sucking disks. Organisms also wedge in furrows on the epithelial surface and lodge in mucus overlying intestinal epithelium. Clearance of the parasite has been associated with intraluminal lymphocyte migration and attachment to the organism (Owen et al. 1979). *Giardia* infects both young and adult mice. In heavy infestations, animals have a rough hair coat and distended abdomen, usually with no evidence of diarrhea. Based on selected surveys, the incidence of positive colonies may be up to 60% (Lindsey 1986; Sparrow 1976).

PATHOLOGY. At necropsy, the small intestine is usually distended, with yellow to white watery contents. Microscopic examination of tissue sections of the small intestine reveals trophozoites that are pear-shaped, with a broadly rounded anterior sucking disk. There may be a reduction in crypt:villus ratio, with increased numbers of inflammatory cells in the lamina propria. Intestinal invasion may occur in immunocompromised mice.

DIAGNOSIS. The demonstration of typical trophozoites on wet mount preparations of feces and on cysts are the basis for the diagnosis of giardiasis.

SIGNIFICANCE. Infections are frequently subclinical in the absence of other predisposing factors. Complications may include significant morbidity and mortality in athymic nude or thymectomized mice (Hsu 1982). Cytokinetic studies have demonstrated a marked increase in cryptal enterocyte turnover in the small intestine of infected mice (MacDonald and Ferguson 1978). Suppression of the immune response to sheep erythrocytes has been observed in infected mice (Belosevic et al. 1985).

Cryptosporidiosis. *Cryptosporidium muris* occurs primarily on the surface of the gastric mucosa of mice. It is relatively nonpathogenic in this species. Similarly, *C. parvum* is a marginally pathogenic inhabitant of the small intestine. The prevalence of infection is not known, but it can be associated with enteritis, probably secondary to viral infections in young mice. It has also been observed to ascend the biliary tract in athymic nude mice, resulting in chronic cholangitis and portal hepatitis.

Coccidiosis. Intestinal coccidiosis due to *Eimeria falciformis* rarely occurs in well-managed facilities. There have been reports of outbreaks of coccidiosis in mice due to more than one species of *Eimeria* (Haberkorn et al. 1983). Intestinal coccidiosis is very common among wild mice, where it causes marked colitis (Fig. 1.54) and runting in juvenile animals. Oocysts can be found in the mucosa of older mice without discernible lesions.

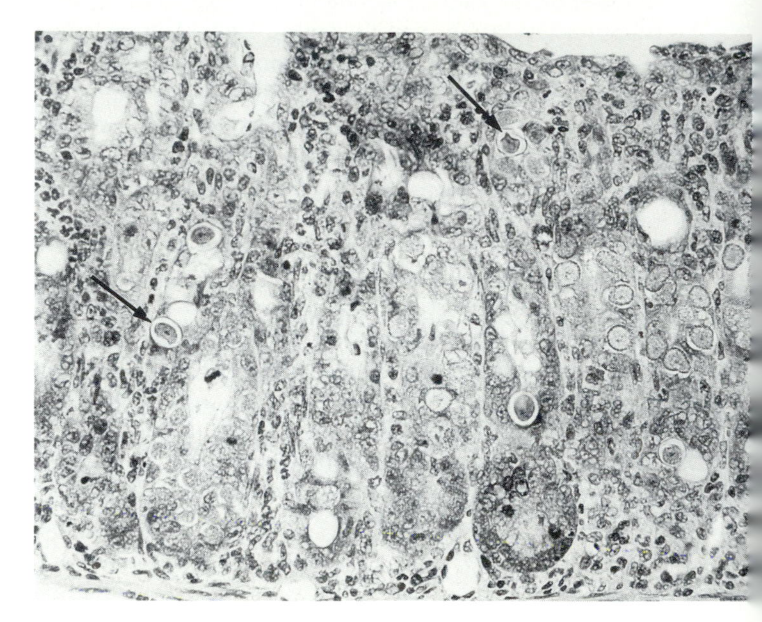

Fig. 1.54. Section of colon from case of intestinal coccidiosis in laboratory mouse. There is a marked hyperplastic colitis. Sloughed cells, leukocytes, and a few developing oocysts (*arrows*) are present on the surface of the gut. Macrogametocytes are evident in some enterocytes.

Renal coccidiosis due to *Klossiella muris* is rarely observed as an incidental finding in wild and laboratory mice. Infection probably occurs by the ingestion of sporocysts, with hematogenous spread to glomerular capillaries and schizogony. Gametogeny and sporogony occur in epithelial cells lining convoluted tubules. On microscopic examination, lesions are usually confined to the convoluted tubules. Organisms appear as eosinophilic spherical structures within the cytoplasm of epithelial cells, with minimal inflammatory response (Fig. 1.55).

SIGNIFICANCE. *Klossiella* infections are rarely detected in laboratory mice and are only occasionally observed in wild house mice. They are considered to be an incidental finding, but if present, they indicate the need for improved sanitation practices. There have been anecdotal reports of transmission of *K. muris* to guinea pigs (Taylor et al. 1979).

***Pneumocystis carinii* Infection.** *P. carinii* is a microorganism that is widespread in the rodent popula-

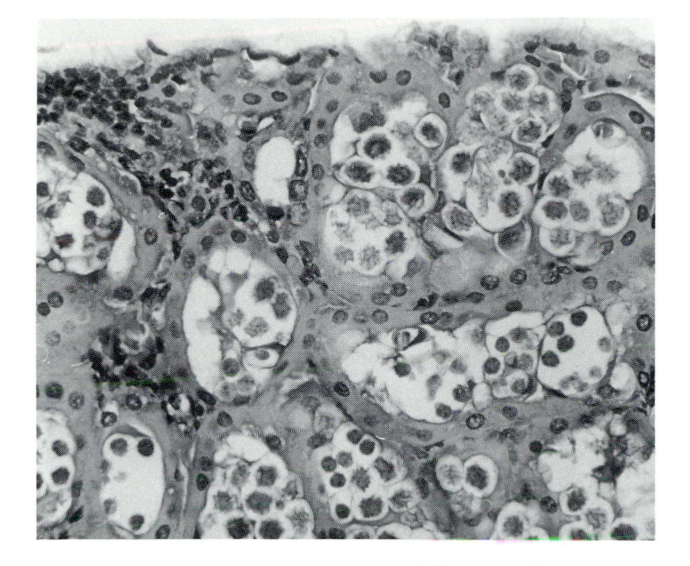

Fig. 1.55. Kidney from mouse with renal coccidiosis due to *Klossiella muris*. Large numbers of sporocysts are present in epithelial cells of renal tubules.

tion. The ongoing issue of whether it should be classified as a protozoan or a fungus has not been completely resolved. Pneumocystis pneumonia is a major complication in AIDS patients during the terminal stages of the disease and is emerging as a significant natural pathogen in athymic and SCID mice.

EPIZOOTIOLOGY AND PATHOGENESIS. *P. carinii* may exist as a saprophytic infection in the lungs of several species, including mice, rats, and humans. Normally, infected animals are asymptomatic. However, experimental immunosuppression of mice may result in pneumocystis pneumonia (Walzer et al. 1979). In addition, spontaneous enzootics of pneumocystis pneumonia have been described in athymic nu/nu mice (Weir et al. 1986). Similarly, there may be a high mortality rate due to pneumocystis infection in older SCID mice (Shultz and Sidman 1987). Clinical signs described include labored respiration, wasting, hunched posture, and dry, scaly skin.

PATHOLOGY. Affected animals are thin, with dry, scaly skin. Lungs collapse poorly and have a rubbery consistency, with pale, patchy areas of consolidation. In the lungs in spontaneous cases of *P. carinii* infection, microscopic examination reveals an interstitial alveolitis, with proteinaceous exudation into the alveolar septa. There is marked thickening of alveolar septa and infiltration with lymphoreticular cells (Fig. 1.56). Finely vacuolated, eosinophilic material and alveolar macrophages are scattered in affected alveoli. In tissue sections from affected lung stained with the PAS or methenamine silver procedures, numerous rounded and irregularly flattened cyst forms 3–5 μm in diameter are present in reactive areas (Fig. 1.57).

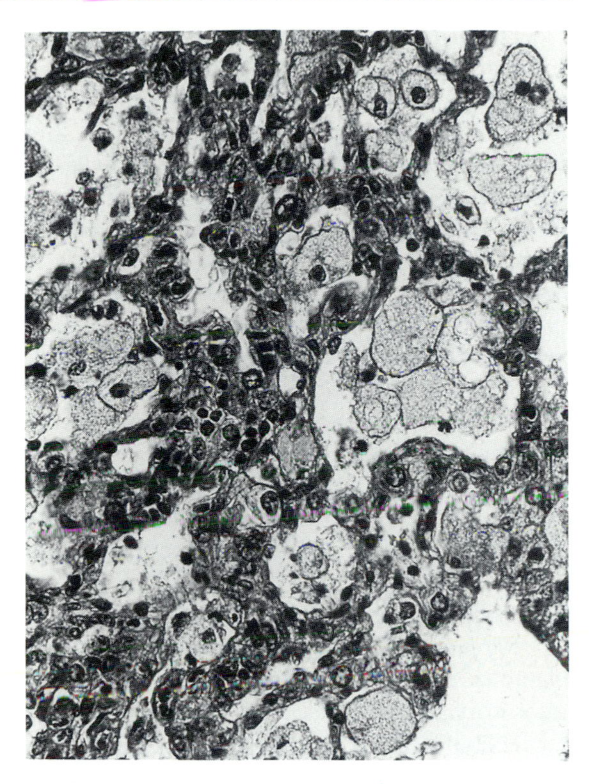

Fig. 1.56. Section of lung from athymic mouse with spontaneous pneumocystis pneumonitis. Alveolar septa are hypercellular, with mononuclear cell infiltration. Foamy proteinaceous exudate and alveolar macrophages are present in many alveoli.

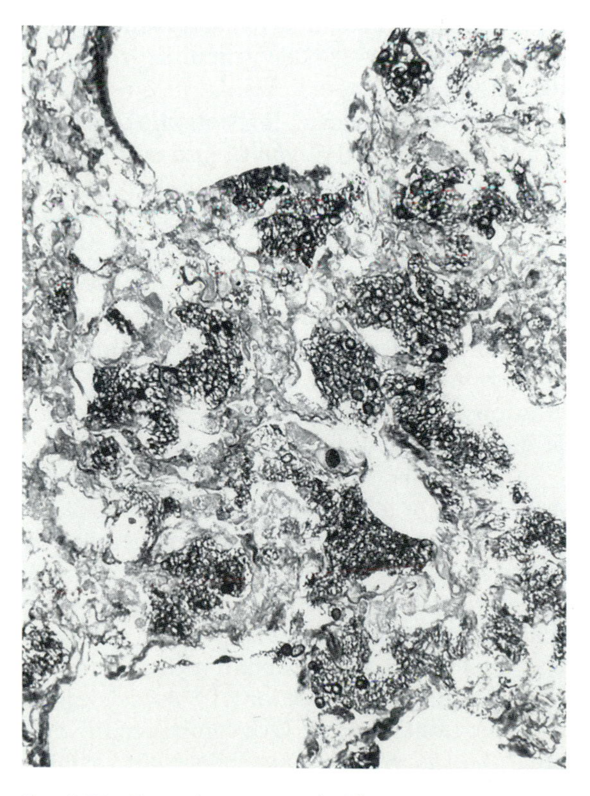

Fig. 1.57. Lung from mouse in Fig. 1.56, stained with methenamine silver method. Large numbers of pneumocystic cysts are present in alveoli.

DIAGNOSIS. The history of experimental procedures leading to immunosuppression or the presence of disease in genetically immunodeficient mice are critical predisposing factors. Organisms should be demonstrated in the typical foamy alveolar exudate, using the methenamine silver or PAS staining procedure. The organisms are best visualized with the methenamine silver staining method. *Differential diagnoses* include viral pneumonitis, particularly Sendai virus and pneumonia virus of mice, and pulmonary edema secondary to congestive heart failure.

SIGNIFICANCE. *Pneumocystis* infections pose a major threat in colonies of immunodeficient mice, with high mortality. Significant antigenic differences have been observed in *Pneumocystis* recovered from different species (Walzer et al. 1989). The issue of possible interspecies transmission has not been resolved.

Helminth Infections: Pinworms (Oxyuriasis)

Syphacia obvelata and *Aspicularis tetraptera* are examples of pinworms that commonly occur in the laboratory mouse (Wescott 1982; Taffs 1976).

EPIZOOTIOLOGY AND LIFE CYCLE. Pinworm infections remain a relatively common problem. In one survey, over 60% of commercial facilities evaluated were positive for *Syphacia* (Lindsey 1986). The life cycle of *Syphacia* is direct and is completed in approximately 12–15 days. Following the ingestion of eggs, larvae emerge and migrate to the cecum. They develop into adults and mate, and females then migrate to the perianal region for egg deposition. Eggs become infective within a few hours. Young mice are particularly susceptible to pinworm infestation. Dual infections with *Aspicularis* and *Syphacia* also occur. The life cycle of *A. tetraptera* is direct and takes approximately 23–25 days. Mature females lay eggs in the terminal colon, which are then passed in the feces. Eggs require incubation at room temperature for 6–7 days in order to become infective and can survive for weeks outside the host. Most mice with pinworm infections are asymptomatic. Clinical signs associated with heavy infestations include rectal prolapse, intussusception, fecal impaction, and diarrhea.

DIAGNOSIS. Visualization and identification of adult worms present in the cecum or colon at necropsy is a standard procedure. These nematodes are frequently found in tissue sections of cecum and colon (Fig. 1.58). Identification of worm eggs will require fecal flotation for *Aspicularis* spp., but cellophane tape applied to the perianal region is the recommended method of collection of *Syphacia* eggs for microscopic identification. Ova can be readily differentiated. *Aspicularis* ova are bilaterally symmetrical, while *Syphacia* ova are banana-shaped.

SIGNIFICANCE. Complications attributed to pinworm infestations include decreased weight gains, behavioral changes, and altered immune responses (Wescott 1982). Athymic nude mice are especially susceptible to heavy infestations. These ascarids are readily treated with anthelmintics such as piperazine preparations, but maintaining colonies free from the parasite is a difficult matter. For additional details on the biology and identification of these parasites, consult Flynn (1973) and Wescott (1982).

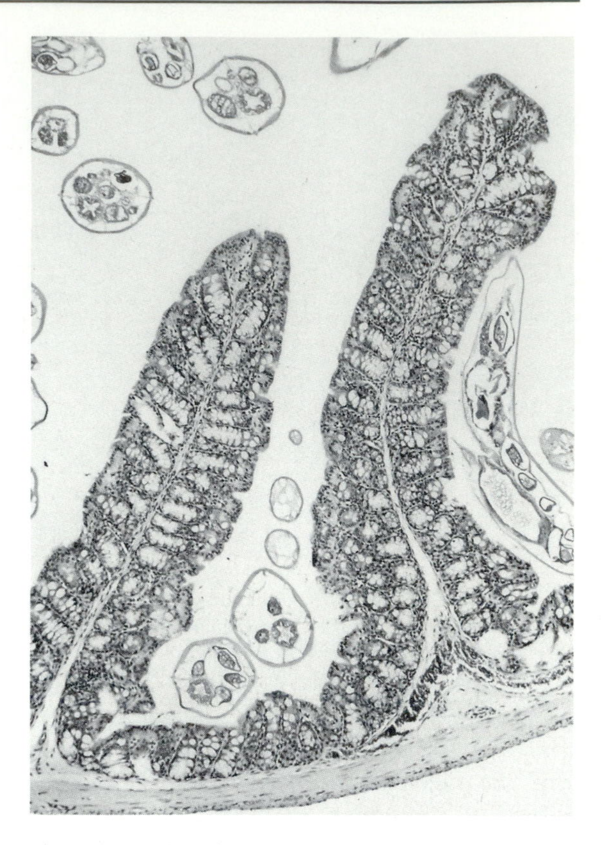

Fig. 1.58. Adult pinworms (*Syphacia obvelata*) present in the cecum of adult mouse.

Helminth Infections: Tapeworms

***HYMENOLEPIS* INFECTIONS.** Wild and laboratory rodents can be infected with three separate species of *Hymenolepis*, including *H. nana* (dwarf tapeworm), *H. diminuta*, and *H. microstoma*. The latter two species are no longer found in laboratory mice, since husbandry conditions preclude the necessary intermediate athropod host.

EPIZOOTIOLOGY, LIFE CYCLE, AND PATHOGENESIS. A variety of species of laboratory animals are susceptible to infection with the dwarf tapeworm, including mice, rats, and hamsters. Its wide host range also includes humans. Husbandry conditions have essentially eliminated *H. diminuta* and *H. microstoma* and have greatly reduced the prevalence of *H. nana* in laboratory mouse populations. These tapeworms all utilize arthropods as intermediate hosts, but *H. nana* can also have a direct life cycle in which onchospheres penetrate the mucosa and develop into the cercocystis stage, subsequently emerging into the lumen as adults. The entire life cycle can occur in the intestine within 20–30 days. Thus, superinfections can occur in the absence of an intermediate host. Immunity develops to these worms, with reduction of parasite numbers over time. Clinical signs associated with heavy infestations include poor weight gains and diarrhea.

PATHOLOGY. *H. nana* adults are threadlike worms in the small intestine. Microscopic findings include the presence of cysticeri within the lamina propria and adults with prominent serrated edges in the lumen. Occasionally, cysticeri can be found in the mesenteric lymph nodes. *H. diminuta* adults are much larger and intermediate forms do not appear in the mucosa. *H. microstoma* adults are the size of *H. diminuta* but often lodge in the bile or pancreatic ducts, inciting inflammatory and atrophic changes in the pancreas and cholangitis.

DIAGNOSIS. The worms can be identified grossly. Only *H. nana* and *H. diminuta* are likely to be encountered in laboratory animals. *H. nana* is typically threadlike (1 mm wide), while the other species are much larger (4 mm wide). The *H. nana* scolex possesses hooks and the ova have polar filaments while those of *H. diminuta* do not.

SIGNIFICANCE. In addition to transient damage to the intestinal mucosa and weight loss, there is danger of interspecies spread, including human infections.

***Taenia taeniaformis* Infection.** Mice can serve as the intermediate host for this cat tapeworm. The larval form, *Cysticercus fasciolaris*, consists of a scolex and segments within a cyst and thus resembles an adult tapeworm (Fig. 1.59). The liver is the most frequent location for cysticerci. The source of the parasite is usually via feed contaminated with cat feces.

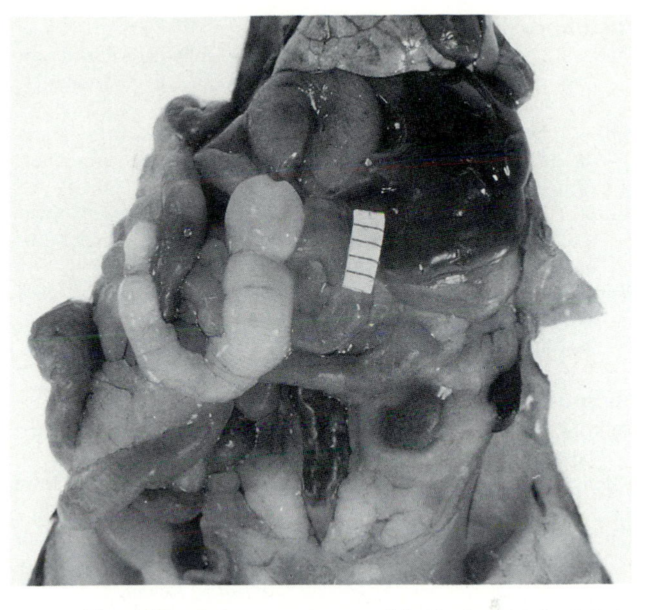

Fig. 1.59. Laboratory mouse with *Cysticercus fasciolaris* infestation of the liver. The lesion has been opened. Note the scolex and identifiable segments of the parasite.

Bibliography for Parasitic Diseases

Ectoparasitic Infections

Csiza, C.K., and McMartin, D.N. 1976. Apparent acaridal dermatitis in a C57BL/6Nya mouse colony. Lab. Anim. Sci. 26:781–87.

Dawson, D.D., et al. 1986. Genetic control of susceptibility to mite-associated ulcerative dermatitis. Lab. Anim. Sci. 36:262–67.

French, A.W. 1987. Elimination of *Ornithonyssus bacoti* in a colony of aging mice. Lab. Anim. Sci. 37:670–72.

Friedman, S., and Weisbroth, S.H. 1975. The parasitic ecology of the rodent mite, *Myobia musculi*. II. Genetic factors. Lab. Anim. Sci. 25:440–45.

Weisbroth, S.H. 1982. Arthropods. In *The Mouse in Biomedical Research. II. Diseases*, ed. H.L. Foster et al., pp 388–90. New York: Academic.

Weisbroth, S.H., et al. 1976. The parasitic ecology of the rodent mite *Myobia musculi*. III. Lesions in certain host strains. Lab. Anim. Sci. 26:725–35.

Weisbroth, S.H., et al. 1974. The parasitic ecology of the rodent mite *Myobia musculi*. I. Grooming factors. Lab. Anim. Sci. 24:510–16.

Wharton, G.W. 1940. Life cycle and feeding habits of *Myobia musculi*. J. Parasitol. 40:29.

Spironucleus (Hexamita) muris Infection

Flatt, R.E., et al. 1978. Hexamitiasis in a laboratory mouse colony. Lab. Anim. Sci. 28:62–65.

Lindsey 1986 (*see* General Bibliography for Protozoal Infections).

MacDonald and Ferguson 1978 (*see* General Bibliography for Protozoal Infections).

Meshorer, A. 1969. Hexamitiasis in laboratory mice. Lab. Anim. Care 19:33–37.

Ruitenberg, E.J., and Kruyt, B.C. 1975. Effect of intestinal flagellates on immune response in mice. Abstract. Parasitology 71:xxx.

Sebesteny, A. 1979. Transmission of*Spironucleus* and *Giardia* spp. and some non-pathogenic intestinal protozoa from infested hamsters to mice. Lab. Anim. 13: 189–91.

Shagemann, G., et al. 1990. Host specificity of cloned *Spironucleus muris* in laboratory rodents. Lab. Anim. 24: 234–39.

Sparrow 1976 (*see* General Bibliography for Protozoal Infections).

Giardiasis

Belosevic, M., et al. 1985. Suppression of primary antibody response to sheep erythrocytes in susceptible and resistant mice infected with *Giardia muris*. Infect. Immunol. 47:21–25.

Hsu 1982 (*see* General Bibliography for Protozoal Infections).

Lindsey 1986 (*see* General Bibliography for Protozoal Infections).

MacDonald and Ferguson 1978 (*see* General Bibliography for Protozoal Infections).

Owen, R.L., et al. 1979. Ultrastructural observations on giardiasis in a murine model. I. Intestinal distribution, attachment, and relationship to the immune system of *Giardia muris*. Gastroenterology 76:757–69.

Sparrow 1976 (*see* General Bibliography for Protozoal Infections).

Coccidiosis

Haberkorn, A., et al. 1983. Control of an outbreak of coccidiosis in a closed colony. Lab. Anim. 17:59–64.

Taylor, J.L., et al. 1979. *Klossiella* parasites of animals: A literature review. Vet. Parasitol. 5:137–44.

Pneumocystis carinii Infection

Shultz, L.D., and Sidman, C.L. 1987. Genetically determined murine models of immunodeficiency. Annu. Rev. Immunol. 5:367–403.

Walzer, P.D., et al. 1989. Outbreaks of *Pneumocystis carinii* pneumonia in colonies of immunodeficient mice. Infect. Immunol. 57:62–70.

Walzer, P.D., et al. 1979. Experimental *Pneumocystis carinii* pneumonia in different strains of cortisonized mice. Infect. Immunol. 24:939–47.

Weir, E.C., et al. 1986. Spontaneous wasting disease in nude mice associated with *Pneumocystis carinii* infection. Lab. Anim. Sci. 36:140–44.

Helminth Infections

Balk, M.W., and Jones, S.R. 1970. Hepatic cysticercosis in a mouse colony. J. Am. Vet. Med. Assoc. 157:678–79.

Flynn, R.J. 1973. *Parasites of Laboratory Animals.* Ames: Iowa State University Press.

Lindsey, J.R. 1986. Prevalence of viral and mycoplasmal infections in laboratory rodents. In *Viral and Mycoplasmal Infections of Laboratory Rodents: Effects on Biomedical Research*, ed. P.N. Bhatt et al., pp. 801–8. New York: Academic.

Sparrow, S. 1976. The microbiological and parasitological status of laboratory animals from accredited breeders in the United Kingdom. Lab. Anim. 10:365–73.

Taffs, L.F. 1976. Pinworm infections in laboratory rodents: A review. Lab. Anim. 10:1–13.

Wescott, R.B. 1982. Helminths. In *The Mouse in Biomedical Research. II. Diseases*, ed. H.L. Foster et al., pp. 373–84. New York: Academic.

General Bibliography for Protozoal Infections

Hsu, C-K. 1982. Protozoa. In *The Mouse in Biomedical Research. II. Diseases*, ed. H.L. Foster et al., pp. 359–72. New York: Academic.

Lindsey, J.R. 1986. Prevalence of viral and mycoplasmal infections in laboratory rodents. In *Viral and Mycoplasmal Infections of Laboratory Rodents: Effects on Biomedical Research*, ed. P.N. Bhatt et al., pp. 801–8. New York: Academic.

MacDonald, T.T., and Ferguson, A. 1978. Small intestinal epithelial cell kinetics and protozoal infection in mice. Gastroenterology 74:496–500.

Sparrow, S. 1976. The microbiological and parasitological status of laboratory animals from accredited breeders in the United Kingdom. Lab. Anim. 10:365–73.

NUTRITIONAL AND METABOLIC DISORDERS

Amyloidosis

Amyloid [from the Greek *amylon,* "starch"] was so named by Virchow because it stained with iodine similar to that seen with cellulose. Amyloidosis is an important disease of laboratory mice, both as a spontaneously occurring, life-limiting disease and as an experimentally induced disease.

EPIZOOTIOLOGY AND PATHOGENESIS. Amyloid is a chemically diverse family of insoluble proteins that are deposited in tissues but have in common a biophysical polymerized conformation known as the β-pleated sheet. It is now known that there are three systemic forms of amyloid. Primary and myeloma-associated amyloid contains immunoglobulin light chains or fragments (called amyloid AL, for amyloid, light chain) (Cohen et al. 1983). Secondary amyloid contains by-products of the acute phase response and occurs in secondary inflammatory processes. Local tissue injury elicits a complex of events in which macrophages release monokines, including interleukin 1 and tumor necrosis factor, which in turn stimulate serum amyloid A (SAA) synthesis in liver. SAAs are polymorphic apoproteins isolated with serum high-density lipoproteins known as acute phase reactants and are a precursor to amyloid A (AA), which is deposited extracellularly in tissues. A third type of systemic amyloid, found in humans as a hereditary trait, is composed of prealbumin. In addition, a number of localized forms of amyloidosis occur, such as in endocrine tumors, ovaries, and the brain (in Alzheimer's disease), each with differing composition. All amyloids possess amyloid P (protein) component, which is a plasma glycoprotein with homology to the acute phase protein, C-reactive protein. It is not known why these native biological products are not catabolized.

Spontaneous amyloidosis is a common event in certain strains of aging laboratory and wild mice (Conner et al. 1983; Dunn 1967). Primary amyloidosis occurs with high prevalence and at a relatively young age in some strains of mice such as A and in SJL and with high prevalence but later onset in strains such as C57BL mice; it can be extraordinarily rare in other strains, such as BALB/c and C3H mice. The patterns of tissue deposition also vary somewhat, depending upon mouse genotype. Secondary amyloidosis can also occur spontaneously in laboratory mice and is related to chronic inflammatory diseases and acariasis. The line of distinction between spontaneous primary and secondary amyloidosis is vague on a morphological basis, but the spleen and liver are only mildly affected in primary amyloidosis and are usually the most severely affected in secondary amyloidosis. The prevalence of amyloidosis can be affected by stress (fighting) and ectoparasitism. Experimental secondary amyloidosis can be readily induced in a variety of laboratory mouse strains with casein injections. The order of susceptibility (in decreasing order) among common mouse strains is CBA, C57BL/6, outbred Swiss, C3H/Hc, BALB/c, and SWR. Localized forms of amyloidosis can also be found in mice. Tumor-associated amyloid can be found, even in the low-amyloid BALB/c strain, and ovarian corpora luteal amyloidosis can be found with frequency in CBA and DBA mice in the absence of systemic disease.

PATHOLOGY. Amyloid has a characteristic hypocellular eosinophilic appearance in H & E–stained sections. When stained with Congo red and subjected to polarized light, amyloid is birefringent. Amyloid deposition occurs in renal glomeruli (Fig. 1.60), renal interstitium, lamina propria of the intestine (Fig. 1.61), myocardium (Fig. 1.62), nasal submucosa (Fig. 1.63), parotid salivary gland, thyroid gland, adrenal cortex, myocardium, perifollicular areas of the spleen, pulmonary alveolar septa, periportal tissue of the liver, tongue, testes, ovary, myometrium, aorta, pancreas, and other tissues. Amyloidosis is often associated with cardiac atrial thrombosis with left- or right-sided congestive heart failure. The mechanism for this association is unknown. Mice with amyloid deposition in the renal medullary interstitium can develop papillary necrosis. Healed lesions give the illusion of hydronephrosis.

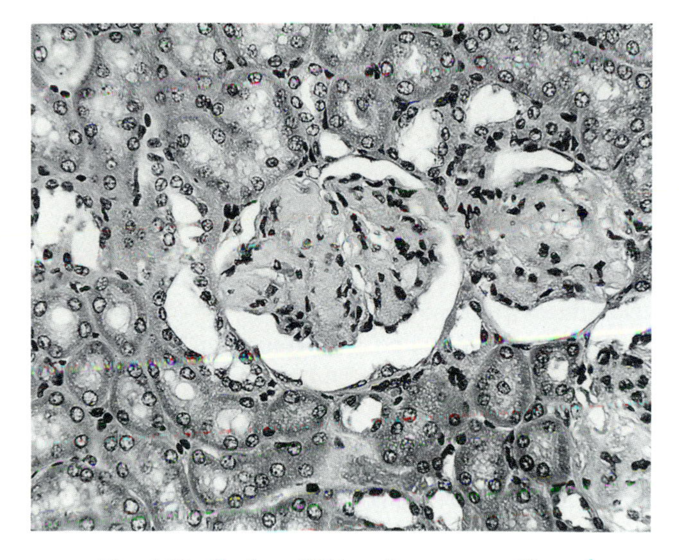

Fig. 1.60. Section of kidney from mouse with renal amyloidosis. There is deposition of amyloid on glomerular basement membranes, with partial obliteration of the normal architecture.

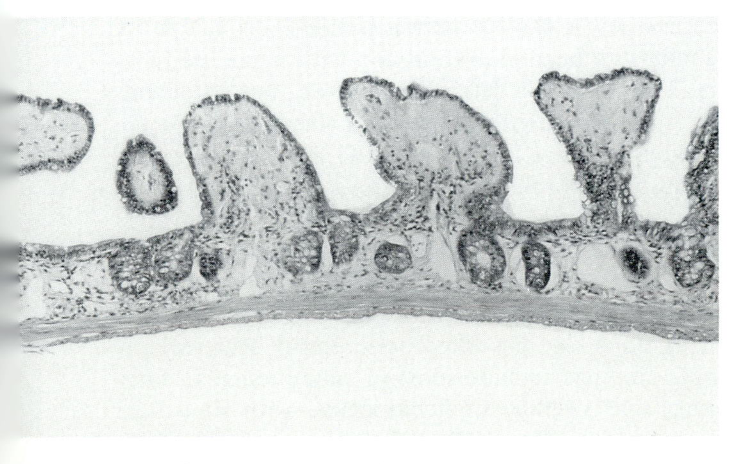

Fig. 1.61. Section of ileum from mouse with intestinal amyloidosis. There is marked deposition of amorphous material in the lamina propria.

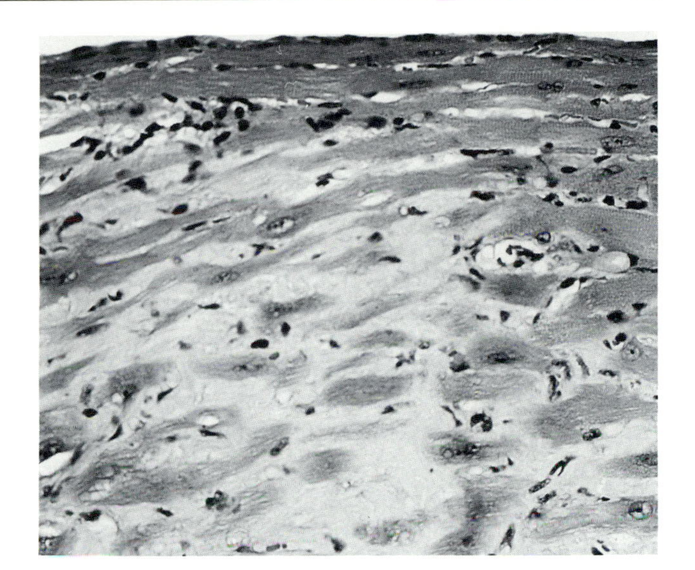

Fig. 1.62. Left ventricle from aged mouse with cardiac amyloidosis. Note the deposition of amorphous material between myofibers.

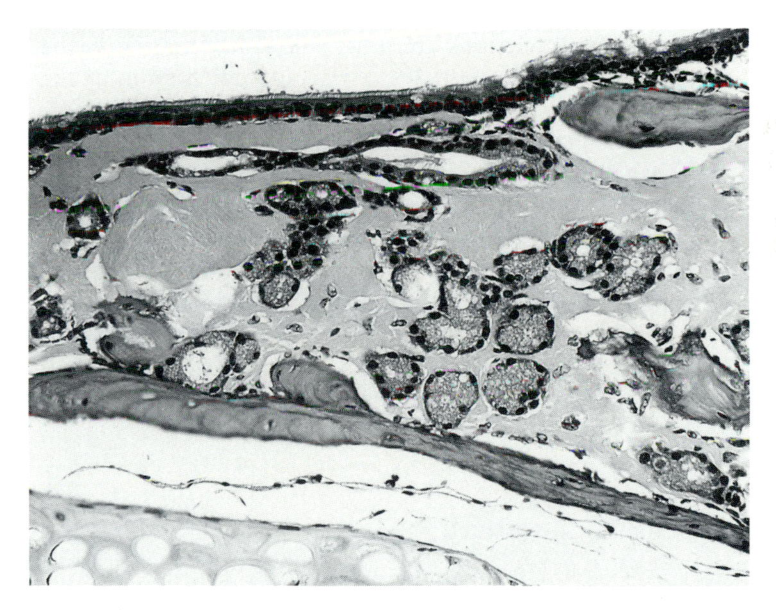

Fig. 1.63. Longitudinal section of turbinates from aged mouse with marked nasal amyloidosis. There is extensive deposition of amyloid around identifiable submucosal glands.

DIAGNOSIS. Diagnosis is confirmed by the typical appearance of amyloid and its staining characteristics. *Differential diagnoses* must include age-related glomerular disease, glomerulonephritis, hydronephrosis, and spontaneous cardiac atrial thrombosis.

SIGNIFICANCE. Amyloidosis is a major life-limiting disease in aging mice and can be exacerbated by stress and other disease states.

Soft-tissue Calcification (Cardiac Calcinosis/ Myocardial Calcification)

EPIZOOTIOLOGY AND PATHOGENESIS. Spontaneous mineralization/calcification of the heart and other soft tissues are a common necropsy finding in strains of mice such as BALB/c, C3H, and DBA mice, most notably in the DBA strain. For example, in the DBA/2 strain, myocardial lesions may be evident at necropsy in up to 100% of males and females by 10 wk of age (Yamate et al. 1987). In DBA mice, calcification may be observed as early as 3 wk of age, with increased incidence in older animals. A variety of factors have been implicated in this condition, including environmental or dietary change, concomitant disease, and elevated levels of corticosteroids (Yamate et al. 1987). Calcification may occur in a variety of tissues, particularly in DBA mice. Focal mineralization most frequently occurs in the myocardium, muscles of the tongue, cornea, and aorta. Calcified lesions at these sites are usually present as an incidental finding at necropsy.

PATHOLOGY. There may be chalky linear streaks evident on the heart, particularly on the epicardium of the right ventricles. On microscopic examination, myocardial lesions are most frequently present in the right and left ventricles, atria, and epicardium of the right ventricle (Fig. 1.64). Changes vary from single mineralized fibers (grade 1) to grade 3 lesions, which are characterized by extensive linear calcification. In recent lesions, there may be interstitial edema. In lesions interpreted to be of some duration, frequently there is concurrent fibrosis and mononuclear cell infiltration. Foci of calcification in the tongue, when present, are often concentrated in the musculature adjacent to the lamina propria, frequently with concurrent granulomatous inflammatory response in the adjacent regions. Lesions may be distributed anywhere along the area from the apex to the root of the tongue. Ulceration of the epithelium overlying affected areas is an infrequent finding. Foci of calcification, when present in the aorta, are characterized by mineralization of the elastic lamina and smooth muscle of the vessel wall. Corneal lesions are characterized by degeneration of Bowman's membrane, with extension into the adjacent collagenous fibers of the corneal stroma.

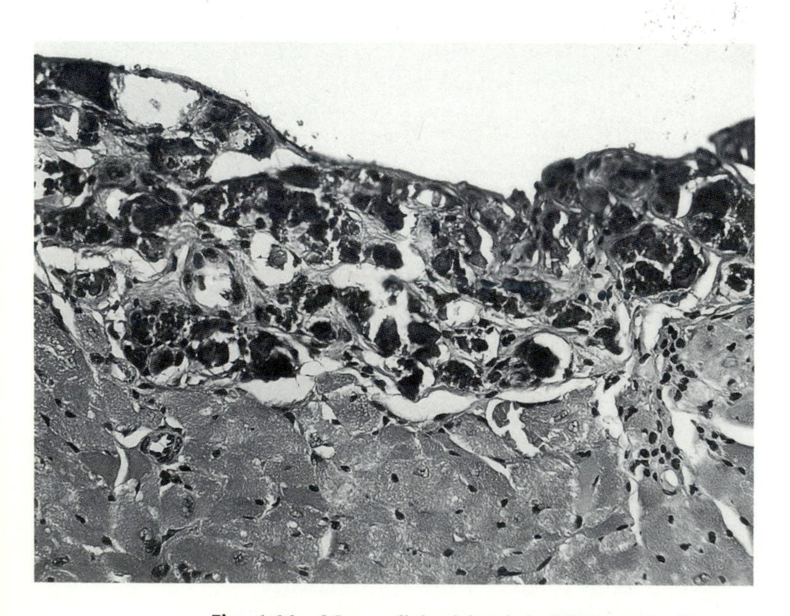

Fig. 1.64. Myocardial calcinosis in DBA mouse, illustrating mineralization and fibrous tissue proliferation in epicardial region.

DIAGNOSIS. The strain of mouse, the demonstration of calcium in lesions by the appropriate staining procedure (e.g., alizarin red), and the nature and distribution of lesions are sufficient to confirm the diagnosis.

SIGNIFICANCE. Soft-tissue calcification commonly occurs in strains such as DBA mice. It is usually an incidental finding at necropsy, although it is likely that there is some impairment of cardiac function, particularly in animals with extensive myocardial lesions.

Reye's-like Syndrome

Reye's syndrome, an important cause of morbidity and mortality among human infants and children, is characterized as encephalopathy and fatty degeneration of viscera. Antecedent viral infections and aspirin therapy are precipitating factors in this disease. Spontaneous Reye's-like syndrome has also been reported in mice (Brownstein 1984) but not in other species.

EPIZOOTIOLOGY AND PATHOGENESIS. Outbreaks of Reye's-like syndrome, although relatively rare, do occur, with high morbidity and mortality. The disease has so far been associated only with BALB/cByJ mice. Precipitating factors have not been defined but may be linked to enterotropic mouse hepatitis virus or other infections. Reye's syndrome in humans is characterized by a rapidly deteriorating encephalopathy secondary to hepatic dysfunction with hyperammonemia. The metabolic defect is unknown, but mitochondrial swelling with dysfunction in hepatocytes is the probable primary lesion. Affected mice become precipitously stuporous and comatose with hyperventilation. Death occurs in most cases within 6–18 hr after onset, but some mice regain consciousness.

PATHOLOGY. Livers are swollen, greasy, and pale, and kidneys are swollen, with pale cortices. Intestines can be fluid- and gas-filled, with empty ceca. Microscopic findings include marked microvesicular fatty change and swelling of hepatocytes, with sinusoidal hypoperfusion (Fig. 1.65). Moderate numbers of fat vacuoles are also present in renal proximal convoluted tubular epithelium. Neurological lesions consist of swelling of protoplasmic astrocyte nuclei (Alzheimer

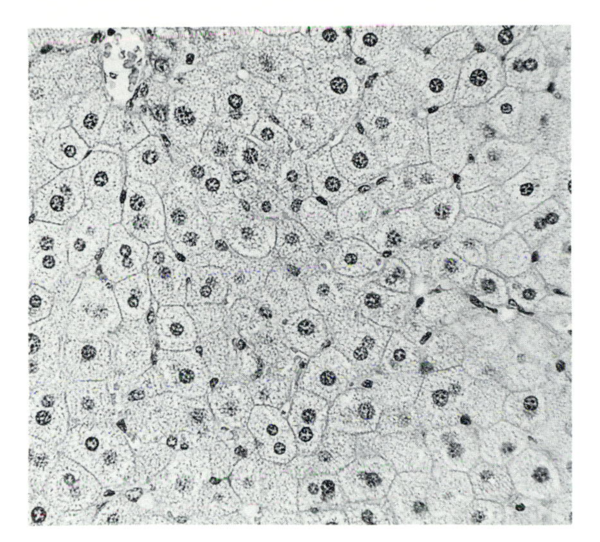

Fig. 1.65. Section of liver from a case of Reye's-like syndrome in BALB/c mouse. Note the increased cytoplasmic volume and the diffuse vacuolation of the hepatocyte cytoplasm.

type II astrocytes) in the neocortex, corpus striatum, hippocampus, and thalamus. Intestinal lesions consistent with enterotropic mouse hepatitis virus are variably present.

DIAGNOSIS. Clinical signs and gross liver lesions are distinctive. Microscopic changes in liver and brain, coupled with hyperammonemia, are diagnostic. *Differential diagnoses* must include other causes of hepatocellular fatty change in BALB/c mice, which normally possess a moderate degree of this change.

SIGNIFICANCE. Although rare, outbreaks of the syndrome can be devastating. If the syndrome could be experimentally reproduced, it would serve as a valuable model of Reye's syndrome in humans.

BIBLIOGRAPHY FOR NUTRITIONAL AND METABOLIC DISORDERS

Amyloidosis

Cohen, A.S., and Shirahama, T. 1980. Amyloidosis, model no. 17. In *Handbook: Animal Models of Human Disease*, ed. C.C. Capen et al. Washington, D.C.: Registry of Comparative Pathology, Armed Forces Institute of Pathology.

Cohen, A.S., et al. 1983. Amyloid proteins, precursors, mediators and enhancers. Lab. Invest. 48:1–4.

Conner, M.W., et al. 1983. Spontaneous amyloidosis in outbred CD-1 mice. Surv. Synth. Pathol. Res. 1:67–78.

Dunn, T.B. 1967. Amyloidosis in mice. In *Pathology of Laboratory Rats and Mice*, ed. E. Cotchin and F.J.C. Roe, pp. 181–212. Edinburgh: Blackwell Scientific, Oxford.

Soft-tissue Calcification

Yamate, J., et al. 1987. Observations on soft tissue calcification in DBA/2NCrj mice in comparison with CRJ:CD-1 mice. Lab. Anim. 21:289–98.

Reye's-like Syndrome

Brownstein, D.G., et al. 1984. Spontaneous Reye's-like syndrome in BALB/cByJ mice. Lab. Invest. 51:386–95.

AGING, DEGENERATIVE, AND MISCELLANEOUS DISORDERS

Ileus in Lactating Mice

A spontaneous disease with relatively high mortality has been recognized in female mice, usually during the second week of their first lactation. Mortality rates may be up 40% and mice of various genetic backgrounds appear to be susceptible to the disease. The condition has been called "paralysis/paresis of peristalsis" (Kunstyr 1986).

PATHOLOGY. At necropsy, the stomach is usually slightly dilated, and filled with watery fluid. The proximal small intestine is distended with fluid contents. Firm, conical fecal plugs are frequently present in the ileum and the tip of the cecum. The ileum caudal to the plug may be empty, or it may contain fecal pellets. The colon and rectum are also usually empty, or they may contain a few fecal pellets or mucous material. Histological findings are usually unremarkable, and pathogenic organisms have not been recovered from the intestine or other tissues.

EPIZOOTIOLOGY AND PATHOGENESIS. Several possible etiologies have been proposed: Clostridial enterotoxemia and exogenous toxins are two suggested causes. *Citrobacter*-induced colitis was associated with the problem in one facility. "Exhaustion" with concurrent imbalance of, for example, calcium, sodium and/or other electrolytes, or glucose, have been proposed as the underlying cause, with subsequent impaired peristalsis, and subsequent impaction (Kunstyr 1986). *Differential diagnoses* include differentiation from advanced autolysis and postpartum bacterial septicemia.

SIGNIFICANCE. Postpartum ileus has been recognized as a significant cause of mortality in primiparous female mice in Europe. It is likely that the problem may occur sporadically in other areas. The etiopathogenesis of the condition is yet to be resolved.

Hypothermia and Hyperthermia

Although mice are highly adaptable to living in different climates, they are inefficiently homeothermic and cannot tolerate sudden and extreme changes in environmental temperature. Hypothermia and hyperthermia are all too common occurrences during shipping, when crates are moved from one environment to another. Water bottle "accidents" often cause mortality from hypothermia. All of these factors can result in high mortality with few, if any, discernible lesions.

Dehydration

Mice require relatively large volumes of drinking water and easily become dehydrated. Hydration can be evaluated at necropsy by skin plasticity, "stickiness" of tissues, pale and contracted spleens, vascular hypovolemia, or elevated hematocrit. Thorough anamnesis will often reveal failure of watering devices. Even if water bottles are full, sipper tubes can become obstructed, or if new, they can contain metal filings that interfere with water flow. Dehydration can also occur when water bottle sipper tubes are too high for young mice to reach or if newly arrived mice are unaccustomed to automatic watering devices.

"Ringtail"

Low ambient humidity can cause skin dryness, which can be a significant problem in infant mice. Dry dermatitis can cause "ringtail," or annular constrictions of the tail and occasionally digits, resulting in edema of the distal extremity and dry gangrene. Hairless strains of mice are also prone to skin problems as adults, which may be manifest as inflammation and gangrene without the classic "ringtail" antecedent stage. These phenomena are probably exacerbated by environmental temperature extremes, particularly low temperatures, and nutritional factors. However, the nature of the predisposing factors and the precise pathogenesis of the condition is yet to be resolved.

Sloughing of Extremities due to Cotton Nesting Material

Necrosis and sloughing of limb extremities in suckling mice have been associated with infarction due to the wrapping of absorbent cotton (cotton wool) nesting material around one or more legs (Rowson and Michaels 1980).

Spontaneous Corneal Opacity

Corneal opacities have been observed in a variety of strains of mice. Opacities are characterized by acute to chronic inflammatory changes of the corneal epithelium and anterior corneal stroma, including acute keratitis with corneal erosion to ulceration, vascularization of the corneal stroma, and mineralization of corneal basement membranes. Van Winkle and Balk (1986) alleviated the problem by more-frequent cage cleaning, and it was concluded that an environmental factor, such as ammonia, may play an important role in the development of the disease.

Suppurative Conjunctivitis/Ulcerative Blepharitis in 129/J Mice

Suppurative conjunctivitis with ulcerations at the mucocutaneous junction have been observed in this strain, and in BALB strains. Suppuration with abscessation of the meibomian glands also occurred. A variety of bacteria were isolated from affected conjunctivae, including *Corynebacterium* and coagulase-negative *Staphylococcus*. These were interpreted to be opportunistic infections and the specific etiopathogenesis is unknown (Sundberg et al. 1991).

Retinal Degeneration

EPIZOOTIOLOGY AND PATHOGENESIS. Retinal degeneration is a very common lesion that can be considered a normal characteristic of certain inbred strains of mice; it also occurs in outbred stocks and wild mice. It may have different modes of inheritance but appears histologically identical between mouse strains. Commonly affected strains include C3H, CBA, and Swiss mice, while strains A, AKR, BALB/c, C57BL, and DBA have normal retinas. The retinal degeneration of C3H/He mice is the best characterized and is inherited as a recessive trait (Sidman and Green 1965). Mice are born with normal-appearing retinas, and the disorder involves both arrested development and subsequent degeneration of photoreceptor cells.

PATHOLOGY. Microscopic changes include absence or degeneration of the rods, outer nuclear layer, and outer plexiform layer (Fig. 1.66). Active degenerative changes can be encountered in young mice, but the lesion evolves rapidly and is nearly complete by weaning age. *Differential diagnosis* should include light-induced retinal degeneration.

SIGNIFICANCE. Mice do not appear to manifest clinical signs. If ophthalmic studies are contemplated, the strain of mouse to be used should be carefully selected.

Aspiration Pneumonia

Accidental inhalation of foreign material can occur under a number of circumstances, but especially when shipping crates containing wood shavings are handled roughly in transit. Foreign plant material can be readily identified in airways (Fig. 1.67).

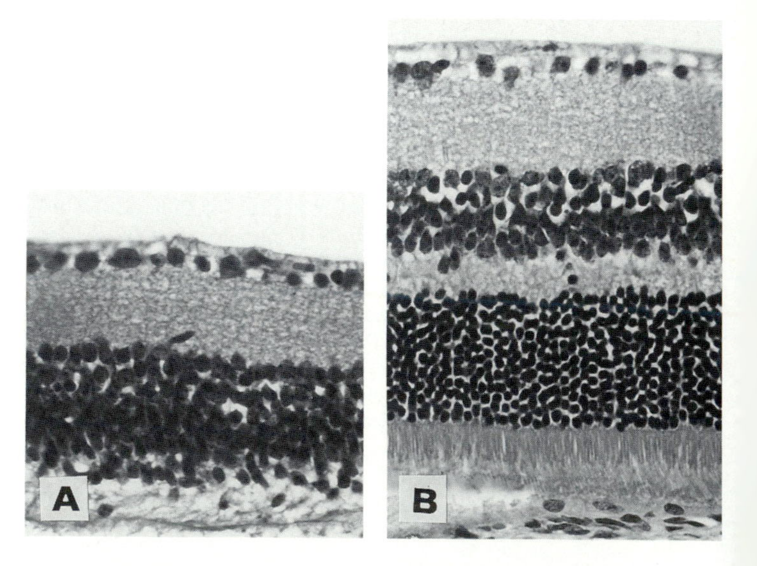

Fig. 1.66. Retinal atrophy. Note the marked reduction in the thickness of the retina and the loss of identifiable outer plexiform and bipolar cell layers in the affected mouse (*A*), compared with the control animal (*B*).

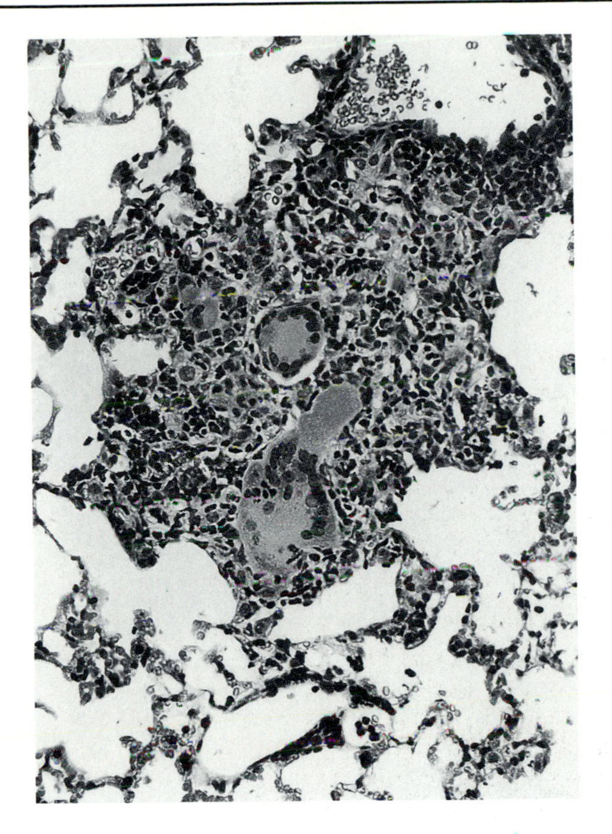

Fig. 1.67. Section of lung from mouse with aspiration pneumonia (from plant fibers). Note the granulomatous inflammatory response and the multinucleated giant cell formation.

SIGNIFICANCE. This is an incidental finding and of no clinical significance.

Behavioral Diseases

The mouse is highly gregarious and thrives in communal groups. Social harmony requires establishment of a dominance hierarchy, which can be easily destabilized. The reproductive cycles of mice are significantly influenced by pheromones, which in turn are intimately associated with the dominance hierarchy and presence of foreign members. Infertility, manifested as altered estrous cycles, fetal resorption, and anestrus, as well as maternal cannibalism, can result from pheromone-driven responses. Mouse behavior and pheromones have been well studied and are recommended subjects for further reading (reviewed in Whittingham and Wood 1976). Adult male mice will fight savagely unless reared as siblings or peers from infancy. Certain strains, such as BALB and SJL, are notorious in this respect. Fight wounds can be diffuse, but they are often oriented around the tail and genitalia (Fig. 1.68). *Barbering*, another common dominance-associated vice of both sexes, can be manifest in different patterns, depending upon genotype or individual idiosyncrasies. Alopecia of vibrissae and other parts of the body is common, often with a well-defined, clipped edge (Fig. 1.69). Typically, the dominant mouse within a group is unaffected. Nasal alope-

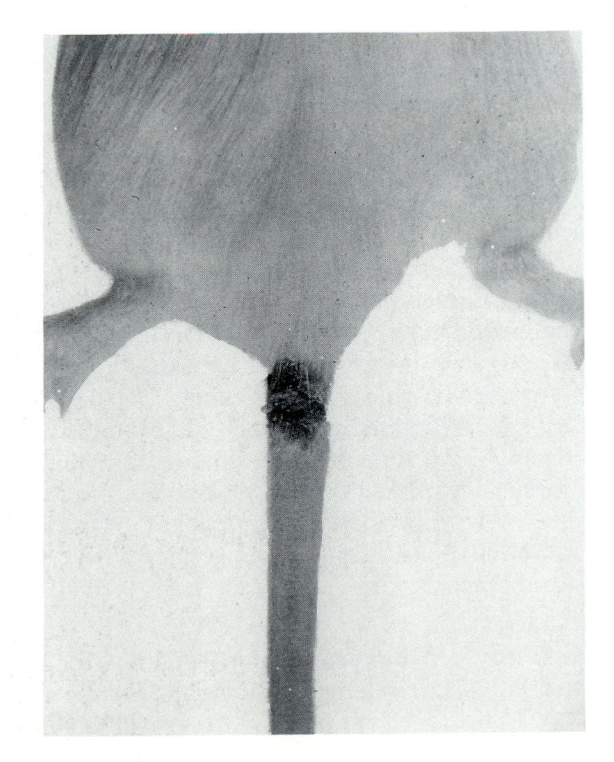

Fig. 1.68. Circumscribed contusions at the base of the tail associated with fighting injuries in sexually mature male mouse.

Nasal Alopecia

Hair loss in the muzzle region occurs occasionally in laboratory mice. *Differential diagnoses* include "barbering" and mechanical denuding due to improperly constructed openings for feeders or watering devices.

Malocclusion

Malocclusion due to improperly aligned upper and lower incisor teeth may result in marked overgrowth, particularly of the lower incisors. This condition has a hereditary basis, and culling is recommended.

Mesenteric Disease

EPIZOOTIOLOGY AND PATHOGENESIS. In this disorder, mesenteric lymph nodes are enlarged and filled with blood. This lesion occurs sporadically in aging mice of various strains, appearing more frequently in C3H mice. Its etiology is unknown.

PATHOLOGY. Mesenteric lymph nodes are grossly enlarged and appear bright red. Microscopically, lymphoid tissue is often atrophic and medullary sinuses are filled with blood. *Differential diagnoses* must include various causes of mesenteric lymphadenomegaly, including *Salmonella*.

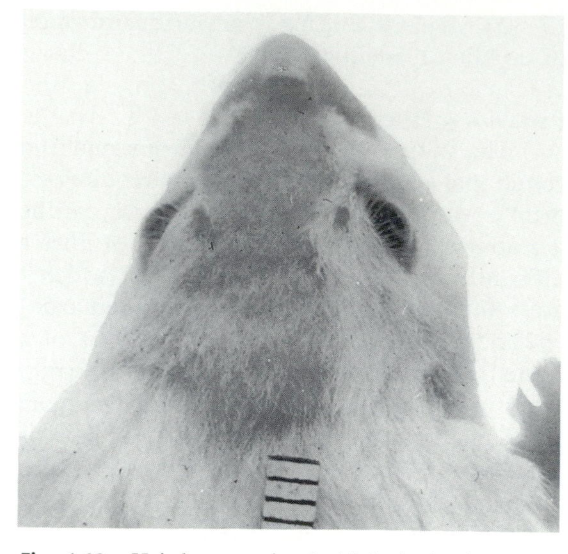

Fig. 1.69. Hair loss associated with barbering by a cagemate.

cia can also be due to mechanical abrasion from feeding devices (Litterst 1974). Individual mice can also display aberrant circling behavior, unrelated to neurological or vestibular disease. Excessive self or peer grooming/barbering, termed "trichotillomania," can result in severe hair loss and pyoderma. C57BL mice are particularly prone to this syndrome (Thornburg et al. 1973).

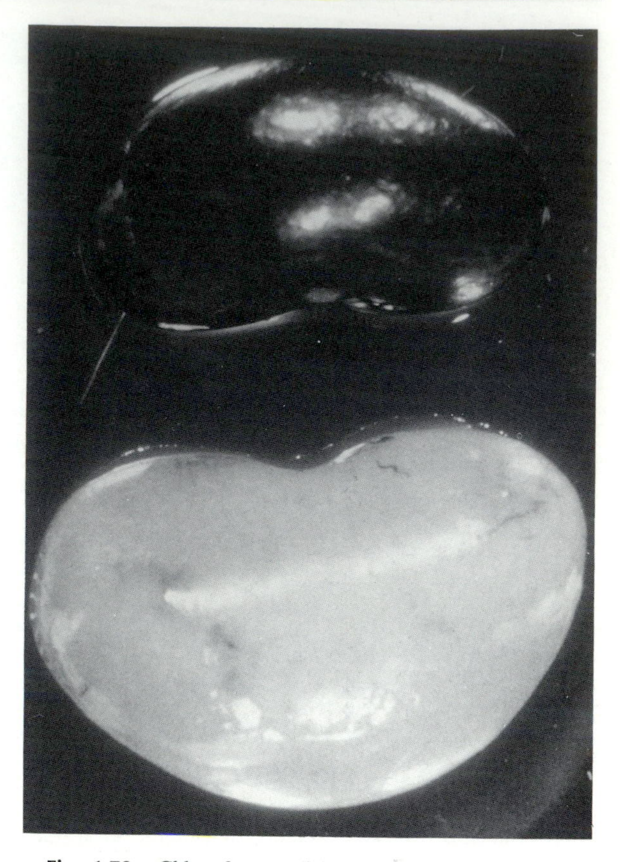

Fig. 1.70. Chloroform toxicity in male mouse. The swollen, pale kidney was collected at necropsy 48 hr post–exposure to chloroform. The control kidney is from normal animal.

Urinary Disorders

Chloroform Toxicity

Mice can develop renal tubular necrosis and mineralization when exposed to chloroform fumes. Mature male mice of certain genotypes such as DBA and C3H are exquisitely sensitive, with high mortality. One factor associated with the sex-related susceptibility appears to be the increased renal binding of chloroform in males, compared with females, and castration of males will eliminate their susceptibility to chloroform nephrotoxicity (Carlton and Engelhardt 1986). Severely affected mice develop swollen, pale kidneys (Fig. 1.70). Microscopic changes are characterized by coagulation necrosis of renal tubules, particularly the proximal convoluted tubules. Surviving mice have residual nephrocalcinosis. "Outbreaks" of mortality, with selective deaths among male mice have been described (Jacobsen et al. 1964; Deringer et al. 1953).

Chronic Glomerulonephritis/Glomerulopathy. Renal lesions are relatively common in certain strains of older mice, such as AKR, BALB/c, and CBA mice. Mice with naturally occurring autoimmune disease, such as the (NZB x NZW) F_1 hybrid, usually have developed extensive glomerular lesions by the time they reach 12 or more mo of age (Sass 1986). However, there are a variety of other factors that may be involved, including viral agents, bacteria or bacterial products, and the deposition of antigen-antibody complexes on glomerular basement membranes. Persistent retroviral infections are an example of a condition that may result in the deposition of antigen-antibody complexes on glomerular tufts.

PATHOLOGY. On gross examination, there may be marked pitting of the cortical surfaces in advanced cases, and small cysts may be evident on cut surface. Microscopic changes are characterized by thickening of glomerular basement membranes due to the deposition of PAS-positive material that does not stain for amyloid. There may be proliferation of mesangial cells, and obliteration of the normal architecture of affected glomeruli in advanced cases. Focal to diffuse mononuclear cell infiltration, and varying degrees of fibrosis in the interstitial regions are other changes that may occur. *Differential diagnoses* include renal amyloidosis and chronic pyelonephritis.

SIGNIFICANCE. In advanced cases of glomerulopathy, there may be manifestations of renal insufficiency, including proteinuria and abnormalities in blood biochemistry. Depending on the strain of mouse affected, investigations into the underlying factors may be warranted.

Interstitial Nephritis. The etiopathogenesis of tubulointerstitial disease in mice is similar to other species and includes sequelae to infections with bacteria such as *Proteus mirabilis*, *Pseudomonas aeruginosa*, or *Staphylococcus aureus*; viruses such as lymphocytic choriomeningitis virus; and following chemically induced nephrosis (Montgomery 1986a). Frequently the etiology cannot be resolved. The nature of the lesions vary, depending on the duration and extent of the disease process. Frequently renal lesions are observed as an incidental finding, although they may be of sufficient magnitude to contribute to the demise of the animal, particularly if there is significant glomerular involvement. In such cases, the carcass is usually pale, with irregular pitting of the renal cortices and often marked ascites. "Metastatic" calcification of target tissues is not a feature of renal failure in the mouse. On microscopic examination, lesions may vary from discrete aggregations of mononuclear cells in perivascular regions in the cortex to segmental to diffuse involvement, with distortion and loss of tubules and to obliteration of the normal architecture in advanced cases. *Differential diagnoses* include renal amyloidosis, glomerulonephritis, and pyelonephritis (Montgomery 1986b).

Hydronephrosis. Unilateral or bilateral hydronephrosis is a common, usually incidental finding. Hydronephrosis can occur in high prevalence among certain strains or lines of mice or be secondary to urinary obstruction or pyelonephritis. *Differential diagnosis* should include renal papillary necrosis due to amyloidosis, creating an *ex vacuo* pelvic enlargement. Presence of the renal papillus should be verified.

Polycystic Disease. Certain strains of mice, such as BALB mice, are prone to congenital cysts of varying size in the kidneys. In some cases, these cysts are quite large and impinge on renal function, resulting in mortality.

Urinary Obstruction/Urologic Syndrome. This spontaneous disease occurs occasionally in male mice and may be manifest as an acute or chronic condition (Bendle and Carlton 1986). Clinical signs may vary from dribbling of urine and wetting of the perineal region to cellulitis with ulceration of the preputial area. In acute cases, animals may be found dead in the cage.

PATHOLOGY. The perineal area may be wet, with varying degrees of swelling and ulceration in the preputial area. Paraphimosis is a variable finding. In acute cases, the urinary bladder is usually markedly distended with urine, and dull white, firm, proteinaceous plugs are often evident in the neck of the urinary bladder and proximal urethra. In chronic cases, the bladder may be distended with cloudy urine and/or calculi. The vesicular glands are sometimes distended with inspissated material, and there may be some evidence of hydronephrosis. On microscopic examination, in acute cases amorphous eosinophilic material containing spermatozoa may be present in the proximal urethra, with minimal to no inflammatory response. In chronic obstruction, there may be varying manifestations of inflammatory response, such as prostatitis, cystitis, urethritis, and balanoposthitis. *Differential diagnoses* include bacterial cystitis and pyelonephritis, and agonal release of secretions from the accessory sex glands at death, which could be confused with a bona fide antemortem urinary obstruction. Urinary obstruction can also occur as a result of fighting injuries to the external genitalia.

Mucometra

Mucometra is a relatively common phenomenon in laboratory mice. It is most commonly encountered among large groups of presumably pregnant mice in which a few never whelp. The abdomen is often distended. Some mice have congenital imperforate lower reproductive tracts, while in others, the cause cannot be determined. One or both uterine horns can be dilated. *Differential diagnoses* must include pyometra (which may occur secondary to mucometra), retained fetuses, and neoplasia. The uterus of mice and rats normally contains small amounts of retained fluid during certain stages of the estrous cycle.

Cystic Endometrial Hyperplasia

Cystic endometrial hyperplasia of endometrial glands is a frequent finding in aged female mice. It may be associated with secondary bacterial pyometras (*Klebsiella oxytoca*).

Vestibular Syndrome

This is a common clinical sign in mice. It is often due to bacterial otitis, and less often, to central nervous system disease. An undescribed but frequent cause of vestibular disease also appears to be necrotizing arteritis of undetermined cause. This lesion occurs in a number of mouse strains in the absence of detectable viral and bacterial pathogens. The internal and middle ear structures are normal, but careful examination of surrounding tissues will reveal active necrotizing change in medium-size arteries. Similar changes will be found in arteries of the heart and mesentery.

Pulmonary Histiocytosis (Crystal Pneumonitis)

Eosinophilic crystals of varying size and shape can occur focally in terminal airways and alveolar sacs

Poxviral Infection

Poxviral infections have been reported to occur in laboratory rats from Eastern Europe and the former Soviet Union (Krikun 1977). The agent (or agents) is Turkmenia rodent poxvirus, which is closely related to cowpox virus and distinctly different from ectromelia virus. Clinical signs resembled mousepox in mice, ranging from inapparent infections to dermal pox and tail amputation with high mortality (Iftimovici et al. 1976). Both dermal and respiratory tract lesions occur. Microscopically, rats with respiratory signs had severe interstitial pneumonia with edema, hemorrhage, and pleural effusion. Focal inflammatory lesions occur in the upper respiratory tract (Kraft et al. 1982). Felids and human contacts are susceptible to infection with this poxvirus (Marennikova and Shelukhina 1976).

RNA Viral Infections

Coronaviral Infections

Coronaviruses, enveloped RNA viruses with characteristic peplomers on the viral surface, replicate in the cytoplasm of the host cell. The two prototype naturally occurring coronaviruses isolated from this species are sialodacryoadenitis virus (SDAV) and Parker's rat coronavirus (PRC). PRC was first isolated from the pooled lungs of specific-pathogen-free CD Fischer laboratory rats. Serological surveys of laboratory rats and wild rats revealed that antibodies to rat coronavirus were common in both groups. Intranasal inoculation of newborn rats with the virus produced interstitial pneumonitis, with focal atelectasis and high mortality (Parker et al. 1970). In intranasal inoculation experiments with young adult CD rats, rhinitis, tracheitis, and focal alveolitis occurred during the acute stages of the disease. Any strain of coronavirus in this species that causes the disease sialodacryoadenitis is termed SDA virus. Although PRC causes similar lesions in the salivary and lacrimal glands (Percy et al. 1990), it has been given a separate name. This is obviously an artificial distinction. Other strains of coronavirus have been isolated from laboratory rats (Hirano et al. 1986; Maru and Sato 1982) but have not been fully characterized. It is likely that additional strains will be isolated and characterized in the future, and there could eventually be a variety of isolates of rat coronaviruses similar to that seen with mouse hepatitis virus.

Sialodacryoadenitis (SDA)

EPIZOOTIOLOGY AND CLINICAL SIGNS. Sialodacryoadenitis is a commonly occurring, transient disease in the laboratory rat. Based on serological surveys for coronaviral antibody, the number of seropositive colonies may be 50% or greater (Caseboldt et al. 1988; Lussier and Descoteaux 1986). Transmission is likely primarily by infected nasal secretions or saliva, and the disease spreads rapidly following the introduction of SDA virus into a susceptible population of rats. In epizootics of SDA, there is a high morbidity and virtually no mortality. Subclinical outbreaks may occur. Typical clinical signs associated with SDA during the acute stages of the disease include sniffling, blepharospasm, and intermandibular swelling. Dark red encrustations may be present around the eyes and external nares. These porphyrin-containing substances are released from damaged Harderian glands and will emit a characteristic pink fluorescence under an ultraviolet light source. Other complications sometimes seen during the convalescent period may include unilateral or bilateral glaucoma/megaloglobus and corneal ulceration. Ocular changes that occur are secondary to the destructive lesions and impaired function of the lacrimal glands. This may result in failure to lubricate the cornea properly, corneal drying, impaired intralocular drainage, and subsequent permanent damage to the eye. Reproductive disorders have also been associated with the disease (Utsumi et al. 1980).

PATHOLOGY. At necropsy, red encrustations may be present around the external nares and eyelids. In animals examined during the acute stages of the disease, there may be marked involvement of the parotid and/or the submandibular (submaxillary) salivary glands. In affected glands,there may be periglandular edema, with variable involvement of the adjacent subcutaneous tissue. In contrast to control rats, there is marked enlargement of affected salivary glands, with separation of the interlobular septa (Fig. 2.6). Similar changes are frequently evident in the exorbital lacrimal glands during the acute stages of the disease. Enlargement, with periglandular and interstitial edema in affected lacrimal glands, is a typical macroscopic finding.

MICROSCOPIC CHANGES. In affected parotid and submandibular salivary glands examined during the acute stages of the disease, there is coagulation necrosis of the ductal structures, with variable involvement of adjacent acini and effacement of the normal architecture. Interstitial edema, with mononuclear and polymorphonuclear cell infiltration, frequently occurs (Fig. 2.7). Similar changes occur in the Harderian and exorbital lacrimal glands during the acute stages of SDA. During the reparative stages of the disease, beginning at 7–10 days postexposure, there is squamous metaplasia of ductal and acinar structures. Cellular infiltrates are primarily lymphocytes, plasma cells, and macrophages (Fig. 2.8). Similar reparative changes occur in affected exorbital glands, including squamous metaplasia of ducts. Squamous metaplasia can be marked in the Harderian glands (Fig. 2.9). In salivary glands, regeneration of acinar and ductal epithelial cells are usually essentially complete by 3–4 wk

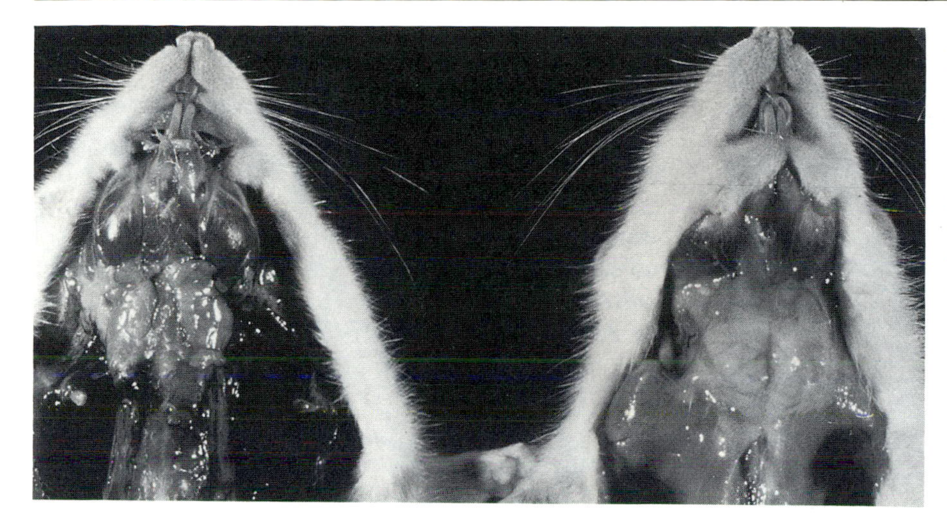

Fig. 2.6. Gross lesions typical of sialodacryoadenitis viral infection when necropsied during the acute stages of the disease (6 days postexposure). There is marked swelling of the submandibular glands, with interlobular and periglandular edema in the animal on the right, compared with the control rat on the left. (Courtesy of Percy and Wojcinski 1986, reprinted by permission)

postexposure. There may be isolated ducts or acini lined by poorly differentiated epithelial cells, with scattered aggregations of mononuclear cells, including mast cells, but usually the salivary glands are essentially normal histologically at this stage postexposure. On the other hand, focal residual inflammatory lesions may persist in the Harderian glands for several weeks (Percy et al. 1989; Jacoby et al. 1975). Reactive foci are frequently associated with interstitial deposition of pigmented material.

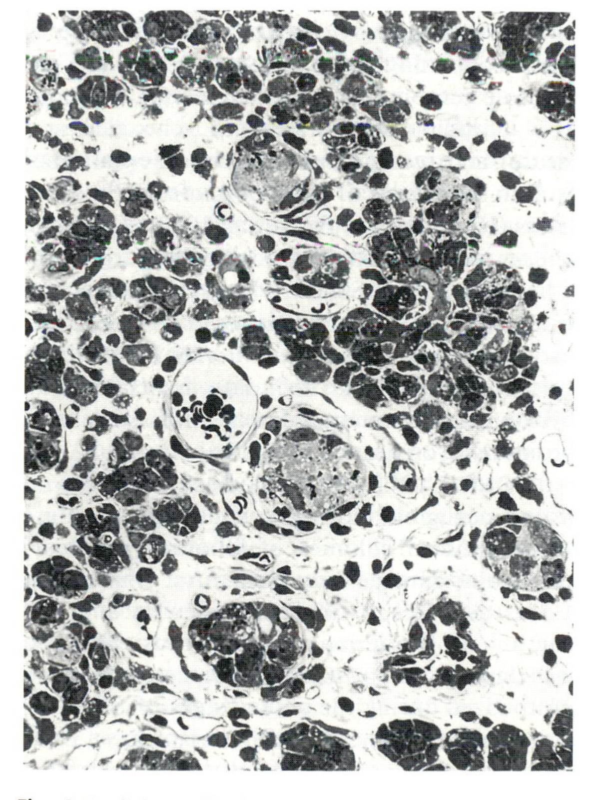

Fig. 2.7. Submandibular salivary gland from Wistar rat collected at 6 days postexposure to SDA virus. Note the marked necrosis of ductal and acinar epithelial cells, with effacement of the normal architecture.

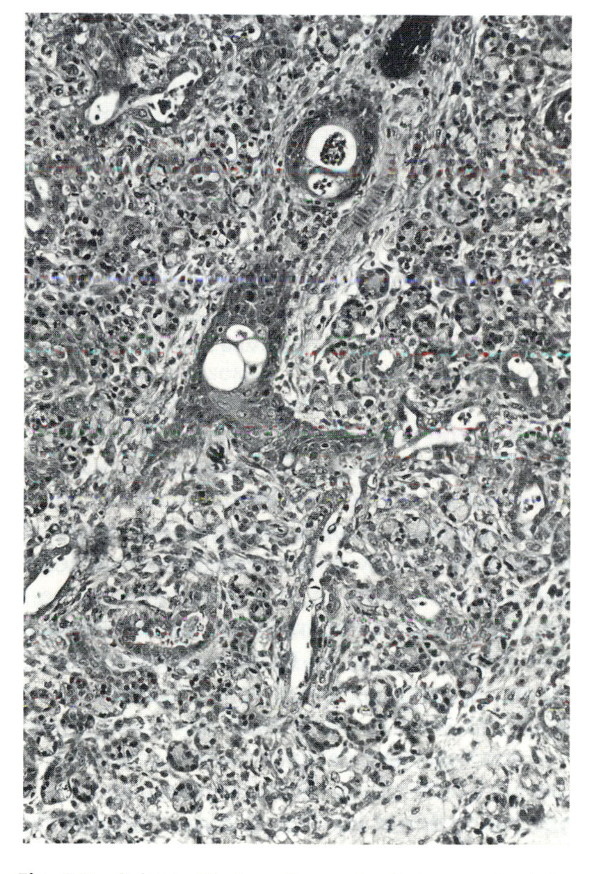

Fig. 2.8. Submandibular salivary gland at approximately 8 days postexposure to SDA virus. Ducts and acini are lined by poorly differentiated epithelial cells, with marked mononuclear cell infiltration.

In the respiratory tract, rhinitis, with mononuclear and polymorphonuclear cell infiltration occur during the acute stages of the disease. Lesions are present in the lower respiratory tract during the first week postexposure. Tracheitis, with focal bronchitis and bronchiolitis, is characterized by mononuclear and polymorphonuclear cell infiltration. There may be hyperplasia of respiratory epithelial cells with flatten-

tions. The abundant polysaccharide capsule is an essential component for bacterial serotyping and also enables the organism to resist phagocytosis by the host cells. Pneumococci are not known to produce soluble toxins. However, several of the recognized serotypes produce tissue damage by activation of the alternate complement pathway (Yoneda and Coonrod 1980). Clinical signs may include serosanguinous nasal discharge, rhinitis, sinusitis, conjunctivitis, and vestibular signs consistent with middle ear infection. In asymptomatic animals infected with the organism, predisposing factors such as concurrent infections or environmental changes may precipitate an outbreak of the disease.

PATHOLOGY. At necropsy, there may be serous to mucopurulent exudate present in the nasal passages, with variable involvement of the tympanic bullae. In the acute systemic form of the disease, there are variable patterns of polyserositis, including pleuritis, peritonitis, pericarditis, periorchitis, and meningitis. There may be fibrinopurulent exudate in the pleural and peritoneal cavities and pericardial sac (Fig. 2.26). Fibrinopurulent lesions may be confined to the leptomeninges in some fatal cases of streptococcosis. There may be consolidation of one or more lobes of the lung. Affected areas are dark red to dull tan and relatively firm and nonresilient. Serous to purulent exudate may be present in the tympanic bullae. Gram-stained touch preparations or smears of exudate should reveal the typical gram-positive encapsulated diplococci (Fig. 2.27).

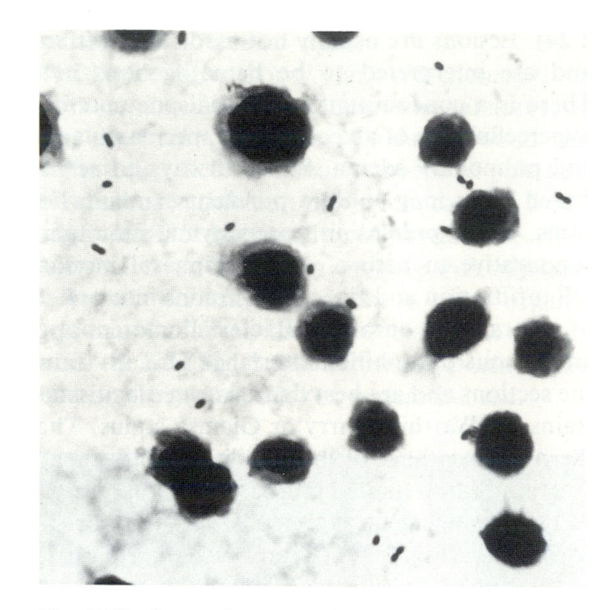

Fig. 2.27. Impression smear from pleura of rat with diplococcal pleuritis (Gram stain). Numerous encapsulated diplococci and mesothelial cells are evident.

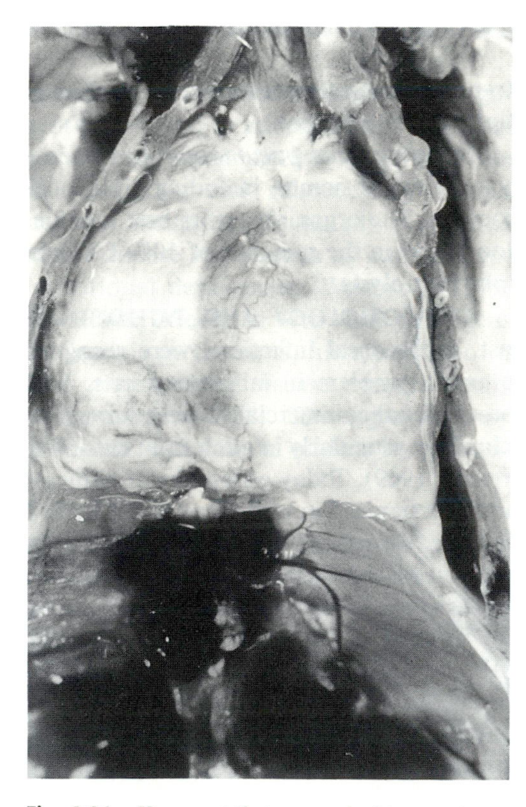

Fig. 2.26. Young rat that succumbed to spontaneous diplococcal (*Streptococcus pneumoniae*) infection. There is an extensive fibrinous pericarditis and pleuritis.

On microscopic examination, in the acute form of the disease fibrinopurulent pleuritis and pericarditis are typical findings. Pulmonary changes vary from localized suppurative bronchopneumonia to acute fibrinopurulent bronchopneumonia, with obliteration of the normal architecture in affected lobes. Fibrinopurulent peritonitis, perihepatitis, and/or leptomeningitis (Fig. 2.28) are frequent findings. Suppura-

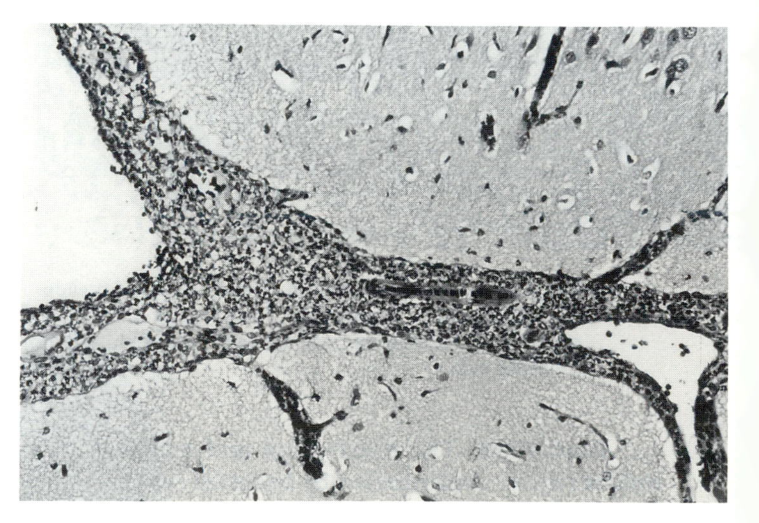

Fig. 2.28. Section of cerebrum from case of spontaneous diplococcal meningitis. Note the marked polymorphonuclear infiltration in the leptomeninges.

tive rhinitis and otitis media may also occur. Embolic suppurative lesions have been observed in organs such as liver, spleen, and kidney. In more chronic, localized disease states, pneumococcal infections have been associated with conditions such as suppurative bronchopneumonia in chronic respiratory disease, and otitis media.

DIAGNOSIS. The demonstration of the typical encapsulated diplococci in Gram-stained smears from typical lesions will provide a provisional diagnosis. Confirmation of streptococcosis requires the collection and culture of material from lesions and identification of the organism. Nasal lavage at necropsy is another recommended procedure to recover the organism for bacterial culture. *Differential diagnoses* include corynebacteriosis, salmonellosis, pseudomoniasis, and pasteurellosis.

SIGNIFICANCE. In previous years, *S. pneumoniae* represented an important cause of disease and mortality, particularly in young rats. Infected animals may harbor the organism in the upper respiratory tract as an inapparent infection (Fallon et al. 1988). Under the appropriate circumstances, these diplococci may become activated to produce a localized infection or an acute systemic disease, sometimes with significant mortality.

Tyzzer's Disease

Bacillus piliformis is a spore-forming, gram-negative rod that multiplies only within host cells and will not grow in cell-free media. Embryonated chick eggs are usually used to grow *B. piliformis* in the laboratory. The organism is best demonstrated in tissue sections with Warthin-Starry, Giemsa, or PAS staining methods.

EPIZOOTIOLOGY AND PATHOGENESIS. The Tyzzer's bacillus has a wide host range and may remain infectious in contaminated bedding for up to 1 yr (Tyzzer 1917). (For additional information on epizootiology and pathogenesis, see the section on Tyzzer's disease in Chap. 6). Outbreaks in laboratory rats usually occur in adolescent rats during the postweaning period. Accidental ingestion of spores is the likely source of the infection. Transplacental transmission has been demonstrated in seropositive rats treated with prednisolone during the last week of pregnancy. Treated carrier dams and affected newborn rats remained asymptomatic throughout the study (Fries 1979). Subclinically infected rats can transmit the organism to naive rats via contaminated bedding. Although seroconversion occurred in rats so exposed, gerbils housed on the contaminated bedding remained asymptomatic throughout the study (Motzel and Riley 1992). However, the organism has been eliminated from a colony of immunocompetent rats by cesarian section and appropriate disinfection techniques (Hansen et al. 1992). In naturally occurring outbreaks of the disease, clinical signs may include depression, ruffled hair coat, abdominal distension, low morbidity, and high mortality in clinically affected animals. Clinical disease with low mortality has also been reported (Fries and Svendsen 1978).

PATHOLOGY. Rats with Tyzzer's disease often develop a unique, marked dilation of the terminal small intestine (megaloileitis). The flaccid ileum may be distended up to 3–4 times the normal diameter, with variable involvement of the jejunum and cecum. Megaloileitis does not always occur in rats with Tyzzer's disease. Enteritis may be evident only at the microscopic level. The mesenteric lymph nodes are swollen and edematous. Disseminated pale foci of necrosis up to several millimeters in diameter are scattered throughout the parenchyma of the liver. There may be circumscribed to linear pale foci present on the heart. Microscopic changes are confined primarily to the ileum, liver, and myocardium. In the intestine, there is frequently a necrotizing transmural ileitis, with segmental involvement of affected areas. There is necrosis and sloughing of enterocytes and edema of the lamina propria and submucosa, frequently with fragmentation and hypercellularity of the muscular layers. Infiltrating inflammatory cells are primarily mononuclear cells, with a sprinkling of neutrophils. In the liver, the histological characteristics vary from foci of acute coagulation necrosis to focal hepatitis, with polymorphonuclear and mononuclear cell infiltration (Fig. 2.29). Hepatic lesions interpreted to be of some duration are characterized by fibrosis, with the presence of multinucleated giant cells and mineralized debris in reparative foci. In the heart, lesions may vary from

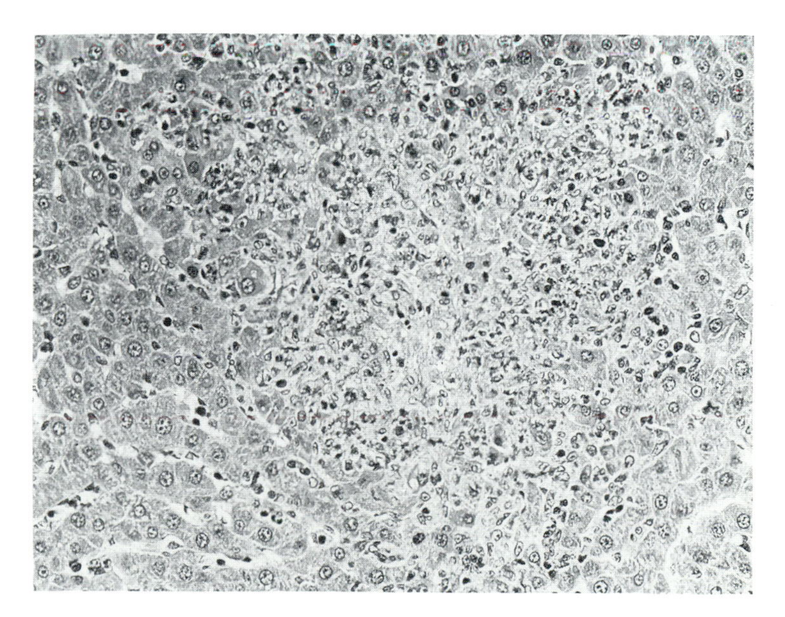

Fig. 2.29. Focal hepatitis in Tyzzer's disease in juvenile rat. There is necrosis of hepatocytes with polymorphonuclear cell infiltration.

necrosis of isolated myofibers to destruction of relatively large segments of myocardium. There is vacuolation to fragmentation of the sarcoplasm, with interstitial edema and mononuclear and polymorphonuclear cell infiltration (Fig. 2.30). Giemsa, Warthin-Starry, or PAS stains are used to demonstrate bundles of slender bacilli in the cytoplasm of enterocytes in ileal lesions, in hepatocytes surrounding necrotic foci, and scattered in the sarcoplasm in myocardial lesions.

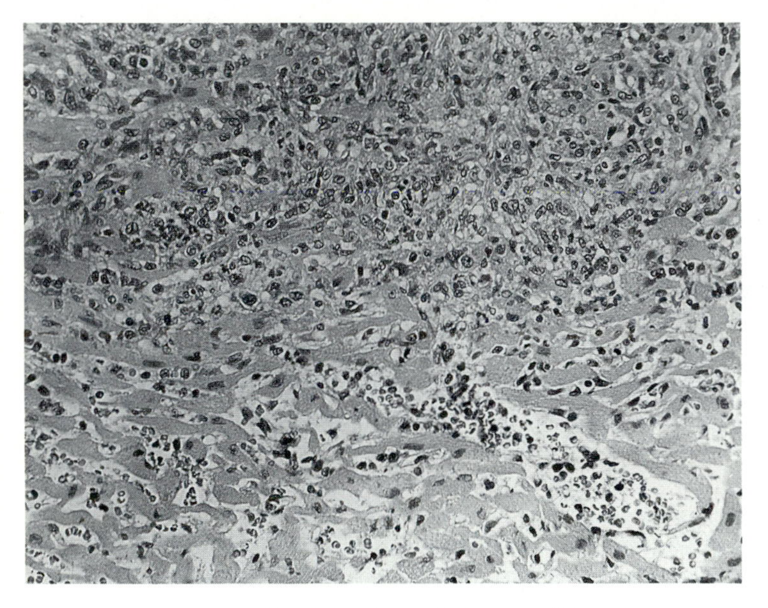

Fig. 2.30. Interstitial myocarditis, with degeneration of myofibers and mobilization of mononuclear cells in spontaneous case of Tyzzer's disease in young rat.

DIAGNOSIS. Confirmation of the diagnosis requires the demonstration of the typical organisms in tissue sections. The triad of organs usually affected in Tyzzer's disease (intestine, liver, and heart) are also useful diagnostic features. Serologic tests, such as the indirect fluorescent antibody or complement fixation procedures, have been used to detect seropositive rats (Fujiwara et al. 1981). *Differential diagnoses* include salmonellosis and ileus following the intraperitoneal administration of chloral hydrate for general anesthesia (Fleischman et al. 1977).

SIGNIFICANCE. Although uncommon in this species, Tyzzer's disease can be a cause of disease and mortality in the laboratory rat. Following confirmation of the diagnosis, investigations should include husbandry and sanitation, as well as the possibility of interspecies transmission from other rodents or rabbits. Clinically normal seropositive rats have been identified in colonies, indicating that inapparent infections may occur.

Salmonellosis

During the early years of this century *Salmonella* infections represented an important infectious disease in laboratory rodents in North America. However, with improved sanitation, health-monitoring methods, and feeding practices, the disease now rarely occurs in laboratory animal facilities.

EPIZOOTIOLOGY AND CLINICAL SIGNS. Of the multitude of serotypes of *Salmonella enteritidis* capable of causing disease in the laboratory rat, serotypes *enteritidis* and *typhimurium* were most frequently implicated. In the period 1895–1910, serotype *enteritidis* was used as a rodenticide to control populations of wild rats in Europe and the United States. Enthusiasm waned when the public health implications became apparent (Weisbroth 1979). Clinical signs include depression, ruffled hair coat, hunched posture, weight loss, and variations in the nature of the feces from softer, lighter, formed stool to fluid contents. There may be porphyrin-containing red encrustations around the eyes and nose in clinically affected animals.

PATHOLOGY. Subclinical infections without discernible lesions are frequent. In clinically affected rats, the ileum and cecum are frequently distended with liquid contents and flecks of blood and there is thickening of the gut wall in affected areas. Focal ulcerations may be present in the mucosa of the cecum and ileum. Splenomegaly frequently occurs. On microscopic examination, lesions in the ileum and cecum are characterized by hyperplasia of crypt epithelial cells, edema of the lamina propria, and leukocytic infiltration with focal ulceration. There is hyperplasia of the mesenteric lymph nodes, spleen, and Peyer's patches, with focal necrosis and leukocytic infiltration. In acute cases, lesions in other viscera are consistent with a gram-negative septicemia. In the spleen, focal necrosis and hemorrhage occur in the red pulp. Sinusoidal congestion and focal coagulation necrosis are frequent findings in the liver. Focal embolization may occur in tissues such as spleen, liver, and lymph nodes. Emboli consist of bacteria, fibrinous exudate, and cellular debris.

DIAGNOSIS. Isolation and identification of the organism from animals with lesions or from inapparent carriers are necessary to confirm the diagnosis of *Salmonella* infection. Salmonellae are intermittently present in the intestine, especially in carrier animals. One recommended procedure is to incubate fecal pellets or macerated tissue in selenite-F plus cystine broth overnight, followed by streaking onto brilliant green to promote the growth of the organism (Weisbroth 1979). Mesenteric lymph nodes may yield *Salmonella* in rats with negative fecal culture. *Differential diagnoses* include pseudomoniasis, rotaviral enteritis, cryptosporidiosis, management-related problems due to failure to provide feed or water, and Tyzzer's disease.

SIGNIFICANCE. Following the diagnosis of salmonellosis, the zoonotic potential should be emphasized. Possible sources of the organism include contaminated feed or fomites and transmission from other species, including human carriers, and from wild rodents. Corrective steps should include the identification of the source of the infection, improved sanitation, and if feasible, immediate slaughter. Repeated fecal samplings may be required in order to detect inapparent carrier animals.

Pasteurella pneumotropica Infection

P. pneumotropica is a gram-negative coccobacillus with bipolar staining properties that grows on conventional media under aerobic conditions.

EPIZOOTIOLOGY AND PATHOGENESIS. The organism may be carried in the nasopharynx, conjunctiva, lower respiratory tract, and uterus as an inapparent infection. *P. pneumotropica* readily colonizes the intestine, where it may be carried for long periods of time. Transmission probably occurs primarily by direct contact or fecal contamination, rather than by aerosols (Weisbroth 1979). The organism is frequently isolated in the absence of disease, and intranasal inoculation has failed to produce lesions in the upper or lower respiratory tract (Burek et al. 1972). On the other hand, it may represent an important secondary bacterial invader and opportunistic infection in primary *Mycoplasma pulmonis* or Sendai virus infections. Interstitial pneumonitis and polymorphonuclear cell infiltration has been observed in pregnant rats with primary Sendai virus and secondary *P. pneumotropica* infection. Fetal death and resorption occurred in approximately 30% of fetuses in infected animals. Abundant growth of *P. pneumotropica* was recovered from the lungs of affected dams (Carthew and Aldred 1988). An outbreak of chronic necrotizing mastitis in Fischer 344 rats was attributed to the organism (Hong and Ediger 1978).

PATHOLOGY. Lesions associated with pasteurellosis include rhinitis, sinusitis, conjunctivitis, otitis media, suppurative bronchopneumonia, subcutaneous abscessation, suppurative or chronic necrotizing mastitis, and pyometra. On the other hand, the organism may be recovered from various tissues in the absence of lesions.

DIAGNOSIS. Recovery of the organism from lesions in pure culture is an important step in confirming the diagnosis. *Differential diagnosis* includes abscessation due to other pyogenic organisms such as staphylococci, *Corynebacterium*, or *Pseudomonas*.

SIGNIFICANCE. *P. pneumotropica* represents a potential opportunistic infection in the laboratory rat, frequently in association with a recognized primary pathogen such as *M. pulmonis*.

Bordetella bronchiseptica Infection

B. bronchiseptica is a small, motile, gram-negative bacillus that grows readily on conventional laboratory media, producing small blue-gray colonies.

EPIZOOTIOLOGY AND PATHOGENESIS. The organism is a relatively common inhabitant of the upper respiratory tract of species such as the guinea pig and domestic rabbit. It is recognized to be an important pathogen in the guinea pig and has been associated with respiratory tract infections in the rabbit. The organism tends to colonize on the apices of respiratory epithelial cells, resulting in impaired clearance by ciliated epithelial cells (Bemis and Wilson 1985).

PATHOLOGY. Aerosol exposure to *B. bronchiseptica* in laboratory rats has resulted in lesions characterized by supurative rhinitis; multifocal bronchopneumonia, with polymorphonuclear cell and lymphocytic infiltration; and peribronchial lymphoid hyperplasia. In animals examined at 2 or more wk postinoculation, there were fibroblast proliferation and mononuclear cell infiltration (Burek et al. 1972). In spontaneous cases of bronchopneumonia associated with *Bordetella* infection, there has been a suppurative bronchopneumonia with consolidation of affected anteroventral areas of the lung. Frequently there is an identifiable concurrent infection, such as rat coronaviral infection.

DIAGNOSIS. The isolation of the organism in large numbers from affected tissues is required.

SIGNIFICANCE. *B. bronchiseptica* is a bone fide opportunistic pathogen in the laboratory rat. When isolated from the respiratory tract in rats with lesions, it is likely that there are concurrent infections with other pathogens, such as mycoplasmal or viral agents.

Haemophilus Infection

A hitherto unknown species of *Haemophilus* has been isolated from the nasal cavity, trachea, lungs, and female genital tract. In rats sampled from one vendor, the organism was recovered from a significant percentage of animals, and antibodies to the *Haemophilus* sp. were detected in close to 50% of animals tested. On microscopic examination, mild inflammatory cell infiltrates were present in the lower respiratory tract. The organism was categorized as a member of the family Pasteurellaceae, *Haemophilus* sp. (Nicklas 1989). Co-infection with other respiratory pathogens was not fully investigated.

SIGNIFICANCE. The prevalence of this organism has not been determined. In view of the sites of colonization and the presence of lesions in the respiratory tract, this unclassified *Haemophilus* sp. represents a possible complicating factor in the laboratory rat under experiment.

Streptococcal (Enterococcal) Enteropathy in Infant Rats

Epizootics of enteric disease due to enterococcal infections, with high morbidity and mortality, have been observed in suckling rats. The causative agent, which had been identified as a streptococcus, has now been identified as *Enterococcus faecium-durans-2*. The disease was reproduced in suckling rats inoculated with pure cultures of the streptococcus (Hoover et al. 1985). A similar disease and agent have been seen in other outbreaks in outbred albino rats. Other strains, including *Enterococcus hirae*, have been associated with diarrhea with low mortality in neonatal rats (Etheridge and Vonderfecht 1992; Etheridge et al. 1988).

PATHOLOGY. At necropsy, animals are stunted, with distended abdomens and fecal soiling in the perineal region. The stomachs are usually distended with milk, and there is dilation of the small intestine. On microscopic examination, large numbers of gram-positive cocci may be present on the surfaces of histologically normal villi of the small intestine, with minimal or no inflammatory response (Figs. 2.31, 2.32).

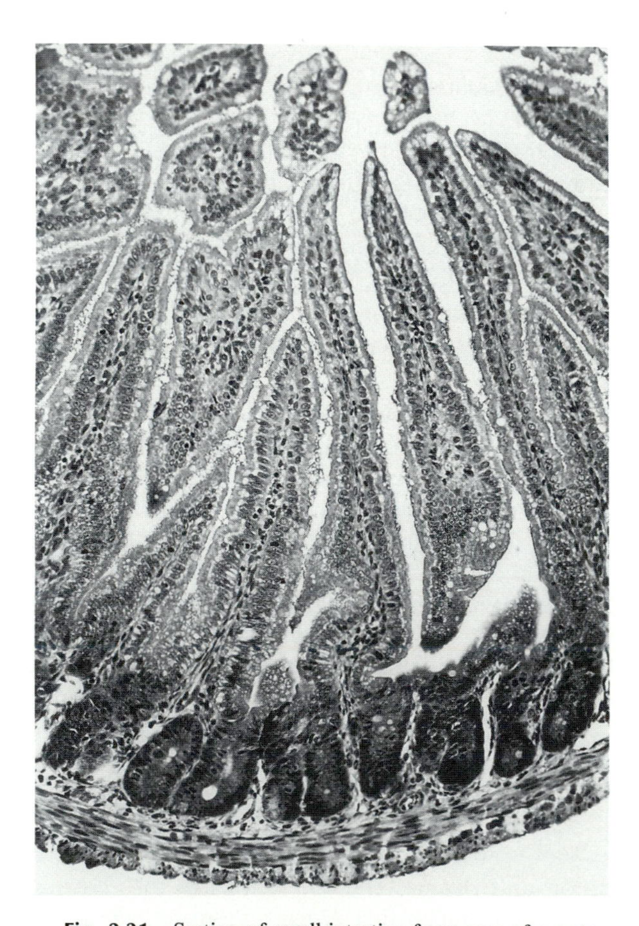

Fig. 2.31. Section of small intestine from case of spontaneous streptococcal enteropathy in suckling rat. The morphology of the villi and enterocytes are essentially normal. Coccoid organisms are adherent to the mucosa. (Courtesy of D.M. Hoover)

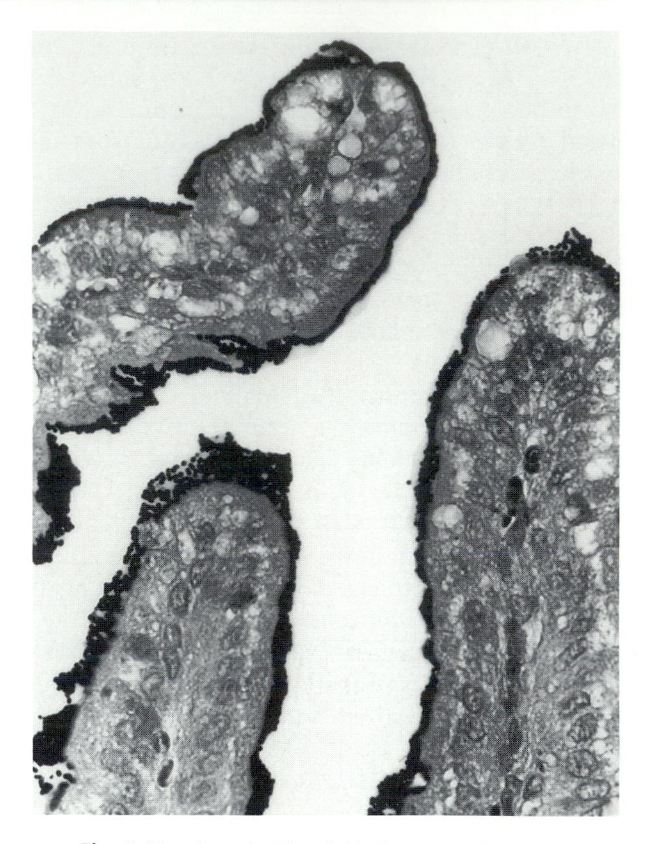

Fig. 2.32. Area in Fig. 2.31 (Brown and Brenn stain). Note the aggregations of cocci on the surface of the villi. (Courtesy of D. M. Hoover)

Ulcerative Dermatitis Associated with Staphylococcal Infections

Ulcerative skin lesions have been observed in Sprague-Dawley and other strains of rats ranging in age from 5 wk to 2 yr old. A relatively small percentage of animals are usually affected.

EPIZOOTIOLOGY AND PATHOGENESIS. The incidence of ulcerative dermatitis may vary from 1–2% to over 20% in certain populations of rats. Lesions are most common in males. Coagulase-positive *Staphylococcus aureus* has been isolated both from lesions and the skin of clinically normal animals (Fox et al. 1977). *S. aureus* has also been isolated from the feces and oropharynx of some affected animals. Toenail clipping or amputation of the toes of the hind feet has resulted in remission of skin lesions, emphasizing the role of self-trauma in the development of the disease (Wagner et al. 1977). Subcutaneous inoculation of *S. aureus* resulted in lesions in 25–40% of inoculated rats, usually at a site other than the inoculation site. Trauma with persistent irritation appears to be an important contributing factor. The possibility of the introduction of staphylococci by human carriers has been proposed (Ash 1971).

PATHOLOGY. Irregular, circumscribed red ulcerative skin lesions occur over the shoulder and rib cage, submandibular regions, neck, ears, and head, with hair loss in the area (Fig. 2.33). In acute cases, microscopic examination of lesions reveals an ulcerative dermatitis, with extension into the underlying dermis. In adjacent areas, there is hyperplasia of the epidermis, with leukocytic infiltration into the underlying dermis, neutrophils predominating (Fig. 2.34). Large colonies of gram-positive coccoid bacteria may be present within the proteinaceous material on the surface of lesions. There may be variable degeneration and leukocytic infiltration of the adnexae in affected areas. In lesions of some duration, dermal sclerosis and mononuclear cell infiltration are prominent features. Healed lesions frequently have dense collagenous tissue in the dermis, with loss of hair follicles and other adnexae.

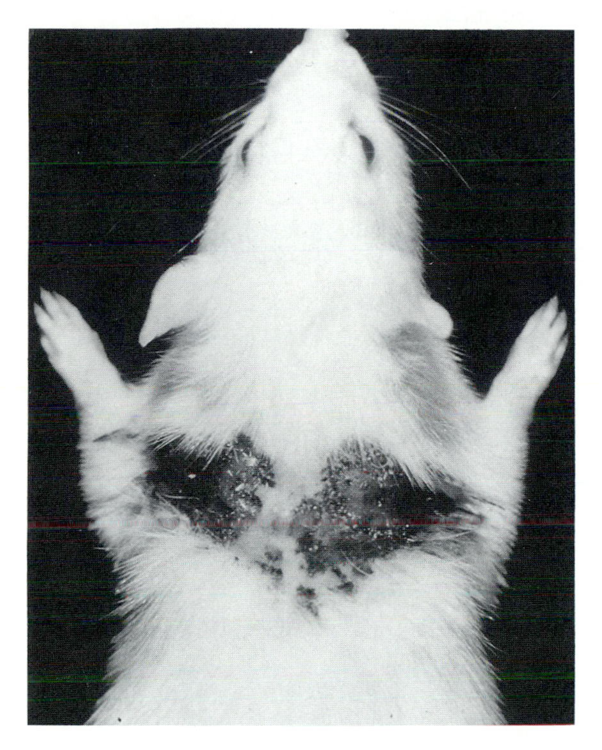

Fig. 2.33. Ulcerative dermatitis in Sprague-Dawley rat, with localization to the dorsal neck and interscapular regions.

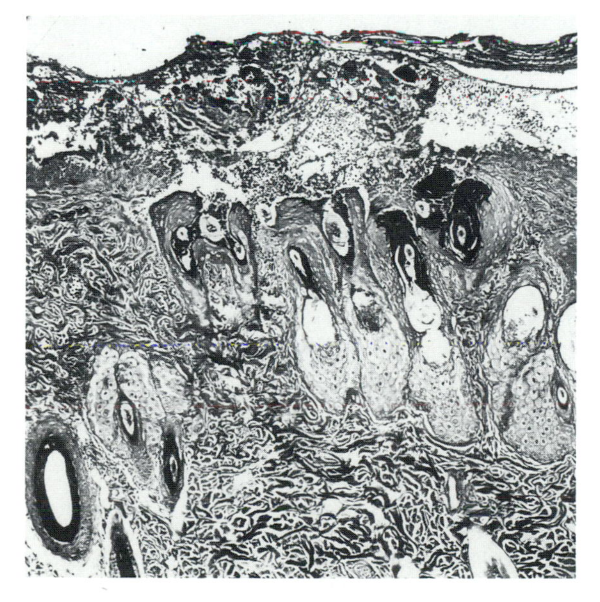

Fig. 2.34. Histological section from rat with ulcerative dermatitis. There is marked epidermal hyperplasia, with ulceration and leukocytic infiltration in the underlying dermis.

DIAGNOSIS. The presence of the typical ulcerative lesion is a useful diagnostic criterion. Bacterial cultures frequently are positive for coagulase-positive *S. aureus*. *Differential diagnoses* include mycotic infections, fighting injuries, and rarely skin lesions associated with epitheliotropic lymphoid tumors (mycosis fungoides).

Suppurative Pyelonephritis/Nephritis

Suppurative renal lesions are more commonly encountered in male rats and may be associated with a concurrent disease process, such as cystitis or prostatitis. A variety of bacteria have been recovered from affected kidneys, including *Escherichia coli*, *Klebsiella*, *Pseudomonas*, and *Proteus* spp. The lower urinary tract is considered to be the most likely portal of entry, particularly in pyelonephritis. However, descending infections may also occur (Duprat and Burek 1986). Lesions are similar to those seen in other species with suppurative disease processes of this system.

Klebsiella pneumonia Infection

K. pneumonia is probably an opportunistic pathogen in the rat, since it can be isolated from feces of normal animals. However, this organism has been associated with abscesses of cervical, inguinal, and mesenteric lymph nodes and of kidney (Jackson et al. 1980). This organism has also been associated with mild suppurative rhinitis in otherwise pathogen-free rats.

Leptospirosis (See the discussion of leptospirosis in Chap. 1.)

Streptobacillus moniliformis Infection

This bacterium is a commensal organism that inhabits the nasopharynx of rats. Contemporary colonies are usually devoid of this organism, but it is common in wild rats and in some populations of laboratory rats. It can be associated with opportunistic respiratory infections and can cause wound infections. Of particular concern is its pathogenicity in humans and mice. In humans, it is the cause of rat bite fever (Anderson et al. 1983). A similar syndrome, called Haverill fever, has been associated with ingestion of rat-contaminated foodstuffs, particularly milk. The bacterium is a gram-negative, pleomorphic rod or filamentous organism that requires blood, serum, or ascities fluid for culture in agar or broth. Another commensal bacterium, *Spirillum muris*, has also been associated with rat bite fever.

Campylobacter Infection

Campylobacter infects a wide range of host species and is emerging as an important intestinal pathogen. *Campylobacter* has been isolated from young rats with mild diarrhea or soft feces, and its frequency may be higher than suspected if it is searched for among colonies of laboratory rats. The lesions resemble those seen in proliferative bowel disease of hamsters and other species, which may not be *Campylobacter* (see Chap. 3).

Erysipelas

Spontaneous erysipelas infections were observed in one outbreak that occurred in laboratory rats in Scandinavia. Lesions included chronic fibrinopurulent polyarthritis, myocarditis, and endocarditis. *E. rhusiopathiae* was isolated from affected joints (Feinstein & Eld 1989).

Rickettsial Disease

Hemobartonella muris infects wild rats, and at one time was common in laboratory rats. The organism, an extracellular parasite of erythrocytes, cannot be cultured in vitro. It is transmitted primarily by *Polyplax spinulosa*, which is no longer found in laboratory rats. It can also be transmitted in utero, but apparently inefficiently, since cesarean section is usually successful at eliminating the organism. It can also contaminate biological products derived from infected rodents and is infectious in both rats and mice. Natural infections are invariably inapparent, with mild transient parasitemia, splenomegaly, and reticulosis of erythrocytes. The reticuloendothelial system, especially the spleen, is critical for clearing the parasitemia. Splenectomy of carrier rats may result in hemolytic anemia with hemoglobinuria and death. Splenectomy has been used as a means of diagnosis in suspect populations of rats. Immunosuppression with corticosteroids is ineffective in activating a subclinical infection (Weisbroth 1979).

BIBLIOGRAPHY FOR BACTERIAL INFECTIONS

Murine Respiratory Mycoplasmosis

Aguila, H.N., et al. 1988. Experimental *Mycoplasma pulmonis* infection of rats suppresses humoral but not cellular immune response. Lab. Anim. Sci. 38:138–42.

Brennan, P.C., et al. 1969. The role of *Pasteurella pneumotropica* and *Mycoplasma pulmonis* in murine pneumonia. J. Bacteriol. 97:337–49.

Broderson, J.R., et al. 1976. Role of environmental ammonia in respiratory mycoplasmosis of the rat. Am. J. Pathol. 85:115–30.

Cassell, G.H. 1982. The pathogenic potential of mycoplasmas: *Mycoplasma pulmonis* as a model. Rev. Infect. Dis. 4:S18–S34.

Cassell, G.H., et al. 1986. Mycoplasmal infections: Disease pathogenesis, implications for biomedical research, and control. In *Viral and Mycoplasmal Infections of Laboratory Rodents: Effects on Biomedical Research*, ed. P.N. Bhatt et al., pp. 87–130. New York: Academic.

Davidson, M.K., et al. 1981. Comparison for methods of detection of *Mycoplasma pulmonis* in experimentally and naturally infected rats. J. Clin. Microbiol. 14:646–55.

Davis, J.K., and Cassell, G.H. 1982. Murine respiratory mycoplasmosis in LEW and F344 rats: Strain differences in lesion severity. Vet. Pathol. 19:280–93.

Lindsey, J.R. 1986. Prevalence of viral and mycoplasmal infections in laboratory rodents. In *Viral and Mycoplasmal Infections of Laboratory Rodents: Effects on Biomedical Research*, ed. P.N. Bhatt et al., pp. 801–8. New York: Academic.

Schoeb, T.R., and Lindsey, J.R. 1987. Exacerbation of murine respiratory mycoplasmosis by sialodacryoadenitis virus infection in gnotobiotic F344 rats. Vet. Pathol. 24:392–99.

Schoeb, T.R., et al. 1985. Exacerbation of murine respiratory mycoplasmosis in gnotobiotic F344/N rats by Sendai virus infection. Vet. Pathol. 22:272–82.

Schreiber, H., et al. 1972. Induction of lung cancer in germ-free, specific-pathogen-free, and infected rats by ^{N}-nitrosoheptamethyleneimine: Enhancement by respiratory infection. J. Natl. Cancer Inst. 49:1107–14.

Tully, J.G. 1986. Biology of rodent mycoplasmas. In *Viral and Mycoplasmal Infections of Laboratory Rodents: Effects on Biomedical Research*, ed. P.N. Bhatt et al., pp. 64–85. New York: Academic.

Respiratory Disease due to CAR Bacillus

Ganaway, J.R., et al. 1985. Isolation, propagation, and characterization of a newly recognized pathogen, cilia-associated respiratory bacillus of rats, an etiological agent of chronic respiratory disease. Infect. Immunol. 47:472–79.

MacKenzie, W.F., et al. 1981. A filamentous bacterium associated with respiratory disease in wild rats. Vet. Pathol. 18:836–39.

Matsushita, S. 1986. Spontaneous respiratory disease associated with cilia-associated respiratory (CAR) bacillus in a rat. Jpn. J. Vet. Sci. 48:437–40.

Matsushita, S. and Joshima, H. 1989. Pathology of rats intranasally inoculated with cilia-associated respiratory bacillus. Lab. Anim. 23:89–95.

Van Zwieten, M.J., et al. 1980. Respiratory disease in rats associated with a filamentous bacterium: A preliminary report. Lab. Anim. Sci. 30:215–21.

Pseudomonas aeruginosa Infection

Flynn, R.J. 1963. Introduction: *Pseudomonas aeruginosa* infection and its effects on biological and medical research. Lab. Anim. Sci. 13:1–6.

Weisbroth 1979 (*see* General Bibliography).

Wyand, D.S., and Jonas, A.M. 1967. *Pseudomonas aeruginosa* infection in rats following implantation of an indwelling jugular catheter. Lab. Anim. Care 17:261–66.

Corynebacterium kutscheri Infection

Ackerman, J.I., et al. 1984. An enzyme linked immunoabsorbent assay for detection of antibodies to *Corynebacterium kutscheri* in experimentally infected rats. Lab. Anim. Sci. 34:38–43.

Barthold, S.W., and Brownstein, D.G. 1988. The effect of selected viruses on *Corynebacterium kutscheri* infection in rats. Lab. Anim. Sci. 38:580–83.

Brownstein, D.G., et al. 1985. Experimental *Corynebacterium kutscheri* infection in rats: Bacteriology and serology. Lab. Anim. Sci. 35:135–38.

Fox, J.G., et al. 1987. Comparison of methods to diagnose an epizootic of *Corynebacterium kutscheri* pneumonia in rats. Lab. Anim. Sci. 37:72–75.

Giddens, W.E., et al. 1969. Pneumonia in rats due to infection with *Corynebacterium kutscheri*. Pathol. Vet. 5:227–37.

McEwen, S.A., and Percy, D.H. 1985. Diagnostic exercise: Pneumonia in a rat. Lab. Anim. Sci. 35:485–87.

Weisbroth 1979 (*see* General Bibliography).

Streptococcosis

Borkowski, G.L., and Griffith, J.W. 1990. Diagnostic exercise: Pneumonia and pleuritis in a rat. Lab. Anim. Sci. 40:323–25.

Fallon, M.T., et al. 1988. Inapparent *Streptococcus pneumoniae* type 35 infections in commercial rats and mice. Lab. Anim. Sci. 38:129–32.

Kohn, D.F., and Barthold, S.W. 1984. *Laboratory Animal Medicine*. New York: Academic.

Weisbroth, S.H., and Freimer, E.H. 1969. Laboratory rats from commercial breeders as carriers of pathogenic pneumococci. Lab. Anim. Care 19:473–78.

Yoneda, K., and Coonrod, J.D. 1980. Experimental type 25 pneumococcal pneumonia in rats. Am. J. Pathol. 99:231–42.

Tyzzer's Disease

Fleischman, R.W., et al. 1977. Adynamic ileus in the rat induced by chloral hydrate. Lab. Anim. Sci. 27:238–43.

Fries, A.S. 1979. Studies on Tyzzer's disease: Transplacental transmission of *Bacillus piliformis* in rats. Lab. Anim. 13:43–46.

Fries, A.S., and Svendsen, O. 1978. Studies on Tyzzer's disease in rats. Lab. Anim. 12:1–4.

Fujiwara, K., et al. 1981. Serologic detection of inapparent Tyzzer's disease in rats. Jpn. J. Exp. Med. 51:197–200.

Hansen, A.K., et al. 1992. Rederivation of rat colonies seropositive for *Bacillus piliformis* and subsequent screening for antibodies. Lab. Anim. Sci. 42:444–48.

Jonas, A.M., et al. 1970. Tyzzer's disease in the rat: Its possible relationship with megaloileitis. Arch. Pathol. 90:516–28.

Motzel, S.L., and Riley, L.K. 1992. Subclinical infection and transmission of Tyzzer's disease in rats. Lab. Anim. Sci. 42:439–43.

Tyzzer, E.E. 1917. A fatal disease of the Japanese waltzing mouse caused by a spore-bearing bacillus (*Bacillus piliformis* n.sp.). J. Med. Res. 37:307–8.

Salmonellosis

Pappenheimer, A.M., and Von Wedel, H. 1914. Observations on a spontaneous typhoid-like epidemic of white rats. J. Infect. Dis. 14:180–215.

Weisbroth 1979 (*see* General Bibliography).

Pasteurella pneumotropica Infection

Brennan, P.C., et al. 1969. The role of *Pasteurella pneumotropica* and *Mycoplasma pulmonis* in murine pneumonia. J. Bacteriol. 97:337–49.

Burek et al. 1972 (*see* General Bibliography).

Carthew, P., and Aldred, P. 1988. Embryonic death in pregnant rats owing to intercurrent infection with Sendai virus and *Pasteurella pneumotropica*. Lab. Anim. 22:92–97.

Hong, C.C., and Ediger, R.D. 1978. Chronic necrotizing mastitis in rats caused by *Pasteurella pneumotropica*. Lab. Anim. Sci. 28:317–20.

Moore, T.D., et al. 1973. Latent *Pasteurella pneumotropica* infection in the intestine of gnotobiotic and barrier-held rats. Lab. Anim. Sci. 23:657–61.

Weisbroth 1979 (*see* General Bibliography).

Bordetella bronchiseptica Infection

Bemis, D.A., and Wilson, S.A. 1985. Influence of potential virulence determinants on *Bordetella bronchiseptica*–induced ciliostasis. Infec. Immunol. 50:35–42.

Burek et al. 1972 (*see* General Bibliography).

Haemophilus Infection

Nicklas, W. 1989. Haemophilus infection in a colony of laboratory rats. J. Clin. Microbiol. 27:1636–39.

Streptococcal (Enterococcal) Enteropathy in Infant Rats

Etheridge, M.E., and Vonderfecht, S.L. 1992. Diarrhea caused by a slow-growing enterococcus-like agent in neonatal rats. Lab. Anim. Sci. 42:548–50.

Etheridge, M.E., et al. 1988. *Enterococcus hirae* implicated as a cause of diarrhea in suckling rats. J. Clin. Microbiol. 26:1741–44.

Hoover, D., et al. 1985. Streptococcal enteropathy in infant rats. Lab. Anim. Sci. 35:653–41.

Ulcerative Dermatitis

Ash, G.W. 1971. An epidemic of chronic skin ulceration in rats. Lab. Anim. 5:115–22.

Fox, J.G., et al. 1977. Ulcerative dermatitis in the rat. Lab. Anim. Sci. 27:671-78.

Wagner, J.E., et al. 1977. Self trauma and *Staphylococcus aureau* in ulcerative dermatitis of rats. J. Am. Vet. Med. Assoc. 171:839–41.

Suppurative Pyelonephritis/Nephritis

Duprat, P., and Burek, J.D. 1986. Suppurative nephritis, pyelonephritis, rat. In *Monographs on Pathology of Laboratory Animals: Urinary System*, ed. T.C. Jones et al. pp. 219–24. New York: Springer-Verlag.

Klebsiella pneumonia Infection

Harwich, J., and Shouman, M.T. 1965. Untersuchen uber gehauft auftretende *Klebsiella*-infectionen bei versuchsratten. Z. Versuch. 6:141–46.

Jackson, N.N., et al. 1980. Naturally acquired infections of *Klebsiella pneumonia* in Wistar rats. Lab. Anim. 14:357–61.

Streptobacillus moniliformis Infection

Anderson, L.C., et. al. 1983. Rat-bite fever in animal research laboratory personnel. Lab. Anim. Sci. 33:292–94.

Erysipelas

Feinstein, R.E., and Eld, K. 1989. Naturally occurring erysipelas in rats. Lab. Anim. 23:256–60.

Rickettsial Disease

Weisbroth 1979 (*see* General Bibliography).

General Bibliography

Burek, J.D., et al. 1972. The pathology and pathogenesis of *Bordetella bronchiseptica* and *Pasteurella pneumotropica* infection in conventional and germ-free rats. Lab. Anim. Sci. 22:844–49.

Kohn, D.F., and Barthold, S.W. 1984. Biology and diseases of rats. In *Laboratory Animal Medicine*, ed. J.G. Fox et al., pp. 91–122. New York: Academic.

Lindsey, J.R. 1986. Prevalence of viral and mycoplasmal infections in laboratory rodents. In *Viral and Mycoplasmal Infections of Laboratory Rodents: Effects on Biomedical Research,* ed. P.N. Bhatt et al., pp. 801–8. New York: Academic.

Shultz, L.D., and Sidman, C.L. 1987. Genetically determined murine models of immunodeficiency. Ann. Rev. Immunol. 5:367–403.

Weisbroth, S.H. 1979. Bacterial and mycotic diseases. In *The Laboratory Rat. I. Biology and Diseases*, ed. H.J. Baker et al., pp. 193–241. New York: Academic.

MYCOTIC INFECTIONS

Dermatophytosis (Dermatomycosis; "Ringworm")

Recognized infections with dermatophytes are now relatively rare in laboratory rats, although it appears to occur more frequently in wild rats. The disease is usually caused by *Trichophyton mentagrophytes* in this species. In confirmed outbreaks of the disease, patterns vary from asymptomatic carriers (Balsari et al. 1981) to rats with florid lesions on the skin.

PATHOLOGY. Lesions, when present, are most frequently observed on the neck, back, and at the base of the tail. There is patchy hair loss, and affected areas of skin are usually raised, erythematous, and dry to moist and pustular in appearance. On microscopic examination, hyperkeratosis, epidermal hyperplasia, and leukocytic infiltration in the underlying dermis with folliculitis are typical findings. Arthrospores investing hair shafts may be seen on H & E–stained tissue sections, but the fungi are better demonstrated with PAS or methenamine silver stains (Fig. 2.35).

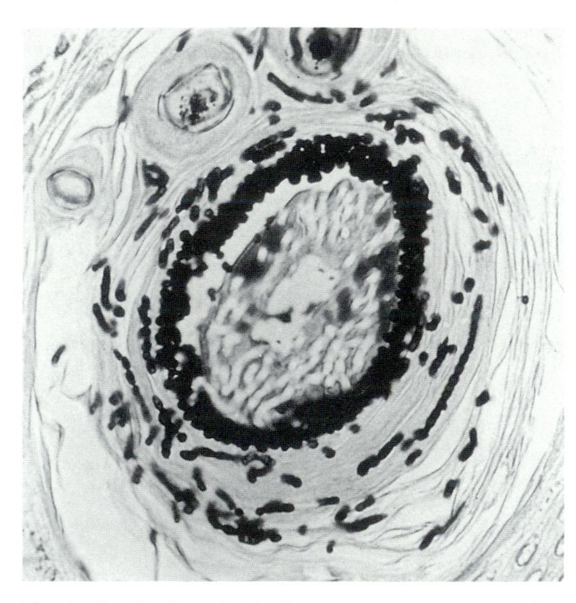

Fig. 2.35. Section of skin from spontaneous case of dermatophytosis in laboratory rat (methenamine silver stain). Numerous arthrospores are investing the hair follicle.

DIAGNOSIS. The skin scrapings with wet mount preparations in 10% KOH under a vaseline-ringed coverslip and fungal culture are recommended procedures. Skin biopsies or histological sections collected at necropsy should enable the pathologist to demonstrate the typical arthrospores microscopically.

SIGNIFICANCE. In view of the possibility of interspecies spread, a slaughter policy and thorough disinfection of premises and fomites are recommended. The source of the infection should be investigated, including the possibility that the organism could have been introduced by wild rodents or human contacts.

BIBLIOGRAPHY FOR MYCOTIC INFECTIONS

Balsardi, A., et al. 1981. Dermatophytes in clinically healthy laboratory animals. Lab. Anim. 15:75–77.

Weisbroth, S.H. 1979. Bacterial and mycotic diseases. In *The Laboratory Rat: I. Biology and Diseases*, ed. H.J. Baker et al., pp. 193–241. New York: Academic.

PARASITIC DISEASES

In addition to the parasitic infestations outlined in this section, there are other parasites that are rarely seen in well-managed facilities. For additional information on the biology and identification of parasites in this species, consult Harkness and Wagner (1989), Hsu (1979), and Flynn (1973).

ECTOPARASITIC INFECTIONS

Ectoparasites are not an important consideration in laboratory rats, although they are relatively common in the wild rat. They are host to two species of lice, *Polyplax spinulosa* (spined rat louse) and *Hoplopleura pacifica* (tropical rat louse), of which only the former has been described in laboratory rats. Polyplax was once an important vector for *Hemobartonella muris* among rats. It can be associated with pruritis, irritability, and anemia, caused directly by feeding and indirectly by *Hemobartonella*. Fleas of several genera, including *Xenopsylla*, *Leptopsylla*, and *Nosopsyllus*, infest wild rats, and rarely laboratory rats. Several different types of mites can infest rats, but all are rare in laboratory rats, except *Radfordia ensifera* (*Myobia ratti*), the fur mite, which can be common in some populations. Pruritis, hair loss, and loss of condition are associated with heavy infestations. The mite may be demonstrated in the pellage and may also be evident in tissue sections of affected skin (Fig. 2.36). *Or-*

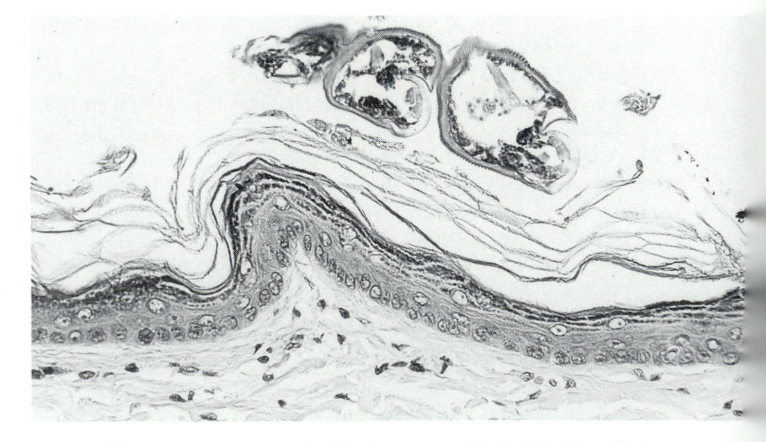

Fig. 2.36. Section of skin from rat with *Redfordia affinis* infestation. Mites are present on the stratum corneum, with minimal reaction in the underlying skin.

nithonyssus bacoti (tropical rat mite) is nonselective in its host range and infests rats on occasion. These mites are associated with rats only while feeding; then they seek refuge in the surrounding environment. Their bites are pruritic, as animal handlers can attest, and they can cause anemia, debility, and infertility. Other mites that reside permanently on the skin or fur of rats include *Demodex* spp., which have been found in follicles as an incidental finding, and *Notoedres muris*, which burrows in the cornified epithelium of the ear and other hairless skin sites. It is frequently referred to as the ear mange mite.

Endoparasitic Infections

Protozoal Infections

Giardia muris and *Spironucleus muris* are flagellates that reside in the small intestine of rats and other rodents. Giardia organisms occur as rounded to crescent-shaped structures situated along the surface of the gut mucosa. *Spironucleus* are visible as pear-shaped organisms in the intestinal crypts approximately 8 x 2.5 μm. These organisms are not considered to be pathogenic in rats under normal conditions, unless there are important predisposing factors such as procedures resulting in immunosuppression (Barthold 1985a,b). Catarrhal enteritis and weight loss are typical signs, particularly with spironucleosis (Hsu 1979).

SIGNIFICANCE. Although there have been reports of aberrations in the immune response in mice with spironucleosis, there is no evidence of a similar effect in the rat (Mullink et al. 1980).

***Pneumocystis carinii* Infection.** *P. carinii*, a microorganism of worldwide distribution, is recognized to be an important cause of disease and mortality in immunocompromised human cases, particularly AIDS patients. Many colonies of rats appear to be naturally infected with *P. carinii*, and there is evidence that laboratory mice and occasionally rabbits may also harbor the parasite (Frenkel 1985). Pneumocystosis has been recognized in other animals, including cats, horses, dogs, and monkeys (Chandler et al. 1976). Pulmonary lesions associated with pneumocystis infection have been readily produced in laboratory rats treated for several weeks with immunosuppressants such as cortisone (Barton and Campbell 1969) and fed a protein-deficient diet. Spontaneous infections have been recognized in athymic rats (Frenkel 1985). In severely affected animals, clinical signs include weight loss, cyanosis, and dyspnea.

PATHOLOGY. There is diffuse to focal consolidation, and lungs are often opaque and pale pink. On microscopic examination, there is alveolar flooding with foamy, eosinophilic material, presenting a honeycomb appearance (Fig. 2.37). In sections stained with preparations such as the Grocott modification of Gomori's methenamine silver technique, numerous black, yeastlike cysts 3–5 μm in diameter are present singly or in groups within alveoli. Intracystic bodies (nuclei) can be demonstrated within cysts by using impression smears of lung stained with the Giemsa method. The classification of the organism has generated considerable discussion, but it has features of both protozoa and fungi.

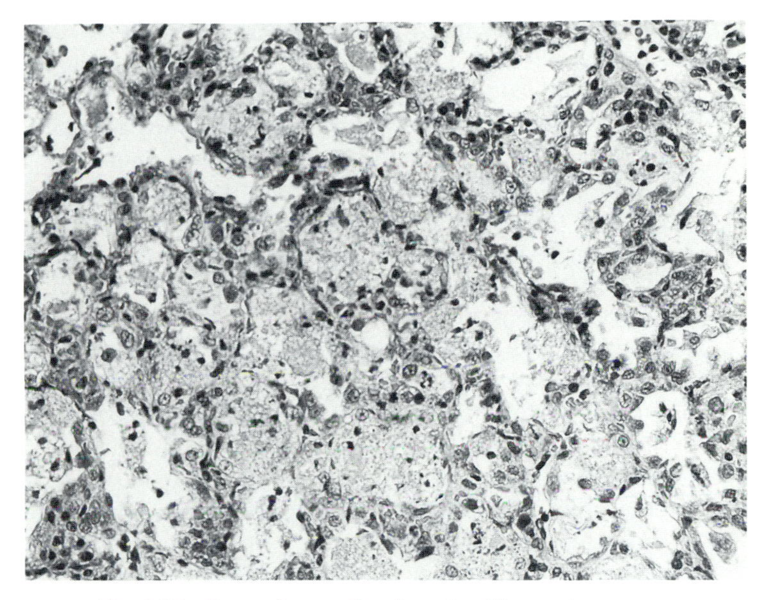

Fig. 2.37. Lung from athymic rat with spontaneous *Pneumocystis carinii* infection. Coccoid organisms are present in the foamy exudate present in alveoli.

DIAGNOSIS. The demonstration of the typical organisms in pulmonary lesions, using silver staining procedures, is the recommended method to confirm the diagnosis.

SIGNIFICANCE. Pneumocystosis normally occurs only in rats subjected to significant predisposing factors, such as immunosuppression or dietary deficiency. Some other species are susceptible under similar circumstances. Spontaneous cases of pneumocystosis may occur in nude mice and in mice with severe combined immunodeficiency (SCID mice) in the absence of immunosuppressive procedures (Walzer et al. 1989).

Cryptosporidiosis. An outbreak of diarrhea and high mortality has been described among infant rats of the Rapp hypertensive strain. Surviving pups were runted and their fur was stained with feces. Lesions in convalescing 21-day-old rats were restricted to the mucosa of the small intestine, primarily jejunum. The mucosa was hyperplastic and villi were shortened and fused, with cryptosporidia attached to the brush borders of enterocytes toward the villus tips. Cryptosporidiosis has also been induced experimentally in rats but is transient and mild unless rats are immunosuppressed or athymic (Moody et al. 1991).

Helminth Infections: Pinworms (Oxyuriasis)

The following oxyurid nematodes are recognized to be infectious for the rat: *Syphacia obvelata*, *S. muris*, and *Aspicularis tetraptera*. They are normally found in the cecum and colon in affected animals. *S. muris* commonly occurs in laboratory and wild rats and is transmissible to the laboratory mouse. These parasites have a direct life cycle. Adults have the characteristic morphologic features of oxyurid worms. Eggs are deposited in the colon or on the perianal area. The eggs embryonate and become infectious within a few hours. Rats may become infected by direct ingestion of embryonated eggs from the perianal region; ingestion of eggs in contaminated food, water, or from fomites; or by direct migration of larvae via the anus to the large intestine. Infected animals are frequently asymptomatic, but younger animals with heavy infections may exhibit various signs, including diarrhea, poor weight gains, impactions, rectal prolapse, and intussusceptions. *Aspicularis tetraptera* frequently occurs in conventional rats and mice. The life cycle is direct. Eggs are passed in the feces and therefore are not found in the perianal region.

DIAGNOSIS. The microscopic demonstration of the characteristic eggs on touch preparations of the anal region (using transparent adhesive tape) is a useful method for *Syphacia*. Eggs may also be demonstrated in stool samples, and adults are visible as small, threadlike worms in the cecum and colon. Adults are also readily visualized in tissue sections of large intestine. Focal submucosal granulomas may be evident in sections of the large intestine on microscopic examination (Figs. 2.38 and 2.39). *S.obvelata* is primarily a pinworm of mice, but rats are also susceptible to infection by this species. The two species can be differentiated by identifying the characteristic morphologic features of the adults and eggs (Hsu 1979). Macroscopic identification of the adults in the large intestine or the identification of the eggs on microscopic examination are the recommended methods to make a positive diagnosis. The perianal adhesive tape method is of little value in making the diagnosis in *Aspicularis* infections.

SIGNIFICANCE OF OXYURIASIS. Except in heavy infections in young rats, clinical signs are usually minimal or absent in affected animals. However, there are concerns regarding possible effects on physiologic processes, such as the immune response. For example, rats infected with *S. obvelata* developed less-severe lesions of adjuvant-induced arthritis than did noninfected animals (Pearson and Taylor 1975). Transmission to other rodents, particularly mice, may also occur.

***Trichosomoides crassicauda* Infection.** *T. crassicauda* infections occur in the urinary bladder of wild

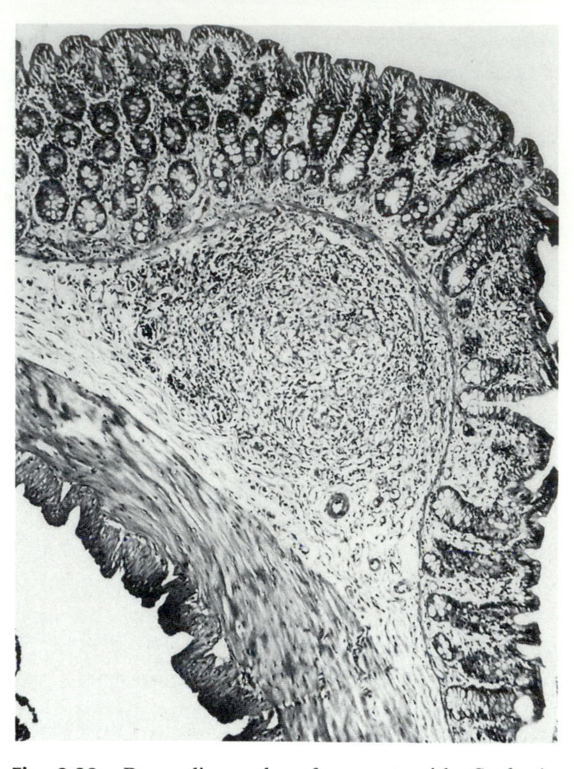

Fig. 2.38. Descending colon from rat with *Syphacia muris* infestation. Note the granuloma in the submucosal region.

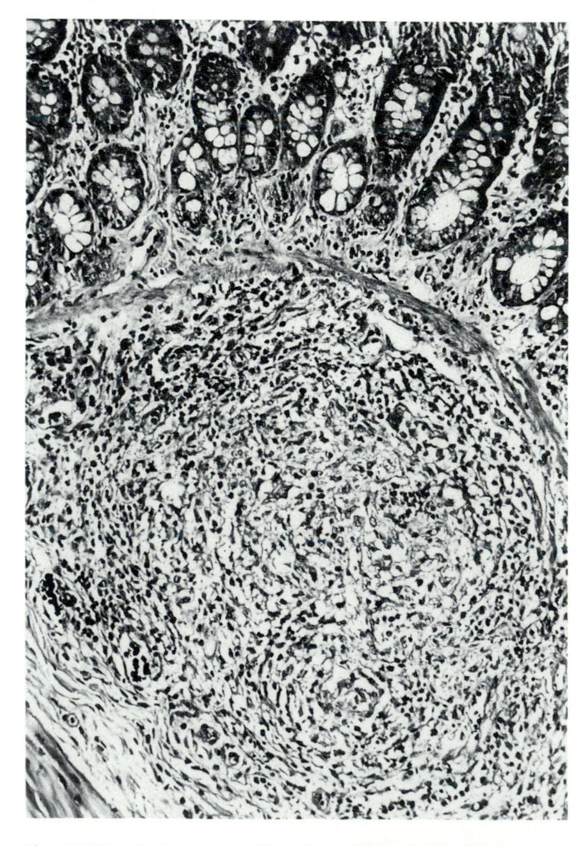

Fig. 2.39. Higher magnification of Fig. 2.38. There were increased numbers of mononuclear cells in the lamina propria, with granulomatous inflammatory response in the submucosa.

rats and, rarely, in laboratory rats. Infected animals are usually asymptomatic, and the threadlike adult worms are found in the lumen of the urinary bladder and renal pelvis at necropsy. In tissue sections examined microscopically, migratory-stage larvae and immature worms may be present in multiple tissues, particularly lungs. Adults reside in the epithelium of the renal pelvis and the urinary bladder, with chronic inflammatory response (Fig. 2.40 and 2.41). Eggs are passed in the urine, and intracage transmission readily occurs. Urinary calculi and bladder tumors have been associated with this parasitic infection, but to date, an unequivocal causal relationship has not been confirmed (Zubaidy and Majeed 1981).

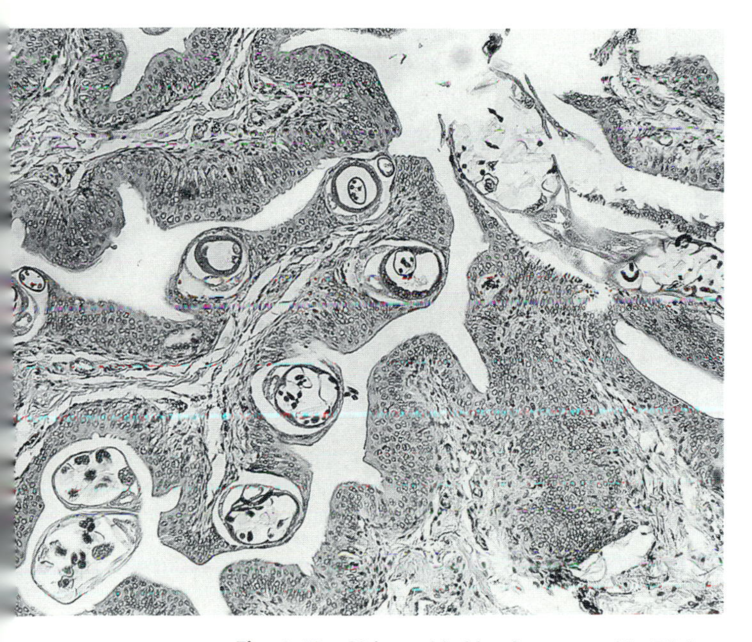

Fig. 2.40. Urinary bladder from rat with *Trichomosoides crassicauda* infestation. Portions of the nematodes are evident in the lumen of the bladder and anterior portions are embedded in the epithelium, with negligible inflammatory cell response.

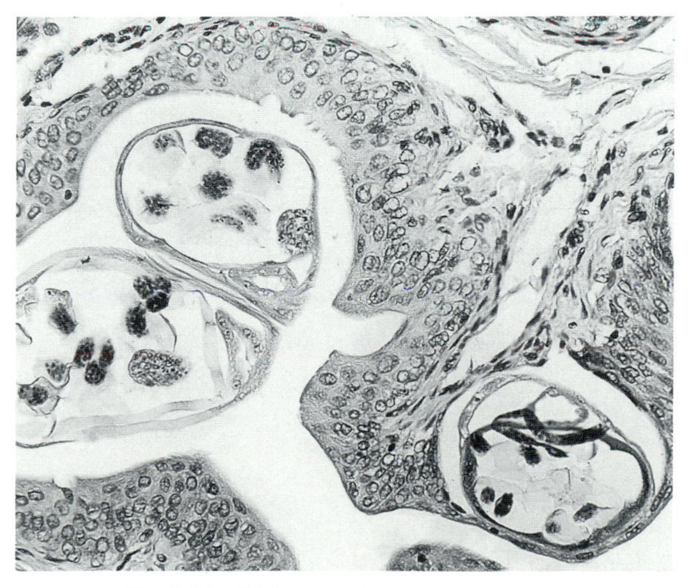

Fig. 2.41. Higher magnification of Fig. 2.40, illustrating the uterine contents of the embedded female worms.

Helminth Infections: Tapeworms

***Hymenolepis* Infections.** *H. nana* and *H. diminuta* may be present in the small intestine of several species, including rats, mice, hamsters, humans, and nonhuman primates. With *H. nana*, the life cycle may be either direct or indirect. In the indirect life cycle, embryonated eggs are ingested by an arthropod host, such as grain beetles or fleas. Ingestion of these arthropods by a susceptible host will then serve as the source of the parasite eggs. In *H. diminuta* infections, an intermediate host, such as beetles or fleas, is essential for the completion of the life cycle.

DIAGNOSIS. Small, flattened white worms in the small intestine at necropsy may be identified as dwarf tapeworms by microscopic examination. Scoleces or cysticercoids may be identified in smears or in tissue sections of small intestine. Eggs can be demonstrated in the feces in order to confirm the diagnosis.

SIGNIFICANCE. Frequently, affected animals are asymptomatic. In a heavy infection, there may be poor weight gain and sometimes catarrhal enteritis. There is a danger of interspecies transmission, including infection in human contacts.

***Taenia taeniaformis* Infection.** *Cysticercus fasciolaris* is the larval stage of *T. taeniaformis*, the cat tapeworm. When eggs of this tapeworm are ingested, they migrate through the bowel and often encyst in the liver of rats, mice, and other rodents. Laboratory rats and mice become infected by contamination of food with cat feces. Usually, only one or two cysts will be found in an infected animal. Parasitism in rats can be associated with the development of sarcomas in the reactive tissue around the cyst (Altman and Goodman 1979).

BIBLIOGRAPHY FOR PARASITIC DISEASES

Ectoparasitic Infections

Flynn, R.J. 1973. *Parasites of Laboratory Animals*. Ames: Iowa State University Press.

Walberg, J.A., et al. 1981. Demodicidosis in laboratory rats (*Rattus norvegicus*). Lab. Anim. Sci. 31:60–62.

Endoparasitic Infections

Altman, N.H., and Goodman, D.G. 1979. Neoplastic diseases. In *The Laboratory Rat. I. Biology and Diseases*, ed. H.J. Baker et al., p. 346. New York: Academic.

Barthold, S.W. 1985a. *Spironucleus muris* infection, intestine, mouse, rat, hamster. In *Monographs on Pathology of Laboratory Animals: Digestive System*, ed. T.C. Jones et al., pp. 356–58. New York: Springer-Verlag.

______. 1985b. *Giardia muris* infection, intestine, mouse, rat, hamster. In *Monographs on Pathology of Laboratory Animals: Digestive System*, ed. T.C. Jones et al., pp. 359–62. New York: Springer-Verlag.

Barton, E.G., and Campbell, W.G. 1969. *Pneumocystis carinii* in lungs of rats treated with cortisone acetate: Ultrastructural observations relating to the life cycle. Am. J. Pathol. 54:209–36.

Chandler, F.W., et al. 1976. Pulmonary pneumocystosis in nonhuman primates. Arch. Pathol. Lab. Med. 100:163–67.

Frenkel, J.K. 1985. Pneumocystosis, lung, rat. In *Monographs on Pathology of Laboratory Animals: Respiratory System*, ed. T.C. Jones et al., pp. 218–23. New York: Springer-Verlag.

Hsu 1979 (*see* General Bibliography).

Moody, K.D., et al. 1991. Cryptosporidiosis in suckling laboratory rats. Lab. Anim. Sci. 41:625–27.

Mullink, J.W.M.A., et al. 1980. Lack of effects of *Spironucleus (Hexamita) muris* on the immune response to tetanus toxoid in the rat. Lab. Anim. 14:127–28.

Pearson, D.J., and Taylor, G. 1975. The Influence of the nematode *Syphacia obvelata* on adjuvant arthritis in rats. Immunology 29:391–96.

Schwabe, C.W. 1955. Helminth parasites and neoplasia. Am. J. Vet. Res. 16:485.

Walzer, P.D., et al. 1989. Outbreaks of *Pneumocystic carinii* in colonies of immunodeficient mice. Infect. Immunol. 57:62–70.

Zubaidy, A.J., and Majeed, S.K. 1981. Pathology of the nematode *Trichosomoides crassicauda* in the urinary bladder of laboratory rats. Lab. Anim. 15:381–84.

General Bibliography

Flynn, R.J. 1973. *Parasites of Laboratory Animals*. Ames: Iowa State University Press.

Harkness, J.E., and Wagner, J.E. 1989. *The Biology and Medicine of Rabbits and Rodents*. Philadelphia: Lea and Febiger.

Hsu, C-K. 1979. Parasitic diseases. In *The Laboratory Rat. I. Biology and Diseases*, ed. H.J. Baker et al., pp. 307–531. New York: Academic.

AGING AND DEGENERATIVE DISORDERS

Chronic Progressive Glomerulonephropathy/Chronic Progressive Nephrosis

The disease has been referred to by a variety of other terms, including glomerulosclerosis, progressive glomerulonephrosis, and "old rat nephropathy."

EPIZOOTIOLOGY AND PATHOGENESIS. The incidence of the disease in older rats varies but may be up to 75% or more in susceptible strains. A variety of predisposing factors may play a role in the development of chronic progressive glomerulonephropathy (PGN). (1) Age: Lesions are usually most extensive in animals at least 12 mo of age. (2) Sex: The disease is considerably more common and more severe in males. (3) Strain: The incidence is usually significantly higher in Sprague-Dawley rats than with most other breeds (Gray et al. 1986). (4) Diet: High-protein diets have been considered to be an important contributing factor (Lalich and Allen 1971), although total dietary restriction, rather than protein content, may be more important in reducing the progression of the disease. (Tapp et al. 1979). PGN has been referred to as a "protein leakage disease," due to increased loss of protein in the glomerular filtrate. The eosinophilic droplets present in epithelial cells lining tubules have been associated with increased lysosomal activity and may be a reflection of functional overload of nephrons (Gray 1986). (5) Immunological factors: Mesangial deposition of IgM has been observed in affected glomeruli consistent with non-complement-fixing immune complexes, but it does not appear to be primarily an immunologically mediated disease (Couser and Stilmant 1975). (6) Endocrine: Prolactin levels have also been implicated as a contributing factor in the development of the disease (Richardson and Luginbuhl 1976). Clinical signs associated with the disease include proteinuria, weight loss, and in advanced cases elevated plasma creatinine levels consistent with renal insufficiency.

PATHOLOGY. The renal cortices are usually pitted and sometimes irregular, with variable degrees of enlargement and pallor in some affected animals (Fig. 2.42). On cut surface, there may be irregularities and linear streaks in the cortex and medulla, with varying degrees of brown pigmentation (Fig. 2.43). Microscopic changes are consistent with a chronic glomerulopathy. Glomerular changes vary from minimal thickening of the basement membranes to marked thickening of glomerular tufts, with segmental sclerosis and adhesions to Bowman's capsule (Fig. 2.44).

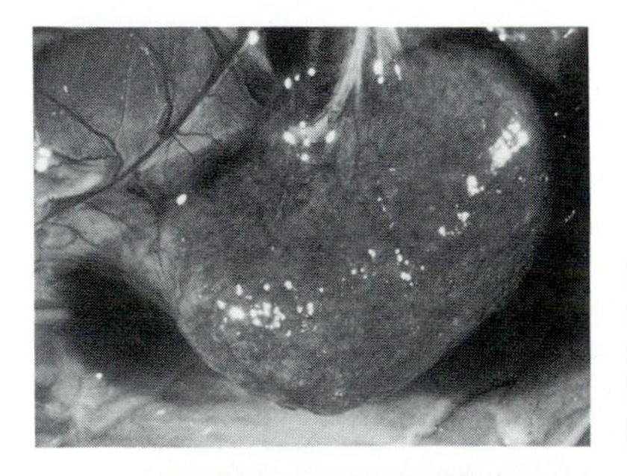

Fig. 2.42. Typical pitting and irregularities of the renal cortical outlines associated with PGN in aged rat.

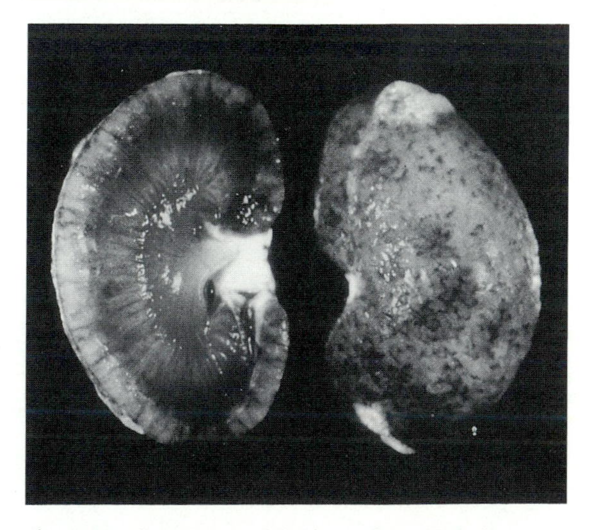

Fig. 2.43. Spontaneous PGN in aged rat. Note the granular appearance to cortical surface and the linear streaks on cut surface.

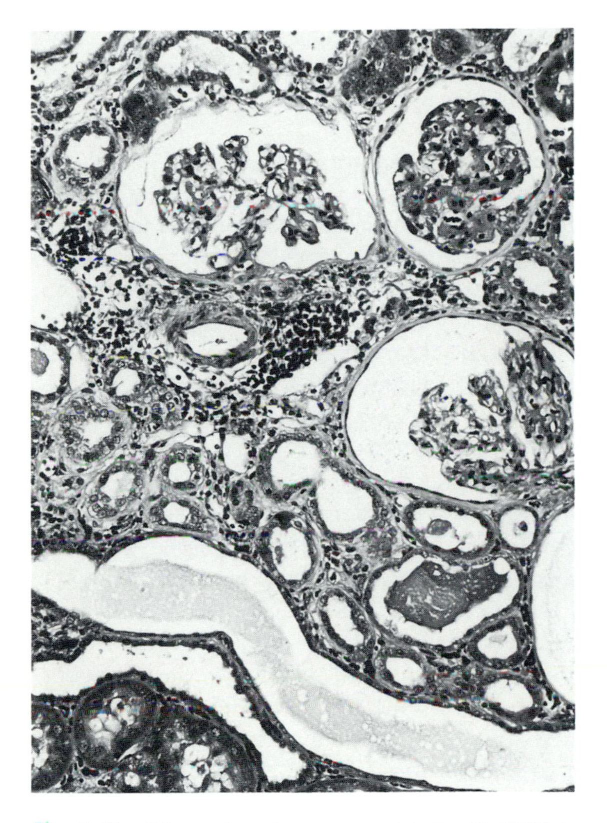

Fig. 2.44. Glomerular changes associated with PGN in aged rat. Note the thickening of glomerular basement membranes and Bowman's capsule. Proteinaceous casts are present in many tubules.

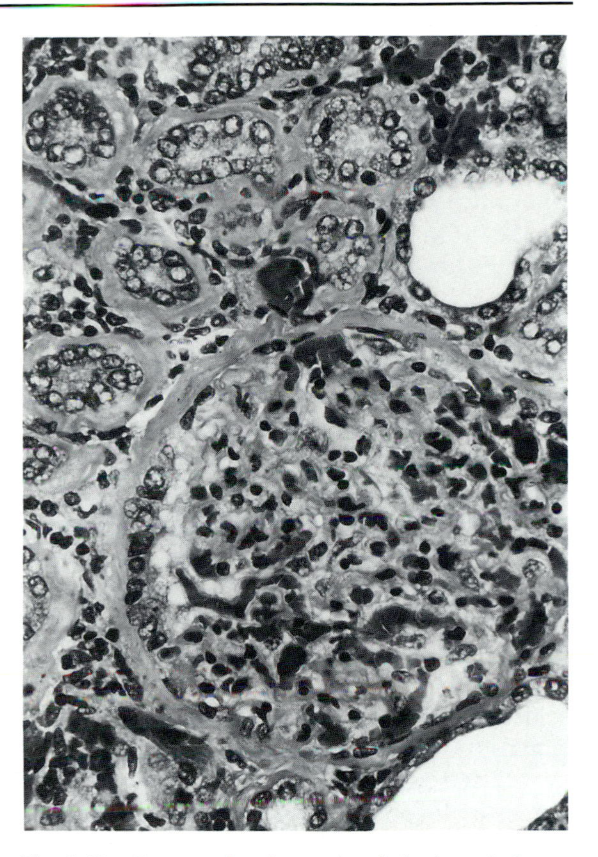

Fig. 2.45. Degenerative changes in tubules in typical case of PGN. There is thickening of basement membranes with flattening of the epithelial cells lining affected tubules.

Proteinaceous casts may be present in tubules in the cortex and medulla. Eosinophilic, PAS-positive resorption droplets are frequently present in epithelial cells lining affected nephrons. The droplets may be iron-positive. Tubules are frequently dilated and lined by flattened epithelial cells; contracted and lined by poorly differentiated, cuboidal, basophilic epithelial cells; or sclerotic. There may be varying degrees of thickening of proximal tubular basement membranes, interstitial fibrosis, and mononuclear cell infiltration (Fig. 2.45). There may be renal secondary hyperparathyroidism, with mineralized deposits in tissues such as kidney, gastrointestinal tract, lungs, and the media of larger arteries. Hypercholesterolemia, hypoproteinemia, and elevated blood urea nitrogen consistent with nephrotic syndrome may be evident in advanced cases. *Differential diagnoses*: PGN must be differentiated from other degenerative nephropathies, such as chronic bacterial pyelonephritis, congenital hydronephrosis, and ischemic injury. Nephrosis associated with toxic insults such as overdosing with aminoglycoside antibiotics is another possible cause of renal injury.

SIGNIFICANCE. Chronic progressive nephropathy is a common disease associated with the aging process in strains such as the Sprague-Dawley rat. Aside from the proteinuria associated with the disease, there may be weight loss and severely affected rats may die due to renal failure. It is a major life-limiting disease in the aged rat. Animals with severe disease appear to cope well but may rapidly decompensate and die. Late-stage disease has been associated with hypertension and polyarteritis nodosa.

Nephrocalcinosis

Renal calcification has been observed on occasion in laboratory rats, including animals on regular commercial diets. The disease has been produced by a variety of dietary manipulations, including those with a low-magnesium content, a high-calcium content, high concentrations of phosphorus, and preparations with a low-calcium-phosphorus ratio (Ritskes-Hoitinga et al. 1989). Lesions are characterized by the deposition of calcium phosphates in the interstitium of the corticomedullary junction, with intratubular aggregations in the same region. In advanced cases, there may be detectable manifestations of renal dysfunction, including albuminuria.

Hydronephrosis

Hydronephrosis is a relatively common incidental finding at necropsy. In some strains, there is a hereditary basis for the disorder. For example, in the brown Norway rat, hydronephrosis appears to be an autosomal polygenetic disorder, with incomplete penetrance (Cohen et al. 1970). In the Gunn rat, it is apparently inherited as a dominant gene and may be lethal when present in the homozygous state (Lozzio et al. 1967). In studies of hydronephrosis in outbred Sprague-Dawley rats, it was concluded that the condition is a highly heritable trait, probably involving more than one gene (Van Winkle et al. 1988). Spontaneous hydronephrosis, particularly of the right kidney, is a well-recognized abnormality, especially in male rats (O'Donaghue and Wilson 1977). Hydronephrosis occurs in a variety of strains and stocks of rats (Maronpot 1986). It has been proposed that the lesion in males may be due to the passage of the internal spermatic vessels across the ureter, resulting in mechanical obstruction to outflow and subsequent hydronephrosis of the affected kidney. However, in one study, sectioning of the right spermatic vessels in young male Wistar-derived rats failed to reduce the incidence of hydronephrosis, compared with control animals (O'Donaghue and Wilson 1977).

PATHOLOGY. At necropsy, there may be varying degrees of involvement. In severely affected animals, the kidney consists of a fluid-filled sac containing clear serous fluid. On microscopic examination, there is marked dilation of the renal pelvis, with excavation of the renal medulla, reduction in the length of the collecting tubules, and absence of an inflammatory response. *Differential diagnoses* include pyelonephritis, polycystic kidneys, and renal papillary necrosis.

SIGNIFICANCE. In most strains, unilateral or bilateral hydronephrosis is often an incidental finding at necropsy. There is a hereditary basis in some strains. The defect may be fatal when bilateral. There may be an increased susceptibility to superimposed renal infections due to urine stasis. Functional abnormalities in renal physiology may occur (Maronpot 1986).

Urinary Calculi

The incidence of spontaneous urolithiasis in nonmanipulated laboratory rats is normally low. Calculi when present in the urinary bladder may be associated with hemorrhagic cystitis, hematuria, and sometimes urinary obstruction. Calculi may also be located at other sites (e.g., renal pelvis, ureter, and urethra). The composition of calculi is variable. Analyses have revealed combinations such as ammonium magnesium phosphate, mixed carbonate and oxalate, and mixed carbonate and phosphate with magnesium and calcium (Paterson 1979).

SIGNIFICANCE. Urinary calculi are frequently sporadic in occurrence. A search for specific contributing factors (e.g., estrogen therapy and water restriction) may be unrewarding. They must not be confused with copulatory plugs, which can form agonally in the urinary bladder and urethra.

Myocardial Degeneration/Fibrosis

Focal to diffuse areas of myocardial degeneration are frequently seen microscopically in conventional and specific-pathogen-free Sprague-Dawley rats, particularly after 1 yr of age. Lesions are more common in male rats. The incidence may be 25% or more in some strains (Anver and Cohen 1979). Endocardial proliferative lesions have also been characterized (Novilla et al. 1991).

PATHOLOGY. At necropsy, there may be moderate to marked ventricular hypertrophy and occasionally pale streaks are evident, but there is usually little evidence of cardiac insufficiency. On microscopic examination, degenerative changes are usually most evident in the papillary muscles of the left ventricle, although the interventricular septum may also be involved. Lesions are characterized by atrophy of myofibers, vacuolation to fragmentation of the sarcoplasm, loss of cross-striations, and mononuclear cell infiltration (Fig. 2.46). Large reactive nuclei are occasionally observed. Interstitial fibrosis, with proliferation of fibrous tissue, is an important microscopic feature of the disease.

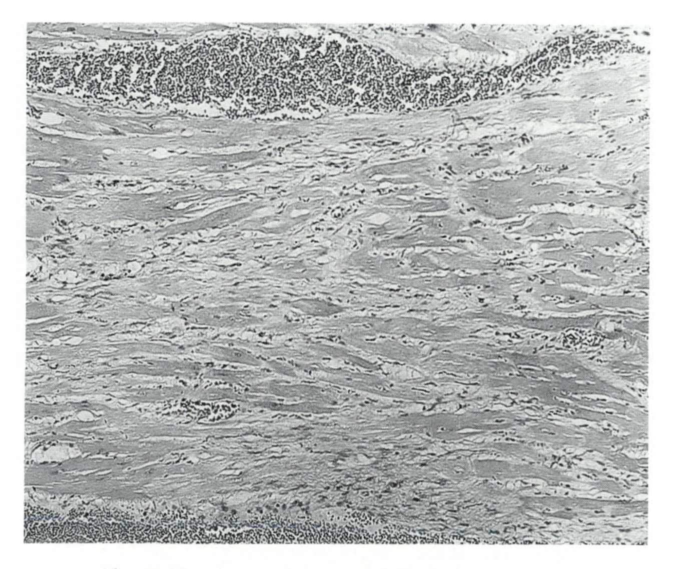

Fig. 2.46. Segmental myocardial degeneration and interstitial fibrosis in aged rat with cardiomyopathy.

SIGNIFICANCE. Although this condition is a frequent microscopic finding in older rats, there may be little or no evidence of cardiac insufficiency. Degenerative myocardial lesions may be secondary to ischemic change (Burek 1978).

Polyarteritis Nodosa

This chronic progressive degenerative disease most frequently occurs in aging rats. The incidence is higher in males. Arterial lesions most frequently occur in medium-size arteries of the mesentery, pancreas, pancreaticoduodenal artery, and testis. The pathogenesis has not been resolved, but an immunologically mediated process is one possible explanation (Bishop 1989; Yang 1965). The spontaneous disease most frequently occurs in the Sprague-Dawley and spontaneous hypertensive rat (SHR) strains, and in rats with late-stage chronic glomerulonephropathy.

PATHOLOGY. At necropsy, affected vessels are enlarged and thickened in a segmental pattern, with marked tortuosity, particularly in the mesenteric vessels (Fig. 2.47). On microscopic examination, vascular lesions may be present in various tissues, such as mesentery, pancreas, testis, kidney, and most other organs except the lung (Bishop 1989). There is fibrinoid degeneration and thickening of the media of affected arteries, with smudging of the normal architecture. Infiltrating leukocytes consist of mononuclear cells, with a few neutrophils. There are marked variations in the size and contours in the lumen of affected vessels, and thromboses, occasionally with recanalization, occur (Fig. 2.48).

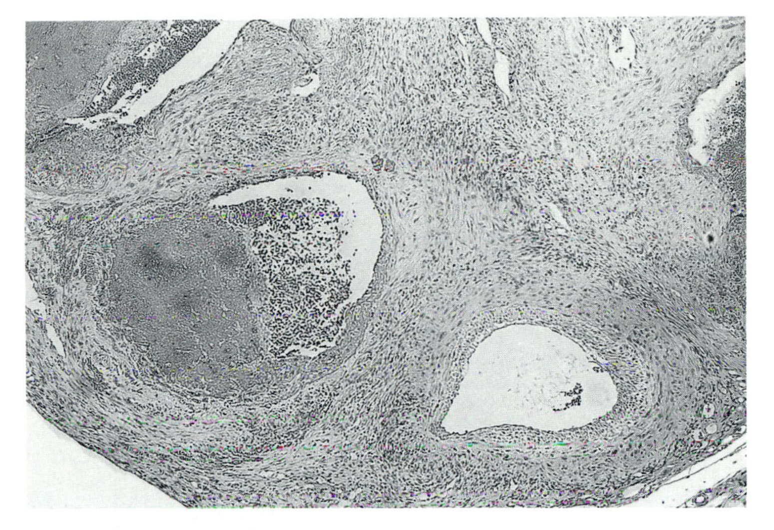

Fig. 2.48. Section of mesenteric arterioles from case of polyarteritis. Note the fibrinoid change in the media of the vessels, with leukocytic infiltration.

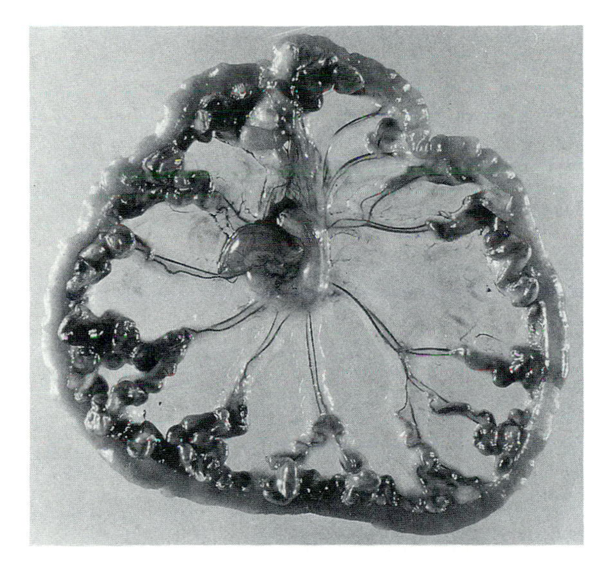

Fig. 2.47. Small intestine and mesenteric attachments in aged Sprague-Dawley rat with polyarteritis nodosa. Note the tortuosity and dilation of mesenteric arterioles, especially along the mesenteric attachments. (Courtesy of Dr. B.S. Jortner)

DIAGNOSIS. The presence of the characteristic lesions, particularly in the Sprague-Dawley and SHR strains, are diagnostic for the disease. Drug-induced vascular lesions may resemble the spontaneous disease (Bishop 1989).

Degenerative Changes in Central Nervous System

Age-related abnormalities seen in aged rats include Wallerian degeneration in focal areas of the spinal cord and segmental demyelination of the peripheral nervous system, particularly in the sciatic nerves (Van Steenis and Kroes 1971). Wallerian degeneration in the cord is characterized by the presence of enlarged axons containing eosinophilic material. In the brain and spinal cord, there may be degeneration of scattered neurons, with astrogliosis. Lipochrome pigment may be present in some neurons in the brain and spinal cord.

Spontaneous radiculoneuropathy is a degenerative disease of the spinal roots, with concurrent atrophy of skeletal muscle in the lumbar region and hind limbs. Vacuolation and demyelination occur primarily in the lumbosacral roots, particularly the ventral spinal regions (Krinke 1988; Berg et al. 1962). This syndrome is manifest clinically as posterior weakness or paresis in the aged rat (Witt and Johnson 1990).

Aging Lesions of the Liver

Rats develop polyploidy, megalokarya, intranuclear cytoplasmic invagination, and intracytoplasmic inclusions of hepatocytes that are similar to but not as striking as in the aging mouse. Foci of sinusoidal dilatation, spongiosis, peliosis, and areas of hepatocellular alteration are all common in the aging rat liver. A striking lesion in the aging rat is bile ductular proliferation. Initially there are increased numbers of bile ductules in portal tracts, which become dilated, lined by atrophic epithelium, and surrounded by collagenous connective tissue, resulting in a cirrhotic appearance (Fig. 2.49). Extramedullary hematopoiesis may be seen in aged rats with conditions such as severe chronic renal disease.

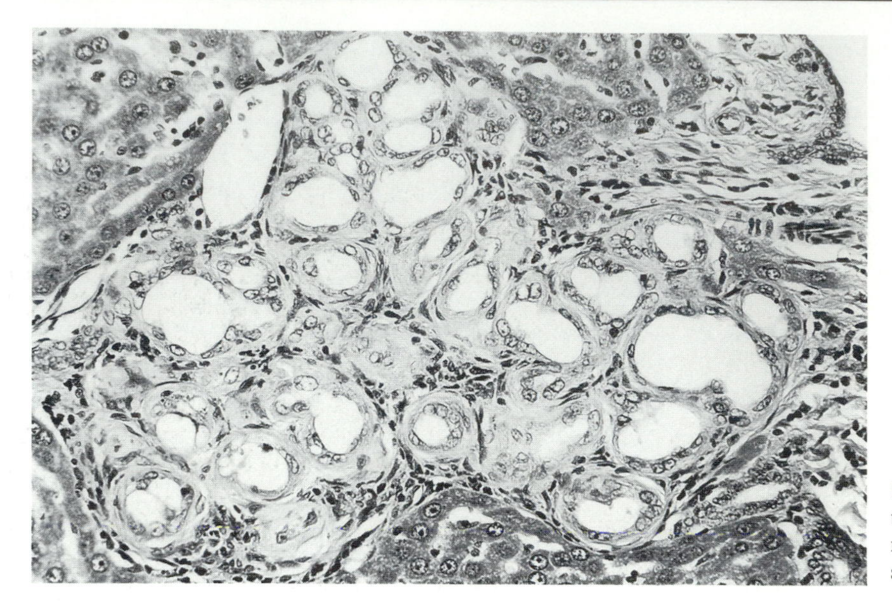

Fig. 2.49. Periportal region from aged laboratory rat, illustrating bile duct hyperplasia, one manifestation of the aging process seen in some strains.

Degenerative Osteoarthritis

Erosion of the articular cartilage occurs in sites such as the sternum and femur in aged rats, with degeneration of cartilaginous matrix, clefting, and cyst formation (Yamasaki and Inui 1985). For additional information on age-related changes seen in the laboratory rat, consult Burek (1978) and Coleman et al. (1977).

Auricular Chondritis

The disease has been observed in several strains, including Sprague-Dawley and Wistar rats. Nodular lesions are present on the pinnae of the ears. Microscopic lesions are characterized by multimodular, granulomatous, inflammatory foci with chondrolysis and invasion mesenchymal cells. Trauma and infectious agents have been considered as possible underlying causes, but it may be an immunologically mediated disease (McEwen and Barsoum 1990).

Malocclusion

Overgrowth of the incisor teeth occurs sporadically in this species. The condition occurs secondary to poor alignment of the upper and lower incisor teeth, with resultant failure to wear in the normal manner. The condition may be secondary to a broken upper or lower incisor tooth but is more often a spontaneous event, and in many cases is considered to be due to genetic factors. Depending on the duration and severity of the problem, affected animals are frequently thin due to their inability to prehend and masticate food normally, and in advanced cases, severely affected lower or upper incisors may penetrate into the soft tissues of the palate or jaw. Cellulitis and increased salivation are frequent sequelae in advanced cases.

Behavioral Diseases

Laboratory rats are usually docile and adapt well to a variety of surroundings. Docility is strain-related. Some strains of rats, such as Long Evans and F344, are somewhat aggressive and can be difficult to handle. They do well singly or in groups, and males seldom fight. Females with litters will tolerate the company of males, but not other females. Rarely, females (usually primiparous dams) will cannibalize their young if stressed.

Environmental Disorders

"Ringtail". Rats adapt to cold, but do not tolerate heat well. Low humidity may predispose young rats to "ringtail," with annular constrictions of the skin of the tail, leading to dry gangrene of the distal tail (Fig. 2.50). This syndrome is most apt to occur in preweaning rats. Traditionally ringtail has been attributed to low environmental humidity (e.g., less than 25%). However, other factors, such as genetic susceptibility, environmental temperatures, degree of hydration, and nutrition may be involved. A detailed histological study may serve to shed light on the etiopathogenesis of this disease.

Other Environmental Disorders. (1) Dehydration: Rats become dehydrated easily, usually from malfunction of water bottle sipper tubes. Dehydration is often accompanied by porphyrin staining around the eyes, a sign of general stress.

(2) High environmental temperature can easily cause infertility in male rats.

(3) The female estrous cycle is sensitive to light cycles. For example, exposure to constant light for as little as three days may induce persistent estrus, hyperestrogenism, polycystic ovaries, and endometrial hy-

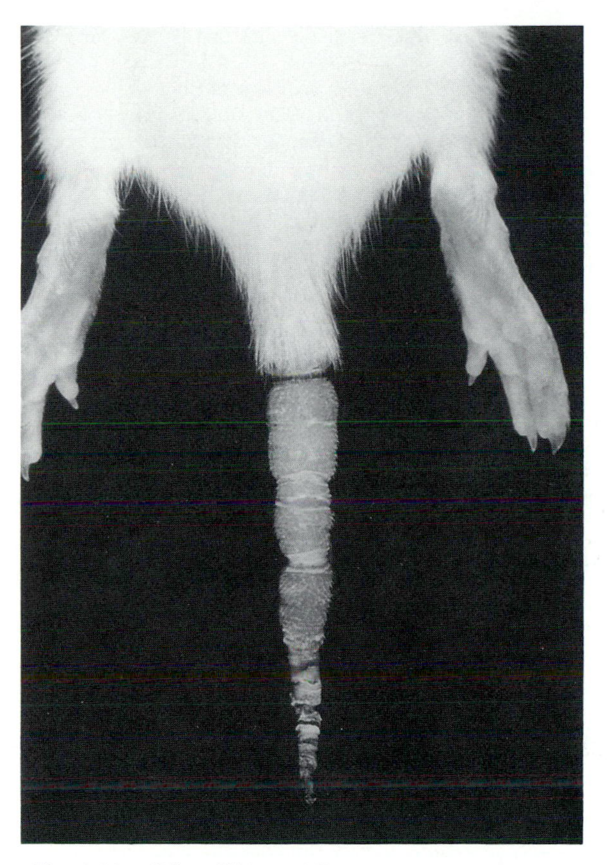

Fig. 2.50. "Ringtail" in suckling rat. There is prominent annular ridging and contraction, a characteristic feature of the disease.

pertrophy. Marked **retinal degeneration** can occur in albino rats subjected to light intensities that would be relatively harmless to animals with pigmented uveal tracts. Retinal changes can occur in rats exposed to cyclic light with an intensity of 130 lux or higher at the cage level. Typically, the changes are most severe in rats housed on the top shelves of racks nearest the ceiling light fixtures. There is a progressive reduction of the photoreceptor cell nuclei in the outer nuclear layer of the central retina. Advanced disease has marked depletion and alteration of the retinal layers, with concomitant cataract formation. This must be differentiated from peripheral retinal degeneration, which occurs in some strains of rats as a genetically inherited disorder.

(4) Dusty bedding material can result in inhalation of foreign material into the lungs, with focal aspiration pneumonia (Fig. 2.51). High mortality has been observed in Sprague-Dawley rats housed on aromatic cedar wood shavings. Reduced weight gains were also observed. The mechanism(s) of the disease was not determined (Burkhart and Robinson 1978).

Chloral Hydrate Ileus

Intraperitoneal injection of chloral hydrate or related compounds causes peritonitis and ileus

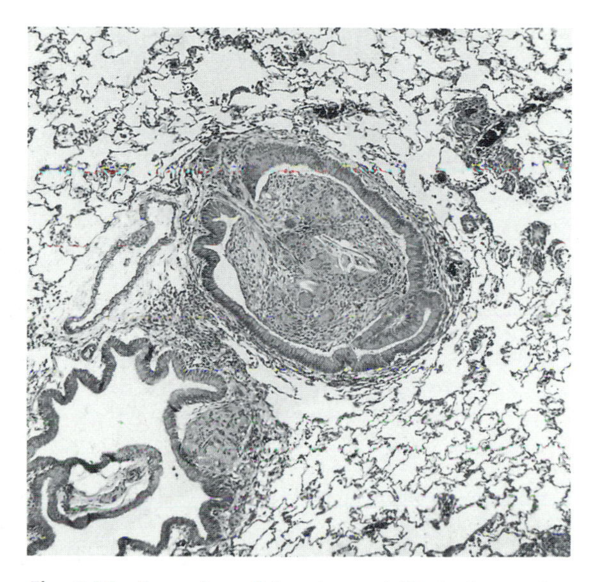

Fig. 2.51. Lung from laboratory rat illustrating aspiration of plant fiber. Note the presence of the aspirated material in the lumen of one bronchus, the chronic inflammatory response, and the subsequent airway obstruction.

(Fleischman et al. 1977). The ileus may not be apparent until up to 5 wk after administration of the drug. Rats develop distended abdomens due to segmental atony and distension of the jejunum, ileum, and cecum (Fig. 2.52). Focal serosal hyperemia can also oc-

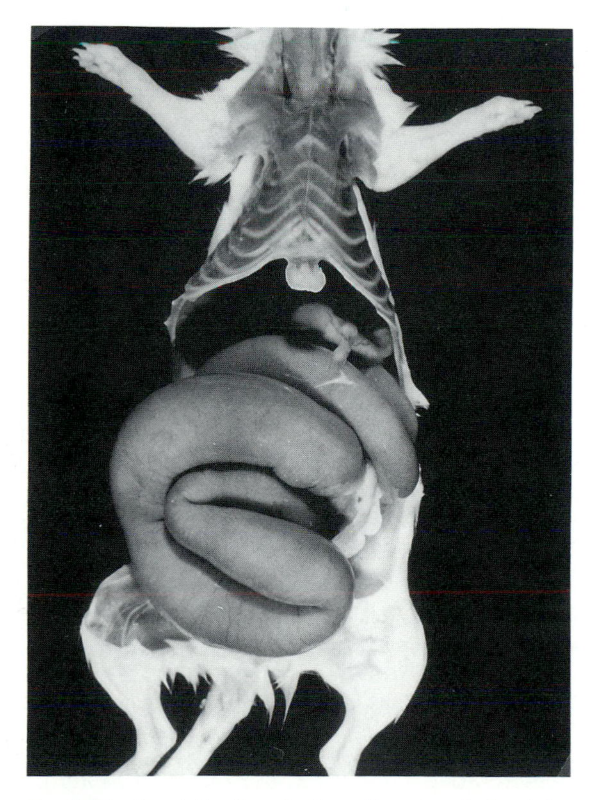

Fig. 2.52. Paralytic ileus in juvenile rat typical of the change associated with the intraperitoneal administration of chloral hydrate as an anesthetic agent. There is marked dilation and ileus of the intestinal tract.

cur. This must be differentiated from ileus/ileitis due to Tyzzer's disease.

Bibliography for Aging and Degenerative Disorders

Renal Disorders

Anver and Cohen 1979 (*see* General Bibliography).

Cohen, B.J., et al. 1970. Veritable hydronephrosis in a mutant strain of brown Norway rats. Lab. Anim. Care 20:489–93.

Couser, W.G., and Stilmant, M.M. 1975. Mesangial lesions and focal glomerular sclerosis in the aging rat. Lab. Invest. 33:491–501.

Gray, J.E. 1986. Chronic progressive nephrosis, rat. In *Monographs on Pathology of Laboratory Animals: Urinary System*, ed. T.C. Jones et al., pp. 174–79. New York: Springer-Verlag.

Gray, J.E., et al. 1982. Early light microscopic changes in chronic progressive nephrosis in several strains of aging laboratory rats. J. Gerontol. 37:142–50.

Lalich, J.J., and Allen, J.R. 1971. Protein overload nephropathy in rats with unilateral nephrectomy. Arch. Pathol. 91:372–82.

Lozzio, B.B., et al. 1967. Hereditary renal disease in a mutant strain of rats. Science 156:1742–44.

Magnusson, G., and Ramsay, C.H. 1971. Urolithiasis in the rat. Lab. Anim. 5:153–62.

Maronpot, R.R. 1986. Spontaneous hydronephrosis, rat. In *Monographs on Pathology of Laboratory Animals: Urinary System*, ed. T.C. Jones et al., pp. 268–71. New York: Springer-Verlag.

O'Donaghue, P.N., and Wilson, M.S. 1977. Hydronephrosis in male rats. Lab. Anim. 11:193–94.

Owen, R.A., and Heywood, R. 1986. Age-related variations in renal structure and function in Sprague-Dawley rats. Toxicol. Pathol. 14:158–67.

Paterson, M. 1979. Urolithiasis in the Sprague-Dawley rat. Lab. Anim. 13:17–20.

Richardson, B., and Luginbuhl, H. 1976. The role of prolactin in the development of chronic progressive nephropathy in the rat. Virchows Arch. Pathol. Anat. 370:13–19.

Ristskes-Hoitinga, J., et al. 1989. Nutrition and kidney calcification in rats. Lab. Anim. 23:313–18.

Tapp, D.C., et al. 1989. Food restriction retards body growth and prevents end-stage renal pathology in remnant kidneys regardless of protein intake. Lab. Invest. 60:184–95.

Van Winkle, T.J., et al. 1988. Incidence of hydronephrosis among several production colonies of outbred Sprague-Dawley rats. Lab. Anim. Sci. 38:402–6.

Weaver, R.N., et al. 1975. Urinary proteins in Sprague-Dawley rats with chronic progressive nephrosis. Lab. Anim. Sci. 25:705–10.

Myocardial Degeneration/Fibrosis

Anver and Cohen 1979 (*see* General Bibliography).

Burek 1978 (*see* General Bibliography).

Novilla, M.N., et al. 1991. A retrospective study of endocardial proliferative lesions in rats. Vet. Pathol. 28:156–65.

Polyarteritis Nodosa

Bishop, S.P. 1989. Animal models of vasculitis. Toxicol. Pathol. 17:109–17.

Yang, Y.H. 1965. Polyarteritis nodosa in laboratory rats. Lab. Invest. 14:81–88.

Degenerative Changes in the Nervous System

Berg, B.N., et al. 1962. Degenerative lesions of spinal roots and peripheral nerves of aging rats. Gerontology 6:72–80.

Krinke, G.J. 1988. Spontaneous radioneuropathology, aged rats. In *Monographs on Pathology of Laboratory Animals: Nervous System*, ed. T.C. Jones et al., pp. 203–8. New York: Springer-Verlag.

Van Steenis and Kroes 1971 (*see* General Bibliography).

Witt, C.J., and Johnson, L.K. 1990. Diagnostic exercise: Rear limb ataxia in a rat. Lab. Anim. Sci. 40:528–29.

Degenerative Osteoarthritis

Burek 1978 (*see* General Bibliography).

Coleman et al. 1977 (*see* General Bibliography).

Yamasaki, K., and Inui, S. 1985. Lesions of articular, sternal and growth plate cartilage in rats. Vet. Pathol. 22:46–50.

Auricular Chondritis

McEwen, B.J., and Barsoum, N.J. 1990. Auricular chondritis in Wistar rats. Lab. Anim. 24:280–83.

Environmental Disorders

Burkhart, C.A., and Robinson, J.L. 1978. High rat pup mortality attributed to the use of cedar-wood shavings as bedding. Lab. Anim. 12:221–22.

Noell, W.K., et al. 1966. Retinal damage by light in rats. Invest. Ophthalmol. 5:450–73.

Semple-Rowland, S.L., and Dawson, W.W. 1987. Retinal cyclic light damage threshold for albino rats. Lab. Anim. Sci. 37:389–98.

Chloral Hydrate Ileus

Fleischman, R.W., et al. 1977. Adynamic ileus in the rat induced by chloral hydrate. Lab. Anim. Sci. 27:238–43.

Kohn, D.F., and Barthold, S.W. 1984. Biology and diseases of rats. In *Laboratory Animal Medicine*, ed. J.G. Fox et al., pp. 91–122. New York: Academic.

General Bibliography

Anver, M.R., and Cohen, B.J. 1979. Lesions associated with aging. In *The Laboratory Rat. I. Biology and Diseases*, ed. H.J. Baker et al., pp. 377–99. New York: Academic.

Burek, J.D. 1978. Pathology of Aging Rats. Boca Raton, Fla: CRC.

Coleman, G.L., et al. 1977. Pathological changes during aging in barrier-reared Fischer F344 male rats. J. Gerontol. 32:258–78.

Goodman, D.G., et al. 1980. Neoplastic and non-neoplastic lesions in aging Osborne-Mendel rats. Toxicol. Appl. Pharmacol. 55:433–47.

Hackbarth, H. 1983. Strain differences in inbred rats: Influence of strain and diet on haematological traits. Lab. Anim. 17:7–12.

Harkness, J.E., and Wagner, J.E. 1989. *The Biology and Medicine of Rabbits and Rodents*. Philadelphia: Lea and Febiger.

Jones, T.C., et al. 1985. *Monographs on Pathology of Laboratory Animals: Digestive System,* ed. T.C. Jones et al. New York: Springer-Verlag.

Lamb, D. 1975. Rat lung pathology and quality of laboratory animals: The user's view. Lab. Anim. 9:1–8.

Losco, P.E., and Troup, C.M. 1988. Corneal dystrophy in Fischer 344 rats. Lab. Anim. Sci. 38:702–10.

Turton, J.A., et al. 1989. Age-related changes in the haematology of female F344 rats. Lab. Anim. 23:295–301.

Van Steenis, G., and Kroes, R. 1971. Changes in the nervous system and musculature of old rats. Vet. Pathol. 8:320–32.

NEOPLASMS

It is beyond the scope of this chapter to give a detailed description of the neoplasms of this species. For additional information, consult Burek (1978),

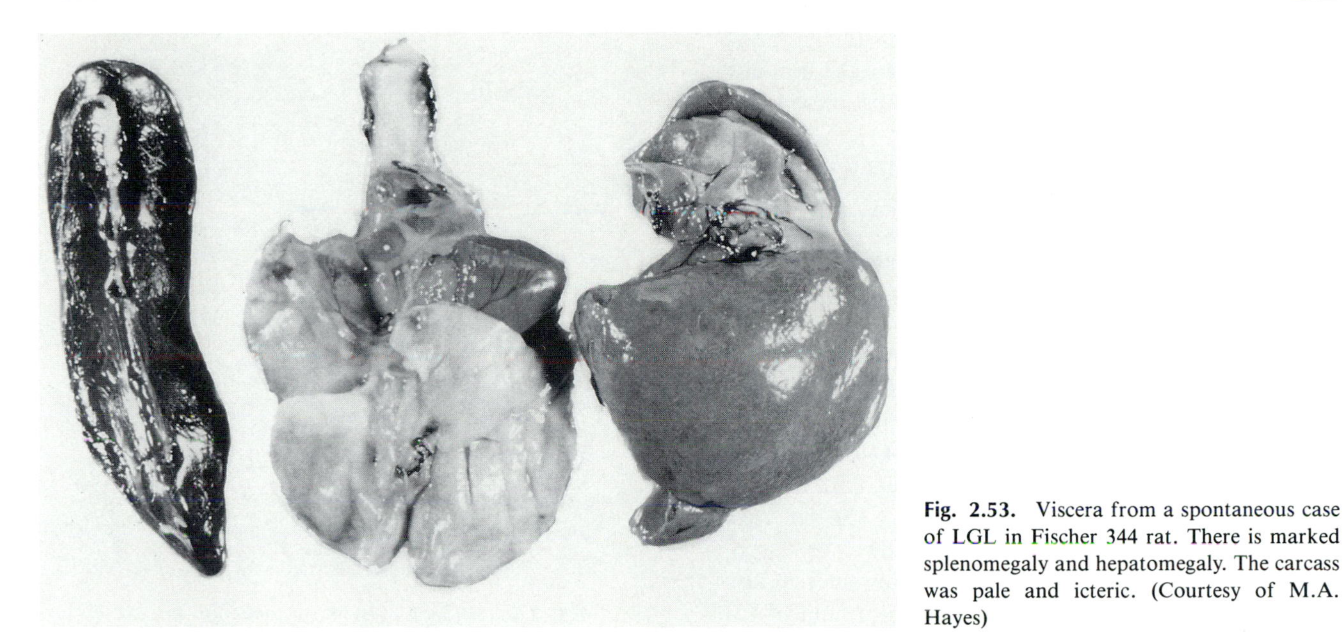

Fig. 2.53. Viscera from a spontaneous case of LGL in Fischer 344 rat. There is marked splenomegaly and hepatomegaly. The carcass was pale and icteric. (Courtesy of M.A. Hayes)

Squire et al. (1978), Turusov 1976), and MacKenzie and Garner (1973).

Lymphoreticular Tumors

Large Granular Lymphocytic (LGL) Leukemia in Fischer 344 Rats. LGL leukemias are a major cause of death in aging F344 rats and occasionally occur in other strains. The neoplastic cells are readily transplantable to rats of the same strain. The malignancy appears to arise in the spleen and then spreads to other organs. Unlike lymphoreticular tumors of mice, retroviruses are not associated with the development of the disease. Although initially considered to be of natural killer (NK) cell origin, studies of cytotoxic activity and surface antigens suggest that these leukemias are of more heterogeneous lymphocytic cell origin (Ward and Reynolds 1983). LGL leukemia is characterized by elevated blood leukocyte counts of up to 180,000/ml. Morphologically, leukemic cells resemble large granular leukocytes. Clinical signs are characterized by weight loss, anemia, jaundice, and depression. There is usually a concurrent, immune-mediated hemolytic anemia, with thrombocytopenia and clotting abnormalities suggestive of disseminated intravascular coagulation (Stromberg et al. 1985).

PATHOLOGY. At necropsy, the carcass is usually pale and icteric. The spleen is markedly enlarged, and there may be moderate to marked enlargement of the liver and lymphadenopathy (Fig. 2.53). Petechial hemorrhages are frequently present on the lung and lymph nodes. Stained impression smears of tissues such as spleen reveal lymphocytes 10–15 μm in diameter, with irregular-shaped, frequently indented nuclei; pale cytoplasm; and prominent, azurophilic cytoplasmic granules (Fig. 2.54). On histological examination of tissue sections, there is diffuse infiltration with malignant lymphocytes in organs such as spleen, lymph nodes, liver, and lung. There is frequently marked depletion of lymphoid follicles in the spleen and diffuse infiltration of leukemic cells in the sinusoids. Hepatocellular degeneration commonly occurs, probably a result of the concurrent anemia and neoplastic infiltrates. Erythrophagocytosis may be evident in the liver and spleen.

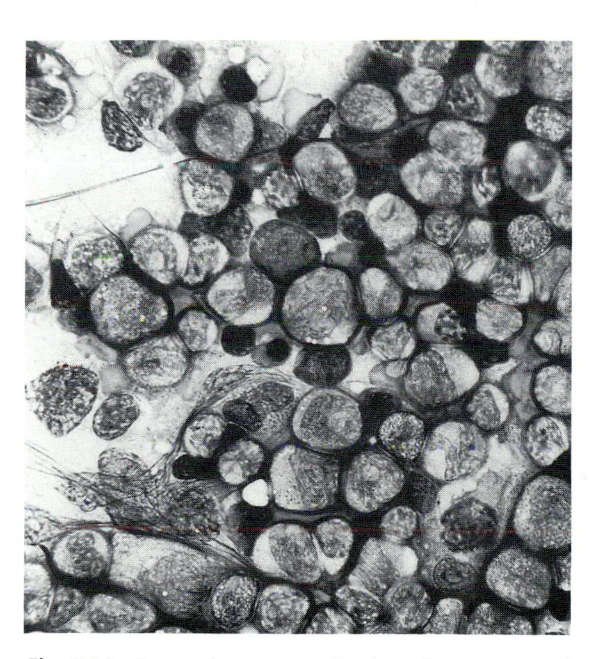

Fig. 2.54. Impression smear of spleen from a case of LGL. There are large numbers of malignant lymphocytes with cytoplasmic granules present in the smear. (Courtesy of M.A. Hayes)

DIAGNOSIS. The presence of the typical clinical and histological pattern, including anemia, splenomegaly, and icterus, and the characteristic lymphocytes in F344 rats are sufficient to confirm the diagnosis. *Differential diagnoses* include lymphosarcoma and histiocytic sarcoma.

SIGNIFICANCE. LGL leukemia is a major cause of disease and mortality in aging F344 rats.

Lymphoma/Leukemia. Spontaneous lymphomas and/or lymphocytic leukemias are relatively uncommon in most strains of rats. They are not associated with retroviral infections. At necropsy, splenomegaly, enlarged lymph nodes, and hepatomegaly are characteristic findings. On microscopic examination, there is frequently diffuse infiltration of neoplastic lymphocytes into organs such as spleen and liver, with obliteration of the normal architecture (Fig. 2.55). On rare occasions, epitheliotropic lymphomas have been recognized in this species. The disease is characterized by the presence of raised plaques on the skin, and histologically by focal lymphocytic infiltration in the dermis, particularly at the dermo-epidermal junction (Fig. 2.56).

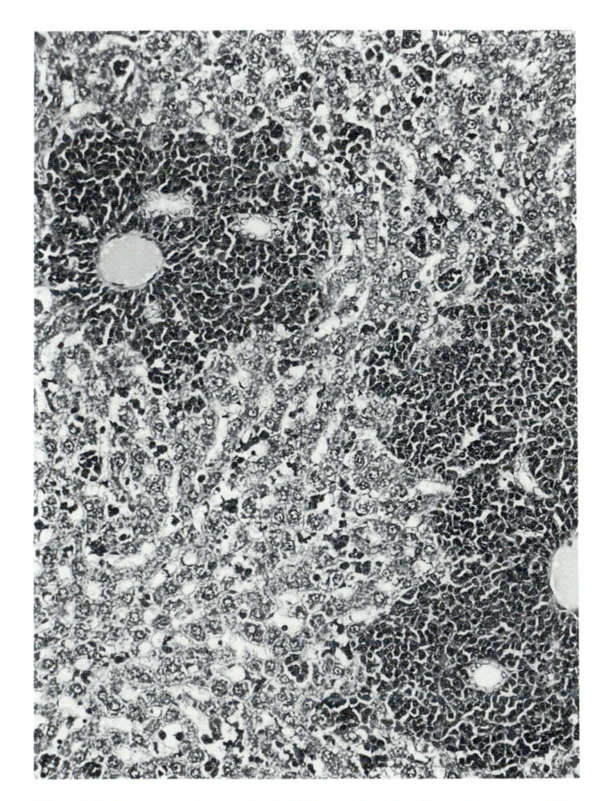

Fig. 2.55. Lymphocytic infiltrate in the liver in a spontaneous lymphosarcoma in mature Wistar rat. Note the marked lymphocytic infiltration in the periportal regions.

Mononuclear Cell Leukemia. Occurs in Wistar-Furth, Wistar, and occasionally other strains of rats (Abbott et al. 1983). Leukemia and diffuse infiltration of organs such as spleen, liver, and lung with mononuclear cells are typical findings.

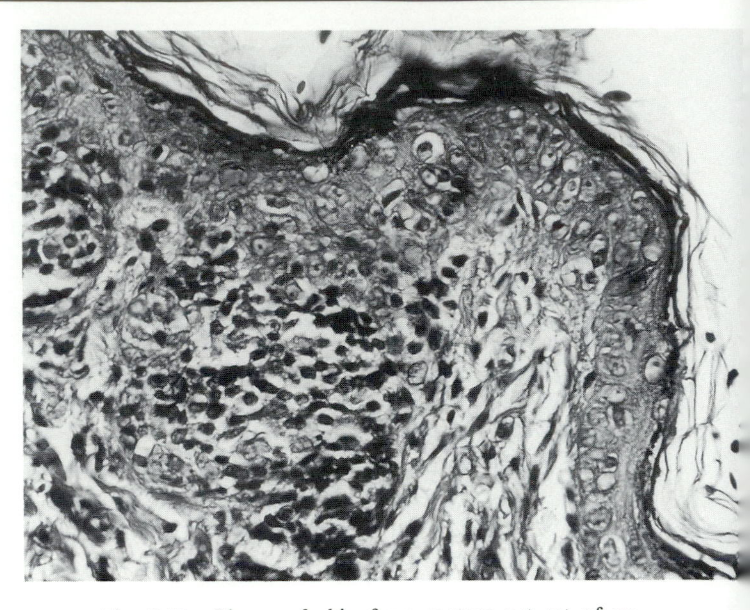

Fig. 2.56. Biopsy of skin from mature pet rat of unknown ancestry, illustrating an example of an epitheliotropic lymphosarcoma. There is an infiltrate of relatively well-differentiated lymphocytes at the dermo-epidermal junction. (Courtesy of M.L. Brash)

Histiocytic Sarcoma. Neoplasms of this nature occur most often in Sprague-Dawley rats (Squire et al. 1981), but they also have been observed in other strains, including Osborne-Mendel and Wistar rats (Barsoum et al. 1984; Goodman et al. 1980). The tumors are present primarily in animals over 12 mo of age, and there is no obvious sex predisposition.

PATHOLOGY. At necropsy, sarcomas of this type may be present in the liver, lymph nodes, lung, spleen, mediastinum, retroperitoneum, and subcutaneous tissue. Neoplasms are pale and moderately firm, and they tend to infiltrate and displace normal tissue. Necrotic areas may be scattered in the mass. On microscopic examination, tumors consist of diffuse sheets of neoplastic cells, varying from elongated, pallisading forms to plump, pleomorphic histiocytic cells. The histiocytic cells have vesicular nuclei, prominent nucleoli, and abundant cytoplasm. Multinucleated giant cells are usually present in tumors with a prominent histiocytic component (Figs. 2.57 and 2.58). Based on electron microscopic and immunohistochemical studies, the histiocytic forms are derived from monocytes or histiocytes, while the origin of the fibrous types remains uncertain (Barsoum et al. 1984; Graves et al. 1982).

DIAGNOSIS. The presence of diffuse to circumscribed sarcomatous tumors with multinucleated giant cells and a spectrum of cell forms varying from histiocytic to pallisading fusiform cells are typical microscopic findings. *Differential diagnoses* include fibrosarcoma, lymphosarcoma, osteosarcoma, and granulomatous inflammatory tissue.

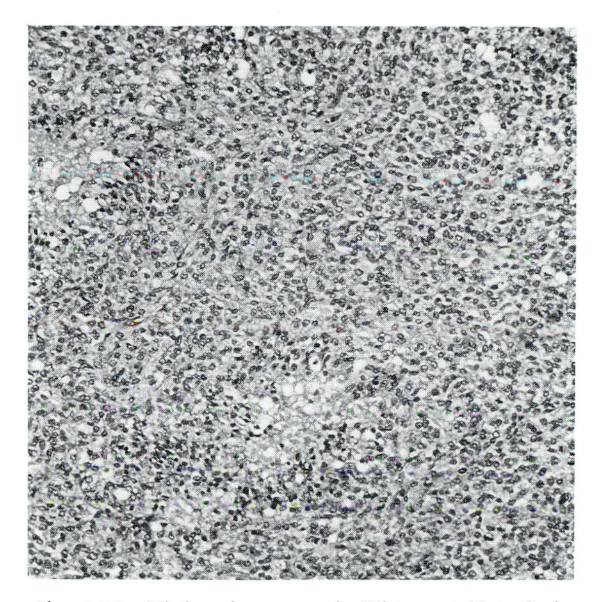

Fig. 2.57. Histiocytic sarcoma in Wistar rat. Note the indistinct cytoplasmic outlines, anisokaryosis, and pleomorphic appearance of the histiocytic cells. (Courtesy of Z.W. Wojcinski)

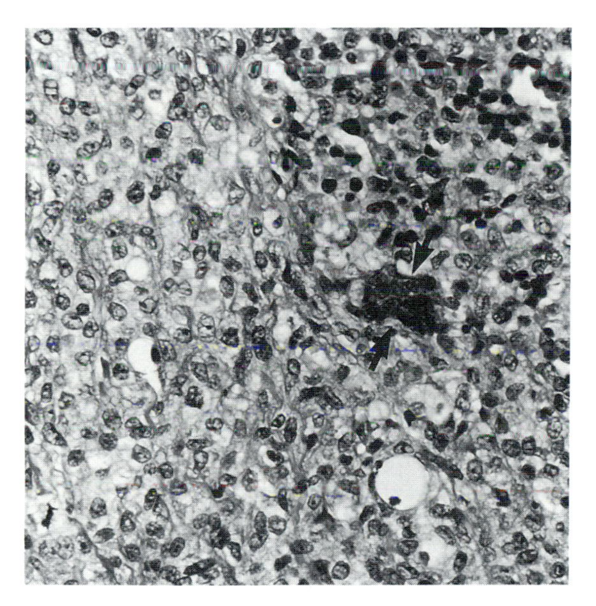

Fig. 2.58. Higher magnification of Fig. 2.57, illustrating anisokaryosis and multinucleated giant cells (*arrows*). (Courtesy of Z.W. Wojcinski)

Mammary Tumors

Mammary tumors are a relatively common occurrence in older female rats, particularly the Sprague-Dawley (S-D) strain.

EPIZOOTIOLOGY AND PATHOGENESIS. The majority of mammary tumors (approximately 85–90%) are the relatively benign fibroadenoma, and most of the remainder are categorized as carcinomas. The benign tumors are particularly common in older S-D females. There are significant variations in the incidence, depending on the source of rats of this strain. Ranges of 7–40% have been recorded in S-D rats studied from different sources (MacKenzie and Garner 1973). This indicates that there is likely to be significant genetic variations over time, although dietary and environmental factors may have also played a role. In one study, restricting the food intake by 20% reduced the incidence of mammary tumors in female S-D rats by 5-fold, compared with controls (Tucker 1979). Prolactin levels have also been identified as a key factor. In one study, the serum prolactin values in females with mammary tumors were over 25 times higher than 6-mo-old virgin females (Okada et al. 1981). There have been attempts to equate mammary tumors with the incidence of pituitary adenomas, but unequivocal correlations have not been made. Unlike mammary tumors in mice, retroviruses do not appear to be involved. Based on current knowledge, it appears that sex, age, genetic, dietary, and endocrine factors may all play a role in the incidence of spontaneous mammary tumors. Mammary fibroadenomas also occur occasionally in male rats.

Regarding their biological behavior, mammary fibroadenomas may become very large and infiltrate locally, but they rarely metastasize. They are transplantable by subcutaneous implantation to recipients of the same strain.

PATHOLOGY. At necropsy, a circumscribed, movable, firm, lobulated mass may be located within any of the 12 mammary glands along the mammary chain. In larger tumors, there may be fixation and ulceration of the overlying skin. The lobulated appearance is readily evident on cut surface (Fig. 2.59).

On microscopic examination, there is distinct interlobular and intralobular connective tissue surrounding relatively well differentiated acinar structures. There are variable proportions of acinar and collagenous tissue. In some cases, the mass may consist primarily of connective tissue, with scattered acinar structures. In other fibroadenomas, epithelial components predominate (Fig. 2.60). Acini are lined by cuboidal epithelial cells, frequently with prominent vacuoles in the cytoplasm (Fig. 2.61).

DIAGNOSIS. The histological pattern and the nature of the tumor, particularly in older female S-D rats, are sufficient to confirm the diagnosis.

SIGNIFICANCE. Mammary fibroadenomas are the most common tumor in female S-D rats. In pet animals, surgical removal is feasible, provided that the mass is not too large for complete excision. However, in both females and males, tumors are likely to recur in another mammary gland. **Mammary carcinomas** represent a relatively small percentage of mammary tumors in the laboratory rat. They occur spontaneously but have been produced experimentally using estrogens (Ito et al. 1984). A variety of patterns may be evident histologically. They have been classified by

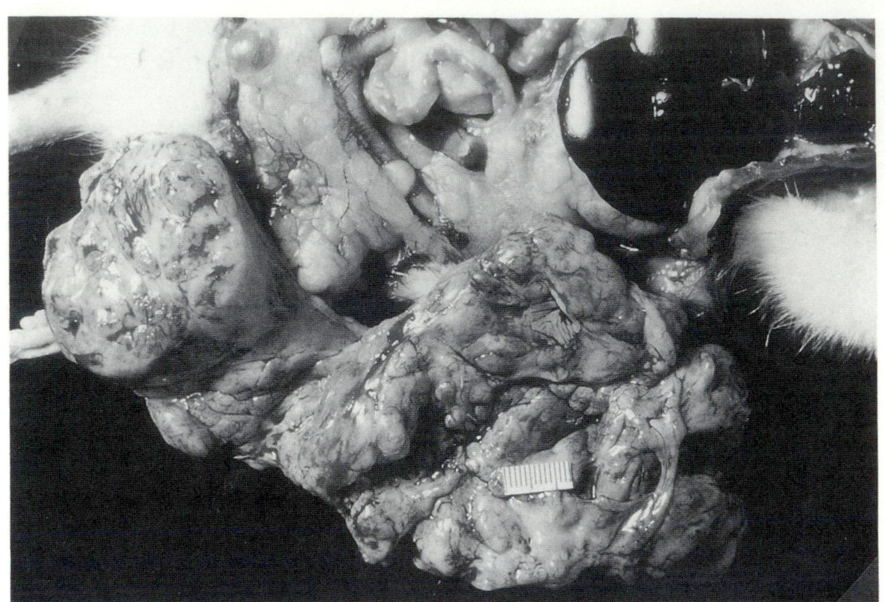

Fig. 2.59. Spontaneous case of fibroadenoma in adult female Sprague-Dawley rat. The prominent lobulations and interlobular fibrous tissue are characteristic gross findings seen with this neoplasm.

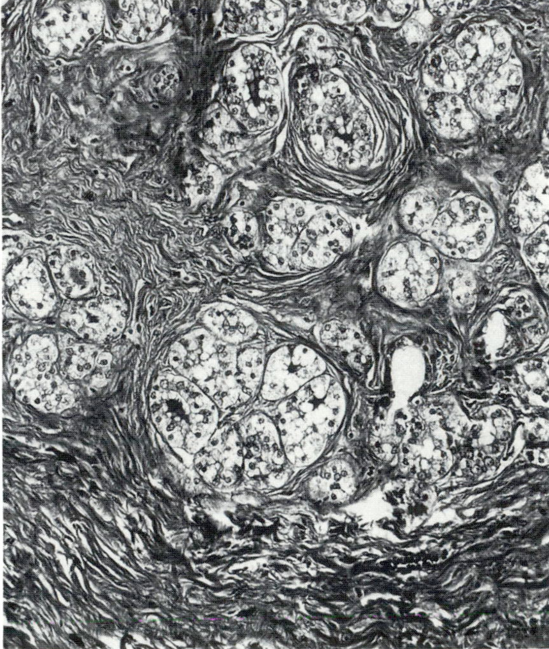

Fig. 2.60. Low-power photomicrograph of fibroadenoma, illustrating the acinar structures and connective tissue components.

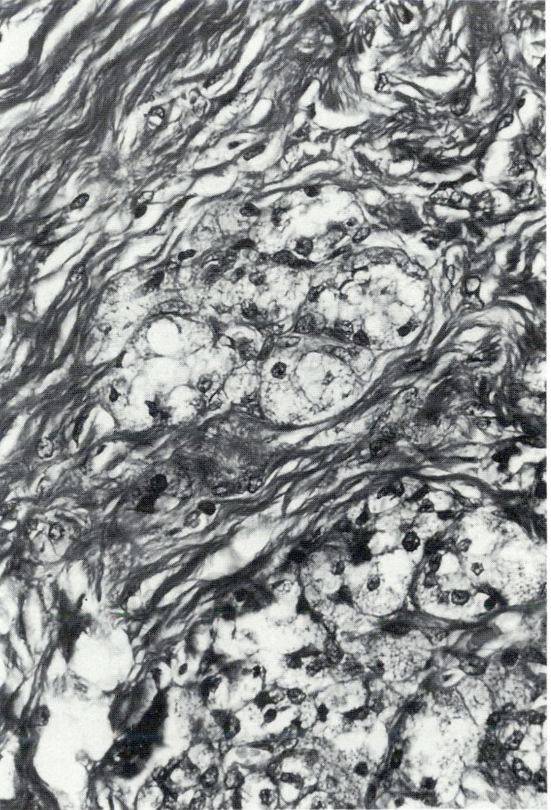

Fig. 2.61. Higher magnification of Fig. 2.60, demonstrating the relatively well-differentiated vacuolated epithelial cells lining acini and the prominent intralobular collagenous tissue.

various terms, including adenocarcinoma (Fig. 2.62) and cribriform, tubular, papillary, and comedo carcinoma (Russo et al. 1989).

Pituitary Gland Tumors

Pituitary adenoma is one of the more common tumors that occur in older animals, particularly in strains such as the Sprague-Dawley and Wistar rat. In addition to age, genetic factors, diet, and breeding history may also play a role. Reduction in food intake reduces the incidence of spontaneous pituitary tumors (Pickering and Pickering 1984b; Tucker 1979), and in another study, mated females had a lower incidence of pituitary tumors than did virgin females (Pickering and Pickering 1984a). In some studies, there is a slightly higher incidence in females, but this is not a consistent finding (McComb et al. 1984). Clinical signs vary, from animals that are asymptomatic to ani-

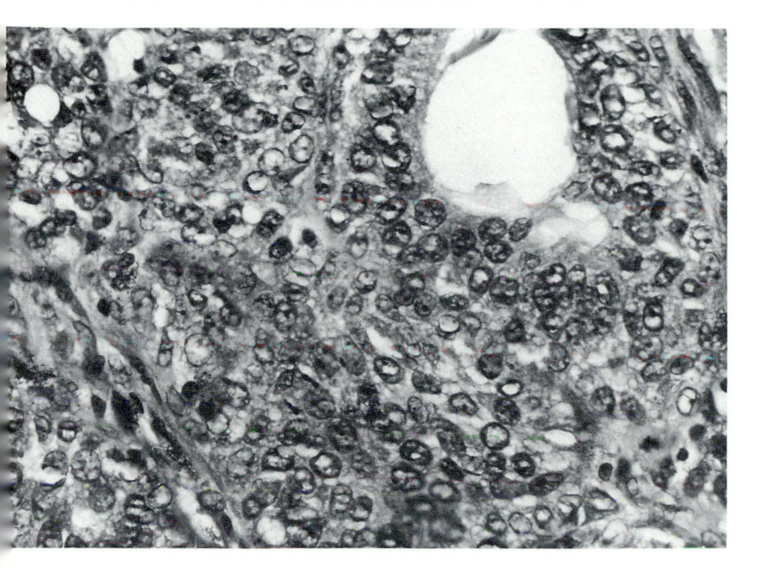

Fig. 2.62. Mammary adenocarcinoma in adult female rat. There are sheets of poorly differentiated cells, with acinar formation (*upper right*).

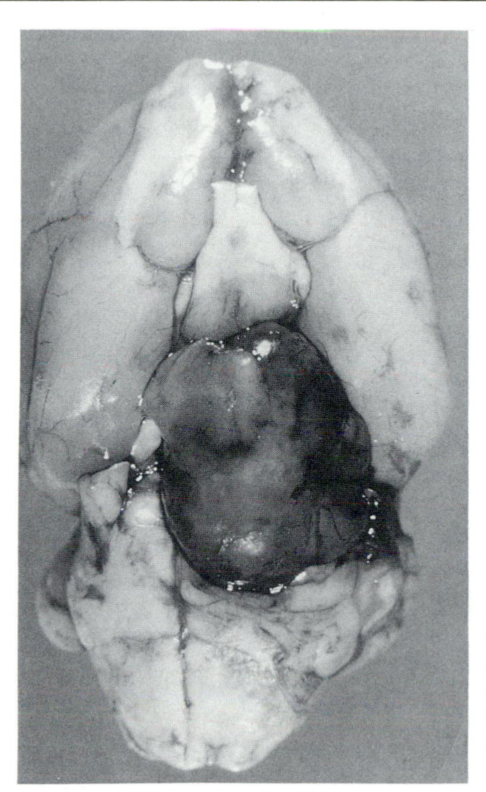

Fig. 2.63. Pituitary gland adenoma in aged female Wistar rat. Note the variable color of the large fleshy mass.

mals with severe depression, frequently with incoordination. The majority of pituitary tumors are interpreted to be chromophobe adenomas. Acidophil and basophil tumors have also been described. However, immunohistochemical techniques are required for positive identification. In pituitary tumors studied by immunocytochemistry, prolactin-producing tumors are the most common type. Most tumors are interpreted to arise from the pars distalis, although tumors of the pars intermedia have also been described (Carlton and Gries 1986). Pituitary carcinomas are relatively uncommon.

PATHOLOGY. At necropsy, the pituitary is enlarged, frequently with prominent lobulations. The tumor is often dark red to brown and hemorrhagic in appearance (Fig. 2.63). In larger tumors, there may be minimal to marked compression of the overlying mesencephalon. On microscopic examination, the anterior pituitary consists of cords or nests of glandular cells bound by strands of connective tissue, with an abundant cavernous vascular capillary network. The cells typically have large nuclei and prominent nucleoli, with abundant, lightly basophilic to amphophilic cytoplasm consistent with a chromophobe adenoma (Fig. 2.64). Giant nuclei may be present in the mass. Mitotic figures are occasionally observed. A pseudocapsule composed of a fine band of connective tissue separates the tumor from the adjacent, nontransformed pituitary tissue. More than one adenoma may be present in an affected gland. Frequently there is evidence of hemorrhage within the mass, and hemosiderin pigment may be present in some tumors.

DIAGNOSIS. The presence of an enlarged pituitary gland composed of nests and cords of large, relatively uniform glandular cells, with an abundant capillary

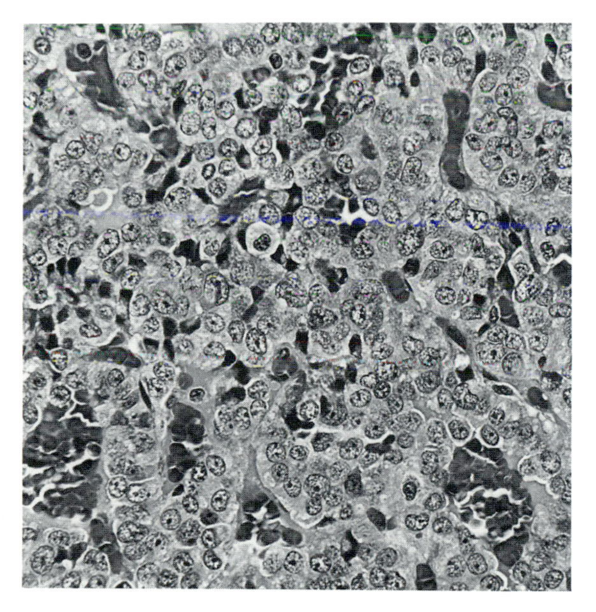

Fig. 2.64. Histological section of spontaneous pituitary adenoma, with prominent cords of epithelial cells interspersed within a vascular stroma.

network and occasionally hemorrhage are typical findings, and consistent with a diagnosis of pituitary adenoma. *Differential diagnoses*: It is necessary to distinguish pituitary adenomas from hyperplastic and hypertrophic lesions. Hyperplastic changes are characterized by the proliferation of cells of normal size,

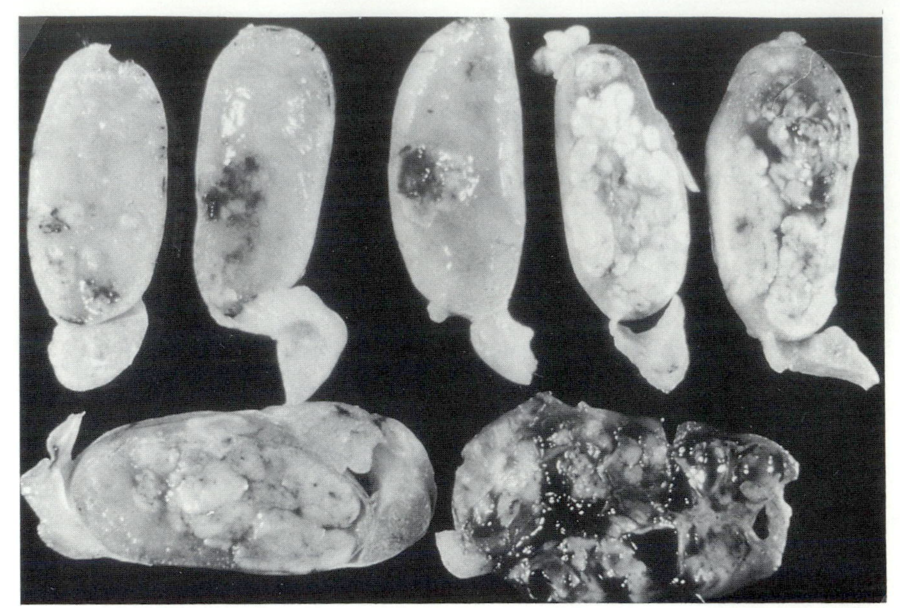

Fig. 2.65. Testes collected from older F344 male rats, illustrating a high incidence of interstitial (Leydig) cell tumors. There are multiple circumscribed masses, with accompanying hemorrhage. (Courtesy of M.A. Hayes)

with no evidence of pseudocapsule formation or marked compression of adjacent pituitary tissue. Nodules of hypertrophic cells may be present in glands and must be differentiated from adenomas. They occur as islands of large cells, sometimes with mitoses, but there is no evidence of encapsulation (McComb et al.1984).

SIGNIFICANCE. Pituitary adenomas are relatively common in aged female and male rats. Attempts have been made to correlate prolactin-producing pituitary tumors with an increased incidence of mammary fibroadenomas, but to date this has not been resolved.

Testicular Tumors

Interstitial Cell Tumors. These tumors most frequently occur in F344 rats, and they are present in most older males of this strain (Goodman et al. 1979). On gross examination, they appear as circumscribed, lobulated, light yellow to hemorrhagic single or multiple masses involving one or both testes (Fig. 2.65). Microscopic changes are consistent with tumors of Leydig cell origin. Masses normally consist of sheets of cells of two types: polyhedral to elongated cells with granular to vacuolated cytoplasm and smaller cells with hyperchromatic nuclei and scanty cytoplasm.

SIGNIFICANCE. In addition to their frequent occurrence in aged male rats, particularly the F344 strain, interstitial cell tumors have been associated with concurrent hypercalcemia (Troyer et al. 1982).

Zymbal's Gland Tumors

These tumors occur in the holocrine gland located at the base of the external ear. On gross examination, there is a circumscribed mass, frequently with ulceration of the overlying skin (Fig. 2.66). On microscopic examination, the mass consists of sheets of epithelial cells with abundant, vacuolated cytoplasm, frequently with foci of necrosis and leukocytic infiltration (Figs. 2.67 and 2.68). Depending on their histological pattern, they are classified as adenomas or adenocarcinomas. The malignant tumors are locally invasive, but not metastatic.

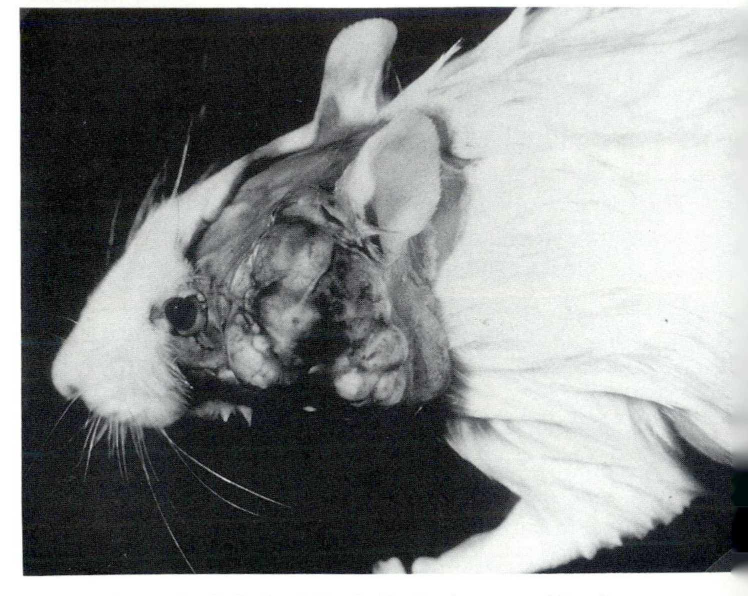

Fig. 2.66. Lobulated Zymbal's gland tumor arising from glands at the external ear canal in adult rat.

Other Neoplasms

Other neoplasms that occur in the laboratory rat include thyroid tumors, particularly parafollicular cell types (Boorman and DeLellis 1983), tumors of the skin and adnexae (Turusov 1976), and neoplastic liver nodules (Stewart et al. 1980).

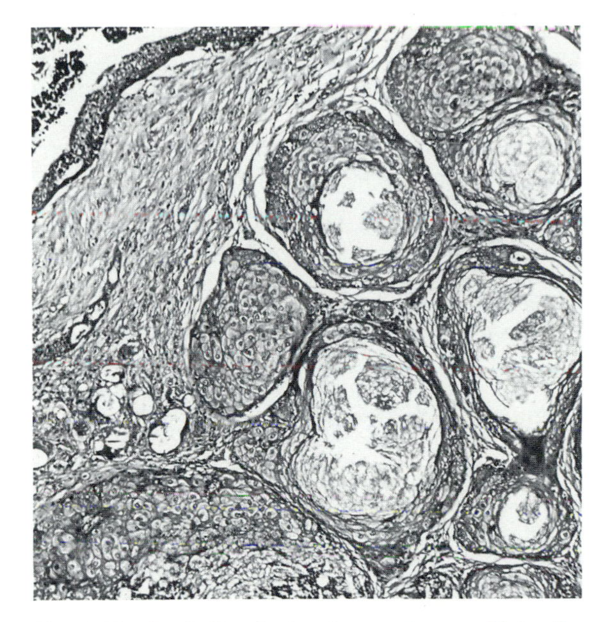

Fig. 2.67. Zymbal's gland adenocarcinoma. Note the polyhedral cells and the formation of acinar-like structures containing keratinized material and debris.

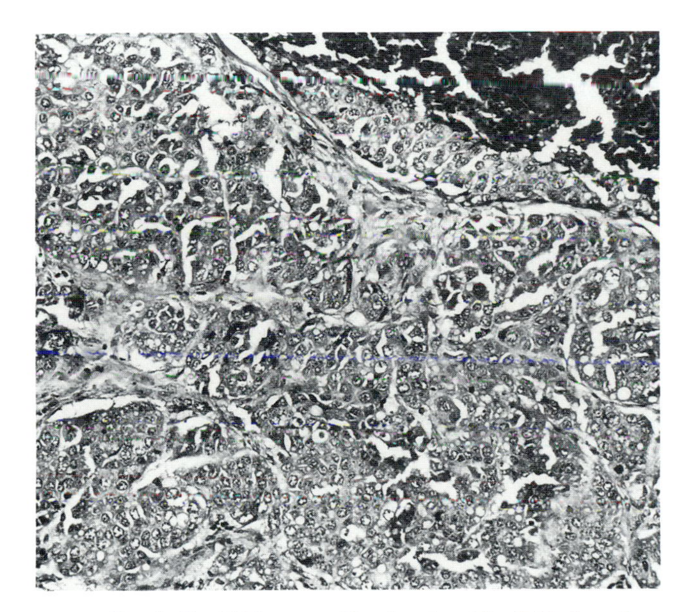

Fig. 2.68. Higher magnification than Fig. 2.67 of another Zymbal's gland carcinoma, illustrating the trabecular and acinar patterns formed by the malignant epithelial cells.

Bibliography for Neoplasms

Lymphoreticular Tumors

Abbott, D.P., et al. 1983. Mononuclear cell leukemia in aged Sprague-Dawley rats. Vet. Pathol. 20:434–39.

Barsoum, N.J., et al. 1984. Histiocytic sarcoma in Wistar rats. Arch. Pathol. Lab. Med. 108:802–7.

Goodman, D.G., et al. 1980. Neoplastic and non-neoplastic lesions in aging Osborne-Mendel rats. Toxicol. Appl. Pharmacol. 55:433–47.

Graves, P., et al. 1982. Spontaneous rat malignant tumors of fibrohistocytic origin: An ultrastructural study. Vet. Pathol. 19:497–505.

Jones, T.C., et al., eds. 1990. *Monographs on Pathology of Laboratory Animals: Hematopoietic System.* New York: Springer-Verlag.

Rosol, T.J., and Stromberg, P.C. Effects of large granular lymphocytic leukemia on bone in F344 rats. Vet. Pathol. 27:391–96.

Squire, R.A., et al. 1981. Histiocytic sarcoma with a granuloma-like component occurring in a large colony of Sprague-Dawley rats. Am. J. Pathol. 106:21–30.

Stromberg, P.C., et al. 1990. Spleen cell population changes and hemolytic anemia in F344 rats with large granular lymphocytic leukemia. Vet. Pathol. 27:397–403.

Stromberg, P.C., et al. 1985. Behavior of transplanted large granular lymphocytic leukemia in Fischer 344 rats. Lab. Invest. 53:200–207.

Ward, J.M., and Reynolds, C.W. 1983. Large granular lymphatic leukemia. A heterogeneous lymphocytic leukemia in F344 rats. Am. J. Pathol. III:1–10.

Mammary Tumors

Ito, A., et al. 1984. Prolactin and aging: X-irradiated and estrogen-induced rat mammary tumorigenesis. J. Natl. Cancer Inst. 73:123–26.

MacKenzie, W.F., and Garner, F.M. 1973. Comparison of neoplasms from six sources of rats. J. Natl. Cancer Inst. 50:1243–57.

Okada, M., et al. 1981. Characteristics of 106 spontaneous mammary tumours appearing in Sprague-Dawley female rats. Br. J. Cancer 43:689–95.

Russo, J., et al. 1989. Classification of neoplastic and nonneoplastic lesions of the rat mammary gland. In *Monographs on Pathology of Laboratory Animals: Integument and Mammary Glands*, ed. T.C. Jones et al., pp. 275–304. New York: Springer-Verlag.

Tucker, M.J. 1979. The effects of long term food restriction on tumors in rodents. Br. J. Cancer 23:803–7.

Pituitary Tumors

Carlton, W.W., and Gries, C.L. 1986a. Adenoma, pars intermedia, anterior pituitary, rat. In *Monographs on Pathology of Laboratory Animals: Endocrine System*, ed. T.C. Jones et al., pp. 145–49. New York: Springer-Verlag.

______. 1986b. Adenoma and carcinoma, pars distalis, rat. In *Monographs on Pathology of Laboratory Animals: Endocrine System*, ed. T.C. Jones et al., pp. 134–45. New York: Springer-Verlag.

McComb, D.J., et al. 1984. Pituitary adenomas in old Sprague-Dawley rats: A histologic, ultrastructural, and immunohistochemical study. J. Natl. Cancer Inst. 73:1143–66.

Nagatani, M., et al. 1987. Relationship between cellular morphology and immunocytological findings of spontaneous pituitary tumours in the aged rat. J. Comp. Pathol. 97:11–20.

Pickering, C.E., and Pickering, R.G. 1984a. The effect of repeated reproduction on the incidence of pituitary tumours in Wistar rats. Lab. Anim. 18:371–78.

Pickering, R.G., and Pickering, C.E. 1984b. The effect of diet on the incidence of pituitary tumours in female Wistar rats. Lab. Anim. 18:298–314.

Sandusky, G.E., et al. 1988. An immunocytochemical study of pituitary adenomas and focal hyperplasia in old Sprague-Dawley and Fischer rats. Toxicol. Pathol. 16:376–80.

Tucker, M.J. 1979. The effects of long term food restriction on tumours in rodents. Br. J. Cancer 23:803–7.

Testicular Tumors

Goodman 1979 (*see* General Bibliography).

Troyer, H., et al. 1982. Leydig cell tumor induced hypercalcemia in the Fischer rat. Am. J. Pathol. 108:284–90.

General Bibliography

Boorman, G.A., and DeLellis, R.A. 1983. C cell adenoma, thyroid, rat. In *Monographs on Pathology of Laboratory Animals: Endocrine System*, ed. T.C. Jones et al., pp. 197–200. New York: Springer-Verlag.

Boorman, G.A., et al. 1990. *Pathology of the Fischer Rat: Reference and Atlas*. New York: Academic.

Burek, J.D. 1978. Pathology of Aging Rats. Boca Raton, Fla.: CRC.

Goodman, D.G., et al. 1980. Neoplastic and nonneoplastic lesions in aging Osborne-Mendel rats. Toxicol. Appl. Pharmacol. 55:433–47.

______. 1979. Neoplastic and nonneoplastic lesions in aging F344 rats. Toxicol. Appl. Pharmacol. 48:237–48.

MacKenzie, W.F., and Garner, F.M. 1973. Comparison of neoplasms from six sources of rats. J. Natl. Cancer Inst. 50:1243–57.

Squire, R.A., et al. 1978. Tumors. In *Pathology of Laboratory Animals*, ed. K. Benirschke et al., pp. 1052–1283. New York: Springer-Verlag.

Stewart, H.L., et al. 1980. Histologic typing of liver tumors of the rat. J. Natl. Cancer Inst. 64:180–206.

Turusov, V.S., et al., eds. 1976. *The Pathology of Tumours in Laboratory Animals. I. Tumours of the Rat*. Lyon, France: IARC Scientific Publications.

3 HAMSTER

A number of different genera and species of hamsters are used in the research laboratory: the Syrian or golden hamster (*Mesocricetus auratus*); Chinese or gray hamster (*Cricetulus griseus*); European or black-bellied hamster (*Cricetus cricetus*); Armenian or migratory hamster (*Cricetulus migratorius*); Dzungarian, Siberian, dwarf, striped hairy-footed hamster (*Phodopus sungorus*); South African hamster or white-tailed rat (*Mystromys albicaudatus*); and others (Clark 1987). However, the majority of those used in research are the Syrian and Chinese hamsters. Most of the Syrian hamsters that are now so commonly used in the research laboratory and kept as pets appear to have originated from a litter captured in Syria in 1930. The original animals were bred in captivity at Hebrew University, and their offspring later served as the foundation stock for *Mesocricetus auratus* in other countries. Although often referred to as outbred, laboratory Syrian hamsters are understandably genetically homozygous. *Cricetulus griseus* was first domesticated in Beijing, China, around 1920. Based on clinical assessment, Syrian hamsters appear to have relatively few serious microbial infections other than those associated with enteritis. Nevertheless, they are remarkably susceptible to a number of xenogeneic viruses and are prone to induction of tumors with many such viruses. The life span of the Syrian hamster varies from 1–3 yr. Female Syrian hamsters are hyperactive during estrus and are capable of travelling a considerable distance during this period (Richards 1966). This chapter deals primarily with diseases of the Syrian hamster. Very little is known about diseases of other types of hamsters, and it must be kept in mind that generalizations cannot be made, since hamsters represent different, distantly related genera.

ANATOMIC FEATURES

The anatomy of the Syrian hamster has been thoroughly reviewed by Bivin et al. (1987) and Magalhaes (1968). Hamsters have a characteristic, compact body, short legs, and very short tails. They possess four digits on the front feet and five on the rear feet. Syrian hamsters have remarkably abundant and loose skin. Adult females are larger than males. The female urethra has a separate opening from the vagina. Both sexes possess paired flank organs, which are most prominent in males. These organs consist of sebaceous glands, pigment cells, and terminal hair. They are darkly pigmented in mature males and appear to play a role in conversion of testosterone to dihydrotestosterone. Hamsters have prominent depots of brown fat beneath and between the scapulae, in the axilla and neck, and around the adrenals and kidneys. The gastrointestinal tract has a number of significant features. As in mice and rats, the incisors (but not the molars) grow continuously. Many, although not all, genera of hamsters possess buccal pouches, which extend dorsolaterally from the oral cavity on either side of the shoulder region. These structures have been utilized as immunologically privileged sites, which allow xenograft transplants to survive. The utility of hamster cheek pouches as an experimental tool has been largely supplanted by the athymic nude mouse. The stomach is divided into nonglandular and glandular segments, which are divided by a muscular sphincter. Paneth cells are a normal constituent of small intestinal crypts. The cecum is divided into apical and basal portions separated by a semilunar valve. In fact, there are a series of four valves in the ileocecocolic region of the hamster. The liver is divided into four lobes, with a gallbladder. As in the mouse and rat, intranuclear cytoplasmic invagination (inclusions) and eosinophilic cytoplasmic inclusions in hepatocytes can be found, particularly in diseased livers. The respiratory tract is similar to that of mice, rats, and guinea pigs, with no respiratory bronchioles. Lungs have a single left lobe and five lobes on the right side (cranial, middle, caudal, intermediate, and accessory). As desert animals, Syrian hamsters have water-conserving kidneys with elongated single papillae that extend into the ureters. The female reproductive tract consists of a duplex uterus with two cervical canals that merge into a single external cervical os. There are seven pairs of mammary glands. The male testes and accessory glands, as with most rodents, are comparatively large and prominent. Adult males develop large adrenal glands, due to enlargement of the zona reticularis to three times the size of females. This enlargement is related to season and sexual maturity. The hamster placenta differs somewhat from the hemochorial placentation of other laboratory rodents and is termed labyrinthine hemochorial. Trophoblastic giant cells of the fetal placenta are in direct contact with the maternal bloodstream and tend to migrate in the maternal tissues. They have a tropism for arterial, not venous, blood and can be found inside uterine vessels and arteries in the mesometrium (Burek et al. 1979). They can persist for up to 3 wk postpartum.

Polychromasia is relatively common in hamster erythrocytes, with moderate anisocytosis. Erythrocyte life spans vary from 50 to 78 days. Life spans are increased during hibernation (Thomson and Wardrop 1987). Leukocyte counts are 5000–10,000/ml. Approximately 60–75% of the circulating leukocytes are lymphocytes. Neutrophils have densely staining eosinophilic granules and thus may be referred to as heterophils.

BIBLIOGRAPHY FOR ANATOMIC FEATURES

Bivin, W.S., et al. 1987. Morphophysiology. In *Laboratory Hamsters*, ed. G.L. Van Hoosier, Jr., and C.W. McPherson. pp. 9–41. New York: Academic.

Burek, J.R., et al. 1979. The pregnant hamster as a model to study intravascular trophoblasts and associated maternal blood vessel changes. Vet. Pathol. 16:553–66.

Clark, J.D. 1987. Historical perspectives and taxonomy. In *Laboratory Hamsters*, ed. G.L. Van Hoosier Jr. and C.W. McPherson. pp. 3–7. New York: Academic.

Magalhaes, H. 1968. Gross Anatomy. In *The Golden Hamster: Its Biology and Use in Medical Research*, pp. 91–109. Ames: Iowa State University Press.

Richards, M.P.M. 1966. Activity measured by running wheels and observations during the oestrous cycle, pregnancy and pseudopregnancy in the golden hamster. Anim. Behav. 14:450–58.

Thomson, F.N., and Wardrop, K.J. 1987. Clinical chemistry and hematology. In *Laboratory Hamsters*, ed. G.L. Van Hoosier, Jr., and C.W. McPherson. pp. 43–59. New York: Academic.

VIRAL INFECTIONS

DNA VIRAL INFECTIONS

Cytomegaloviral Infection

Cytomegaloviruslike lesions have been observed in salivary glands of subclinically infected Chinese hamsters (Lussier 1975). Intranuclear and intracytoplasmic inclusions with cytomegaly occur in the acinar epithelium of the submaxillary salivary glands. Cytomegaloviruses are generally considered to be host-specific, but the relationship of the hamster agent to cytomegaloviruses of other species has not been defined.

Parvoviral Infection

An epizootic of high mortality with malformed and missing incisors has been observed among suckling and weaning pups in a breeding colony of Syrian hamsters. Necrosis and inflammation of the dental pulp with mononuclear leukocytic infiltration of the dental lamina and osteoclasis of alveolar bone was noted. This epizootic was associated with seroconversion to Toolan H-1 virus, a parvovirus of rats that has previously been shown to cause facial and dental deformities in experimentally infected neonatal hamsters (Gibson et al. 1983). Seroconversion to both rat virus and H-1 virus occurs without disease in hamsters.

Hamster Papovaviral Infection

Hamster papovavirus (HaPV) belongs to the polyomavirus subgroup of Papoviridae. It is structurally and biologically very similar to polyomavirus of mice but is a distinctly different virus (see Chap. 1). HaPV has suffered a longstanding history of misunderstanding. It is the cause of transmissible lymphoma, which can occur in epizootics among young hamsters, keratinizing skin tumors of hair follicle origin, or of subclinical infections. This agent has erroneously been called hamster papillomavirus because of its ability to induce papillomalike skin lesions. It clearly does not belong to the papillomavirus subgroup of Papoviridae, and thus the name should not be used.

EPIZOOTIOLOGY AND PATHOGENESIS. HaPV is not common, but infections of Syrian and European hamster colonies have been reported on several occasions in the United States and Europe. The origin of these infections has not been definitively determined. HaPV is probably one of few truly hamster-origin agents. It was probably introduced to laboratory hamsters in Eastern Europe through acquisition of wild European hamster stocks and mixing with laboratory Syrian hamsters. The biology of HaPV closely parallels that of polyomavirus of mice, but there are unique features peculiar to the hamster, especially their sensitivity to the oncogenic effects of DNA viruses in general. HaPV is probably spread by environmental contamination with infected urine. Like polyomavirus of mice, it causes a multisystemic infection with persistence in the kidney and shedding in the urine. This behavior is typical of many viruses of the polyomavirus subgroup, which infect a number of different species. Like polyomavirus of mice, HaPV is also oncogenic, but tumor formation is a side effect of infection and not critical to the virus life cycle. HaPV infection can result in the formation of lymphomas and hair follicle epitheliomas in hamsters (Barthold 1991; Barthold et al. 1987). Other types of tumors have not been described. Typical of polyomaviruses, HaPV can infect cells lytically with virus replication or transform cells without virus replication. Thus, lymphomas do not have detectable infectious virus. On the other hand, HaPV epitheliomas have HaPV replication in keratinizing epithelium, similar to the behavior of papillomaviruses. Polyomavirus of mice can cause similar virus-replicating skin tumors in mice as well, but unlike mice, hamsters are susceptible to the oncogenic effects of this virus (and other DNA viruses) beyond the neonatal period.

With the above brief synopsis, the epizootiology of HaPV can be understood. When first introduced to a naive population of breeding hamsters, HaPV can result in epizootics of lymphoma, with attack rates as high as 80% among young hamsters within 4–30 wk postexposure, which is a diagnostic phenomenon, since lymphomas normally occur in very low incidence and only among aged hamsters. Infected hamsters may also have a variable incidence of epitheliomas, usually around the face and feet. Although the epitheliomas contain infectious virus, they are not necessary for virus transmission, which occurs primarily through the urine. Lymphomas do not contain infectious HaPV, but HaPV nucleic acid can be detected in their genome. Hamster leukemia virus particles also occur in these tumors, as they do in other tumors and normal tissues as an incidental finding. Once enzootic, the incidence of lymphoma declines to much lower levels because young hamsters are presumably protected from infection and the virus infects only older hamsters, which tend to resist the oncogenic effects. Infection of older hamsters results in a clinically silent infection with persistent viruria, typical of polyomaviruses. Enzootically infected hamsters, however, tend to develop a higher incidence of HaPV skin tumors than do hamsters during the epizootic form. These complex features have led to considerable confusion as to the etiology of transmissible lymphoma, including claims that it is caused by a DNA viroid-like agent. These claims have been refuted and the etiological role of HaPV confirmed. Barthold (1991) presents a thorough review of HaPV biology.

PATHOLOGY. Affected hamsters appear thin, often with palpable masses in their abdomen. Lymphomas usually arise in the mesentery without involvement of the spleen, but they can arise in axillary and cervical lymph nodes. Infiltration of liver, kidney, thymus, and other organs can also occur (Fig. 3.1). Tumors vary cytologically. They are usually lymphoid, but erythroblastic, reticulosarcomatous, and myeloid types have been described. Lymphoid tumors are variably differentiated, usually immature, although sometimes they have plasmacytoid features. Lymphomas of the abdomen have been shown to possess B-cell markers, and those of the thymus T-cell markers. Mesenteric masses involve the intestinal wall and lymph nodes, with necrosis of the central region. Infiltration of hepatic sinusoids is also common (Fig. 3.2). Affected hamsters may also present few to many nodular masses involving nonglabrous skin. These lesions consist of keratinizing follicular structures reminiscent of trichoepitheliomas (Figs. 3.3 and 3.4).

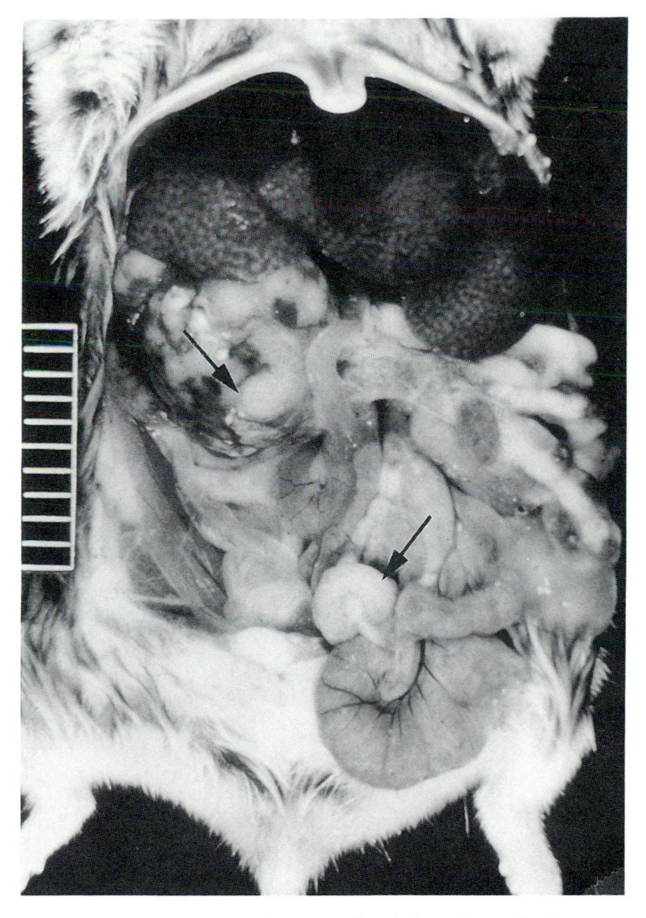

Fig. 3.1. Papovavirus-associated lymphoma in hamster. Note the marked enlargement of lymph nodes (*arrows*) in the abdominal cavity. (Courtesy of Barthold et al. 1987, reprinted by permission)

DIAGNOSIS. Epizootic HaPV is unmistakable. Lymphoid tumors are otherwise rare in hamsters, and when they occur, it is usually in aged hamsters. As stated earlier, electron microscopy of lymphoid tu-

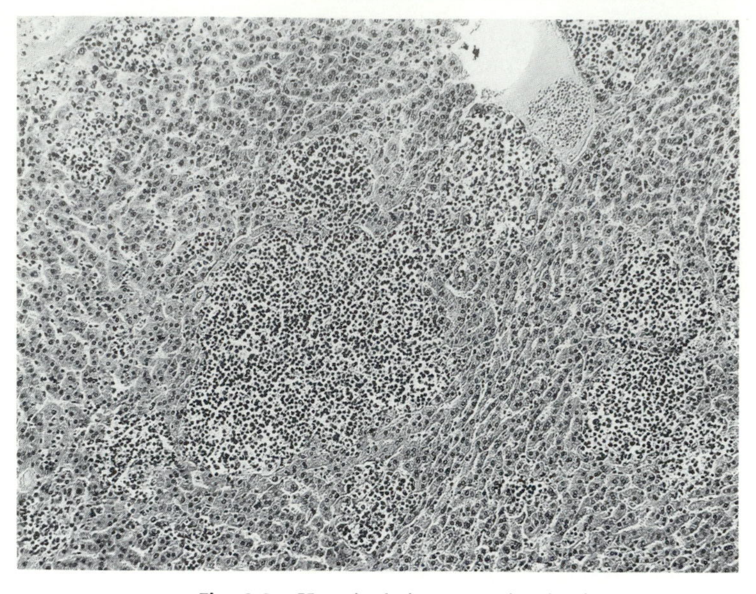

Fig. 3.2. Hepatic lesions associated with hamster papovavirus infection. Note the marked infiltration of neoplastic lymphoid cells in sinusoids.

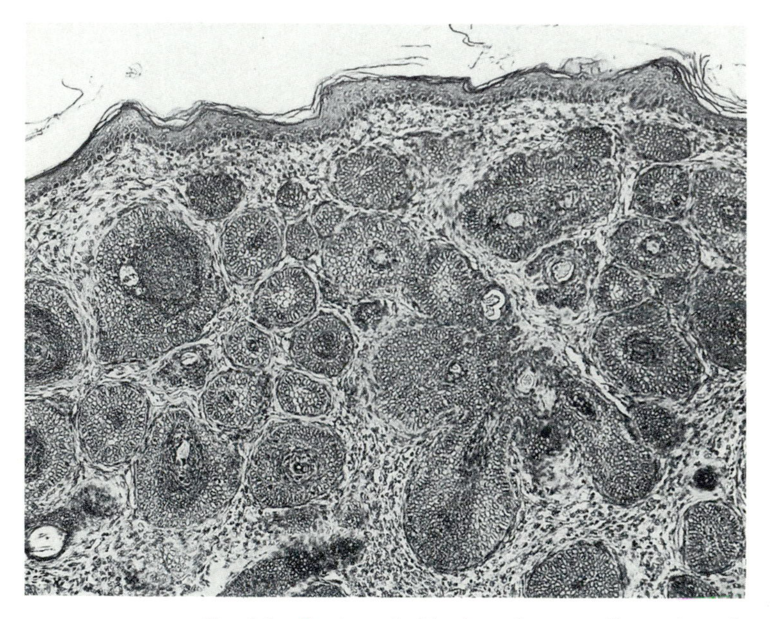

Fig. 3.3. Section of skin from hamster, illustrating trichoepithelioma associated with HaPV infection.

mors in an effort to visualize HaPV is a vacuous exercise. Trichoepitheliomas have not been described in hamsters unless associated with HaPV. If present, they offer the opportunity to visualize HaPV crystalloids in the nucleus of keratinizing epithelial cells. A serological test for this virus is not available. *Differential diagnoses* must include transmissible ileal hyperplasia, which can cause palpable enlargement of the terminal ileum, spontaneous lymphoid tumors, and skin lesions such as *Demodex* folliculitis.

SIGNIFICANCE. HaPV can cause devastating epizootics, which have caused the total loss of some in-

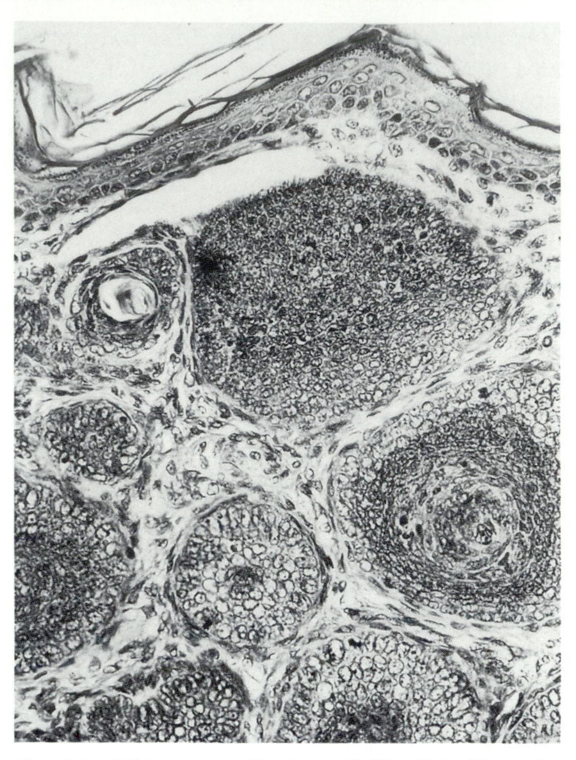

Fig. 3.4. Higher magnification of Fig. 3.3. Note the prominent hair follicle formation.

bred strains of Syrian hamsters. Once enzootic, the virus cannot be effectively eliminated without slaughter of the entire population and thorough decontamination of the premises. Even under these circumstances, repeated outbreaks have been known to occur, possibly because of the resistance of the virus to environmental decontamination.

Adenoviral Infection

Adenoviral intranuclear inclusion bodies have been observed in ileal enterocytes in tissues collected from hamsters during the first few weeks of life. In addition, antibodies to the K87 strain of mouse adenovirus are commonly present in hamsters from commercial suppliers in the United States.

PATHOLOGY. Large, amphophilic, intranuclear inclusions may be present in the enterocytes lining villi and goblet cells of the jejunum and ileum, and rarely, in the cryptal epithelial cells. In the typical case, animals are asymptomatic and there is no evidence of intestinal tract damage or inflammatory response. To date, typical adenoviral inclusions have been observed only in hamsters less than 4 wk of age (Gibson et al. 1990).

DIAGNOSIS. The presence of adenoviral infection may be confirmed by electron microscopy and serology.

SIGNIFICANCE. Infected hamsters are considered to be asymptomatic, and the significance of the infection in this species is currently unknown.

RNA Viral Infections

Sendai Viral Infection

Based on serological surveys, Sendai virus infections are relatively widespread in some colonies of hamsters. However, there are few reports of confirmed clinical disease due to Sendai virus infections in this species. Seronegative hamsters introduced into a facility housing infected rodents may seroconvert, but it is unlikely that any clinical signs will be observed (Profeta et al. 1969), although there are reports of mortality in newborn Syrian and Chinese hamsters (Parker et al. 1987). Currently mice, rats, and hamsters are regarded as the natural hosts for Sendai virus. Respiratory lesions resemble Sendai viral lesions in rats and in resistant strains of mice and are characterized by mild necrotizing bronchiolitis and focal interstitial pneumonia (Fig. 3.5).

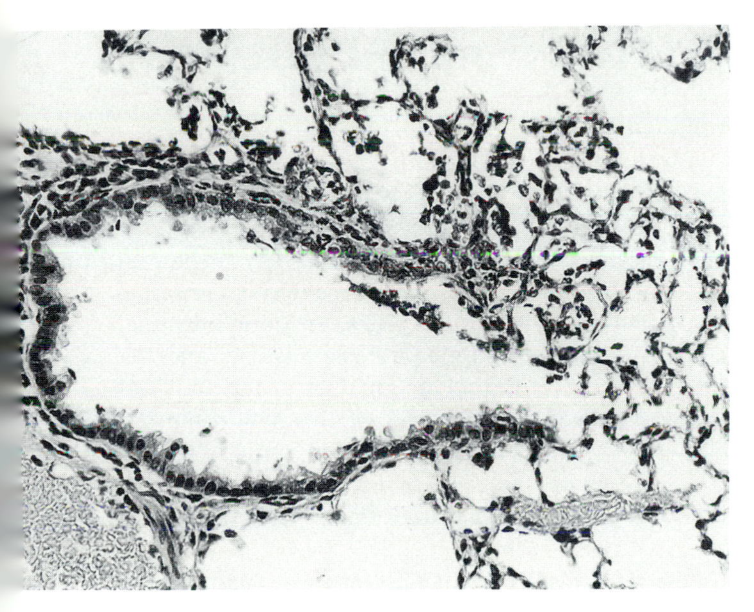

Fig. 3.5. Lung from spontaneous case of Sendai virus pneumonitis in Syrian hamster. There is a bronchoalveolitis with mobilization of alveolar macrophages.

Pneumonia Virus of Mice (PVM) Infection

Hamsters, mice, and rats are considered to be the natural hosts for this paramyxovirus. In an early report, interstitial pneumonitis with consolidation was observed in hamsters inoculated with an infectious agent interpreted to be contaminated with PVM, but there are few details of the morphologic changes (Pearson and Eaton 1940). Conventional colonies of hamsters may be seropositive, usually in the absence of clinical disease. Based on current information, it is evident that PVM infections in this species normally go unrecognized as a subclinical event (Parker et al. 1987). Thus the significance of PVM infection in hamsters is currently unknown; it does, however, represent a potential complicating factor, particularly in research related to respiratory function.

DIAGNOSIS. In suspected cases of pneumonitis due to Sendai virus or PVM, the demonstration of seroconversion is the most practical method to confirm the diagnosis. *Differential diagnoses* would include alveolar changes secondary to congestive heart failure.

Lymphocytic Choriomeningitis (LCM)

LCM virus is an arenavirus with a wide host range, including rodents and human and nonhuman primates. Its principal natural reservoir host is the wild mouse (see Chap. 1).

EPIZOOTIOLOGY AND PATHOGENESIS. Infection with LCM virus may occur by exposure to saliva or urine from animals shedding the virus. Portals of entry include the oronasal route and skin abrasions. Cage-to-cage transmission via aerosols does not appear to play an important role in spread. Congenital infections also occur in hamsters. Cell cultures or transplantable tumors contaminated with the virus are an important source of the virus in the laboratory (Bhatt et al. 1986). The patterns of disease that occur in hamsters postexposure will depend on the age of the animal, strain and dose of the virus, and route of administration. In one study, newborn hamsters were inoculated subcutaneously with LCM virus. Approximately half of the recipients cleared the virus, with minimal to moderate lymphocytic infiltration in the viscera. In the remaining inoculated animals, viremia and viruria persisted for approximately 3 and 6 mo, respectively. In addition, there was chronic wasting, and lymphocytic infiltration was observed in tissues such as liver, lung, spleen, meninges, and brain. Vasculitis and glomerulitis were present in hamsters examined histologically at 6 or more mo postinoculation. Antigen-antibody complexes were demonstrated in arterioles and glomerular basement membranes (Parker et al. 1976).

SIGNIFICANCE. LCM virus–infected hamsters are recognized to be the primary source of the virus in human patients. Thus, the public health aspects are an important consideration. Epidemics of LCM have occurred in laboratory personnel exposed either to hamsters shedding virus or to infected cell lines. Pet hamsters are also the recognized source of virus in some human cases. In one outbreak reported in Europe, there were approximately 200 reported cases of LCM in humans after contact with subclinically infected pet hamsters. It was suggested that there may have been up to 4000 additional cases of exposure in pet owners after hamsters were distributed to homes nationwide from the supplier of infected animals (Biggar et al. 1975). In human cases, sequelae postexposure may vary from subclinical infections to influenzalike symptoms (Biggar et al. 1975; Bowen et al. 1975). On rare occasions, viral meningitis or encephalomyelitis may occur.

DIAGNOSIS. Serology is the recognized method for confirming the diagnosis. Sera collected from hamsters infected early in life may have a high percentage of samples with anticomplementary activity, thus complicating interpretations if the complement fixation test is used. Hamsters that acquire LCM infections as adults usually seroconvert early in the disease and remain seropositive for a long period of time. The indirect fluorescent antibody test is one procedure recommended for serological testing.

Viral Infections of Uncertain Significance

Laboratory Syrian hamsters seroconvert to mouse encephalomyelitis virus, reovirus 3 and SV5, a paramyxovirus, and parvoviruses of the rat (rat virus and H-1 virus). Except for a single epizootic associated with H-1 virus (see Parvoviral Infection), disease has not been associated with these infections. Hamsters are also host to an endogenous retrovirus, which is expressed in tissues and cells as C-type particles without evidence of oncogenicity.

Virus-associated Neoplasia

Newborn hamsters are recognized to be a sensitive in vivo test system to screen for potentially oncogenic viruses isolated from other mammalian species (Trentin et al. 1987).

BIBLIOGRAPHY FOR VIRAL INFECTIONS

DNA Viral Infections

Barthold, S.W. Hemolymphatic tumors. In The *Pathology of Tumours in Laboratory Animals, III. Tumours of the Hamster*, ed. U. Mohr et al. Lyon, France: IARC Scientific Publications.

Barthold, S.W., et al. 1987. Further evidence for papovavirus as the probable etiology of transmissible lymphoma of Syrian hamsters. Lab. Anim. Sci. 37:283–88.

Gibson, S.V., et al. 1990. Naturally acquired enteric adenovirus infection in Syrian hamsters (*Mesocricetus auratus*). Am. J. Vet. Res. 51:143–47.

Gibson, S.V., et al. 1983. Mortality in weanling hamsters associated with tooth loss. Lab. Anim. Sci. 33:497.

Kuttner, A.G., and Wang, S. 1934. The problem of the significance of the inclusion bodies in the salivary glands of infants, and the occurrence of inclusion bodies in the submaxillary glands of hamsters, white mice and wild rats (Peiping). J. Exp. Med. 60: 773–91.

Lussier, G. 1975. Murine cytomegalovirus (MCMV). Adv. Vet. Sci. Comp. Med. 19:223–47.

RNA Viral Infections and General Bibliography

Bhatt, P.N., et al. 1986. Contamination of transplantable murine tumors with lymphocytic choriomeningitis virus. Lab. Anim. Sci. 36:136–39.

Biggar, R.J., et al. 1975. Lymphocytic choriomeningitis outbreak associated with pet hamsters: Fifty-seven cases from New York state. J. Am. Med. Assoc. 232:494–500.

Bowen, G.S., et al. 1975. Laboratory studies of a lymphocytic choriomeningitis virus outbreak in man and laboratory animals. J. Epidemiol. 102:233–40.

Carthew, P., et al. 1978. Incidence of natural virus infections of laboratory animals 1976–1977. Lab. Anim. 12:245–46.

Parker, J.C., and Richter, C.B. 1982. Viral diseases of the respiratory system. In *The Mouse in Biomedical Research, Vol. II., Diseases,* ed. H.L. Foster et al., pp. 107–55. New York: Academic.

Parker, J.C., et al. 1987. Viral diseases. In *Laboratory Hamsters*, ed. G.L. Van Hoosier, Jr., and C.W. McPherson, pp. 95–110. New York: Academic.

Parker, J.C., et al. 1976. Lymphocytic choriomenigitis virus infection in fetal, newborn, and young adult Syrian hamsters (*Mesocricetus auratus*). Infect. Immunol. 13:967–81.

Pearson, H.E., et al. 1940. A virus pneumonia of Syrian hamsters. Proc. Soc. Exp. Biol. Med. 45:677–79.

Profeta, M.L., et al. 1969. Enzootic Sendai infection in laboratory hamsters. Am. J. Epidemiol. 89:316–24.

Reed, J.M., et al. 1974. Antibody levels to murine viruses in Syrian hamsters. Lab. Anim. Sci. 24:33–38.

Trentin, J.J. 1987. Experimental biology: Use in oncological research. In *Laboratory Hamsters*, ed. G.L. Van Hoosier, Jr., and C.W. McPherson. pp. 201–25. New York: Academic.

BACTERIAL AND MYCOTIC INFECTIONS

Proliferative Ileitis (Transmissible Ileal Hyperplasia)

Proliferative ileitis is among the most commonly recognized diseases in the Syrian hamster; it usually results in high morbidity and mortality. This specific condition has been referred to by a variety of names, including regional ileitis, hamster enteritis, terminal enteritis, atypical ileal hyperplasia, enzootic intestinal adenocarcinoma, proliferative bowel disease, and wet tail. It seems that each individual or group that has become involved in the study of this syndrome has endowed it with its own epithet. The term "wet tail" should not be used because it includes virtually all the numerous conditions that may cause diarrhea in hamsters.

ETIOLOGY AND PATHOGENESIS. The many quests toward identifying the etiology of proliferative ileitis in hamsters have incriminated a number of apparently secondary and possibly contributory agents, including *Escherichia coli* (Frisk et al. 1978), *Campylobacter* (Humphrey et al. 1986), and *Cryptosporidium* (Davis and Jenkins 1986). Certainly, *E. coli* isolates from cases of proliferative ileitis have been shown to be enteropathogenic in naive hamsters but did not induce proliferative disease (Frisk et al. 1981). Inoculation of young, naive hamsters with homogenates of proliferative ileal lesions results in a progressive disease process. Hyperplasia of crypt cells occurs, with migration of mitotically active, immature epithelium onto the villi, with villus elongation, distortion, and fusion. This is followed by downward extension of hyperplastic crypts into Peyer's patches and the submucosa, with variable invasion of the muscular layers of the ileal wall (Amend et al. 1976). However, an organism has recently been isolated and identified as a new species of *Chlamydia*, with close but distinctly different

16S ribosomal RNA homology to *C. trachomatis* and *C. psittaci* (Stills 1991). Koch's postulates seem to have been fulfilled, but its true classification as a *Chlamydia* remains equivocal. Chlamydiae are obligate intracellular organisms with a cell wall and with DNA and RNA similar to conventional bacteria, but which are incapable of autonomous vegetative growth and are transmitted as small, environmentally resistant elementary bodies. Vegetative organisms are found abundantly in the apical cytoplasm of ileal enterocytes. Organisms of this type have been a recognized feature of this syndrome since its original description and thought for some time to be variants of *Campylobacter*. There is still some evidence that *Campylobacter*-like organisms may play a role in this disease, but whether they are an essential component has not been resolved.

CLINICAL SIGNS AND PATHOLOGY. Epizootics of the disease are usually confined to younger animals, particularly during the postweaning period. Hamsters are normally resistant to the experimental disease by 10–12 wk of age. Overcrowding, transport, diet, and experimental manipulations have also been identified as predisposing factors. In epizootics of the disease, there may be a morbidity rate of up to 60%, and mortality rates in affected animals may approach 90%. Clinical signs include lethargy; unkempt hair coat; anorexia; weight loss; foul-smelling, watery diarrhea; and dehydration. Rectal prolapse or intussusceptions frequently occur.

Affected hamsters are runted and emaciated, with soiling of the perineum with diarrhea. The ileum is segmentally thickened, often with prominent serosal nodules and fibrinous peritoneal adhesions to adjacent structures (Fig. 3.6). The opened bowel reveals an abrupt transition of the craniad, normal ileum, and the caudal cecum with the affected, hyperplastic mucosa. Microscopic lesions, as noted, consist of marked crypt and villous epithelial hyperplasia, villous elongation, villous fusion, varying degrees of necrosis and hemorrage, crypt invasion of underlying structures, crypt abscessation, and granulomatous inflammation (Figs. 3.7 and 3.8). With silver or PAS stains, numerous and characteristic small bacteria can be seen in the apical cytoplasm of proliferating enterocytes, and macrophages in the lamina propria and submucosa contain abundant granular PAS-positive material in their cytoplasm.

DIAGNOSIS. The demonstration of the typical lesions should be sufficient to confirm the diagnosis. *Differential diagnoses* would include Tyzzer's disease, salmonellosis, antibiotic-associated enterocolitis, coliform enteritis, and giardiasis.

SIGNIFICANCE. Sporadic outbreaks of transmissible ileitis are an important cause of disease and mortality in hamsters from pet stores and represent a po-

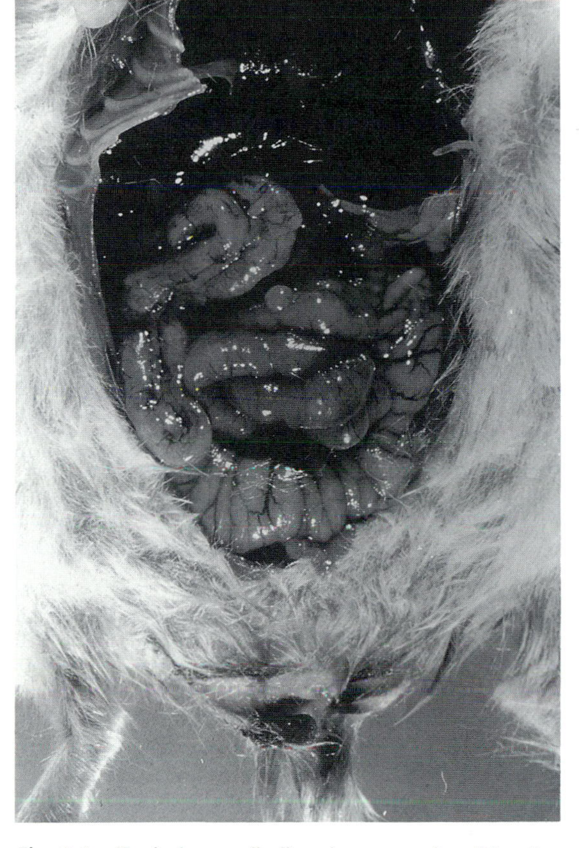

Fig. 3.6. Typical gross findings in a case of proliferative ileitis in young hamster. There is soiling of the perineal region and marked thickening and edema of the terminal small intestine.

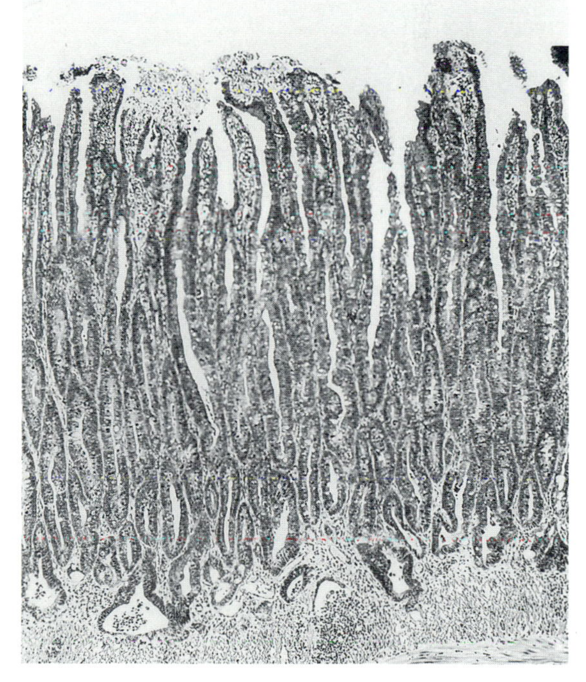

Fig. 3.7. Section of ileum from a spontaneous case of proliferative ileitis. There is marked hyperplasia of enterocytes lining crypts and villi, with extension into the submucosa. Some crypts are dilated and contain cellular debris.

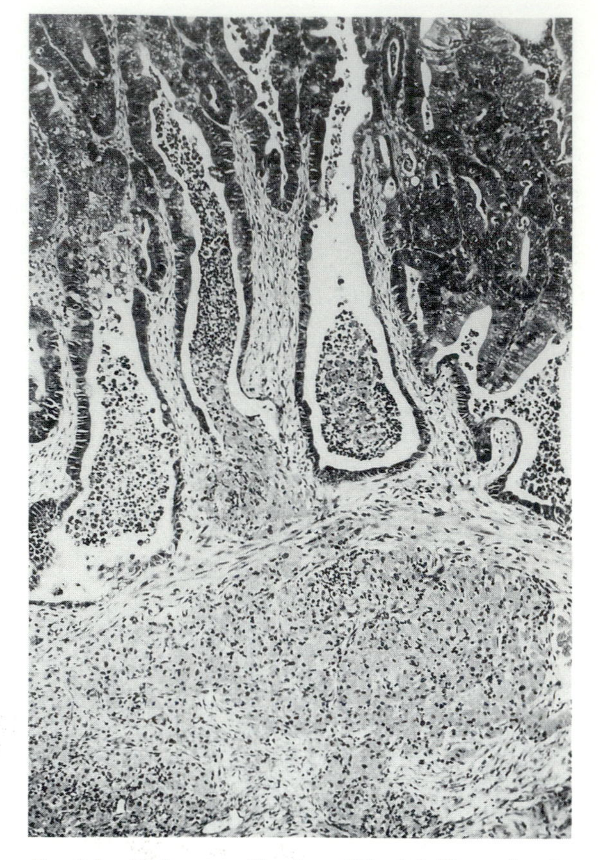

Fig. 3.8. Higher magnification of Fig. 3.7, illustrating the cryptal diverticular, with concurrent granulomatous inflammatory response in the submucosa and adjacent muscle.

tential complicating factor in hamsters used in research facilities.

Campylobacter Infection

Aside from their association with some outbreaks of proliferative ileitis, *C. fetus* ssp. *jejuni* has been isolated on numerous occasions from clinically normal hamsters. In one study, 24 of 30 hamsters acquired from pet stores were positive for *Campylobacter*. A few animals had watery diarrhea (Fox et al. 1983). Hamsters have been shown to be relatively resistant to experimental disease. Manipulations may be required in order to produce clinical disease consistently in inoculated animals. Subclinically infected hamsters may shed the organism in the feces for up to several months (Fox et al. 1986).

SIGNIFICANCE. *Campylobacter*-infected hamsters represent a zoonotic threat to both pet owners and laboratory animal personnel. The organism has been associated with some cases of proliferative ileitis and occasionally enteritis in this species.

Escherichia coli Infection

EPIZOOTIOLOGY AND PATHOGENESIS. Isolates of *E. coli* from hamsters with enteritis have proven to be pathogenic when inoculated into susceptible recipient animals. Strains 1056, 1126, and 4165 were isolated from naturally occurring cases of hamster enteritis. When inoculated into ligated intestinal loops to test for enteropathogenicity, the pathogenic strains produced changes in most of the inoculated weanling hamsters and in some of the adult animals (Frisk et al. 1978).

One isolate of *E. coli*, strain 1056, has been recovered from ground ileal suspension prepared from a hamster with proliferative ileitis. Many of the weanling Syrian hamsters inoculated orally with this strain developed acute enteritis within 2 wk postinoculation. Animals inoculated with a nonenteropathogenic strain remained asymptomatic throughout the study (Frisk et al. 1981).

PATHOLOGY. At necropsy, the small intestine may contain yellow to dark red, fluid material. On microscopic examination, blunting and fusion of villi are frequently observed. Affected villi were lined by cuboidal epithelial cells. Degeneration and sloughing of enterocytes and polymorphonuclear cell infiltration in the lamina propria commonly occurs. In the mesenteric lymph nodes, changes may vary from lymphoid hyperplasia to diffuse polymorphonuclear cell infiltration. Focal coagulation necrosis in the liver, with polymorphonuclear cell infiltration, and gastric ulcers are other variable findings. Colitis and/or typhlitis may be present in some affected animals, sometimes with concomitant colonic intussusception. Ultrastructural studies of sections of ileum have revealed bacilli in the cytoplasm of enterocytes and blunting and irregularities in microvilli. *Differential diagnoses* include antibiotic-associated enterocolitis, proliferative ileitis, salmonellosis, and *Campylobacter* infections.

SIGNIFICANCE. Enteropathogenic strains of *E. coli* may cause acute enteritis in weanling hamsters. It appears to be a different syndrome from ileal hyperplasia, although concurrent *E. coli* infections may also play a role in this disease entity.

Tyzzer's Disease

Epizootics of Tyzzer's disease have been observed in Syrian hamsters in various parts of the world (Zook et al. 1977; Takasaki et al. 1974). The causative agent, *Bacillus piliformis*, is a gram-negative, spore-forming bacillus that appears to multiply only within cells. The organism has a wide host range. (For additional details, see Tyzzer's disease in the rabbit, Chap. 6.)

EPIZOOTIOLOGY AND PATHOGENESIS. Hamsters may become infected by contact with affected animals or by contaminated bedding. Predisposing factors, such as poor sanitation, intestinal parasitism, and inappropriate feeding practices, may play a role in precipitating clinical outbreaks of the disease (Motzel and Gibson 1990). In hamsters inoculated with infected liver homogenates, organisms and lesions were detectable in the mucosa of the small and large intestine by 3 days postinoculation, and multiple lesions and bacilli were present in the liver by days 6–8 postexposure (Nakayama et al. 1976). Weanling hamsters are most often affected.

PATHOLOGY. At necropsy, there is a variable distribution of lesions. In some epizootics, lesions may be confined to either the liver or the intestinal tract. Multifocal hepatic necrosis is evident in some cases. Intestinal lesions, when evident grossly, usually involve the lower ileum, cecum, and colon and are associated with soiling of the perineum. Affected areas are edematous and dilated, with fluid contents. Microscopically, hepatic lesions are characterized by foci of necrosis with leukocytic infiltration. Intracellular bundles of bacilli are usually best demonstrated at the periphery of hepatic lesions. When lesions are present in the intestinal tract, there is edema of the lamina propria, with leukocytic infiltration and effacement of the mucosal architecture. There may be extension of the inflammatory process into the underlying muscular layers. Typical bacilli are usually demonstrable within enterocytes in the region (Fig. 3.9) and in hepatocytes adjacent to necrotic foci (Fig. 3.10). Focal granulomatous myocarditis, with conspicuous pale bulging nodules, has been associated with Tyzzer's disease in this species.

DIAGNOSIS. Confirmation of the diagnosis entails the demonstration of the typical intracellular bacilli in affected cells, using Warthin-Starry or Giemsa stains. *Differential diagnoses* include proliferative ileitis, salmonellosis, coliform enteritis, antibiotic-associated enterocolitis, and *Campylobacter* infections.

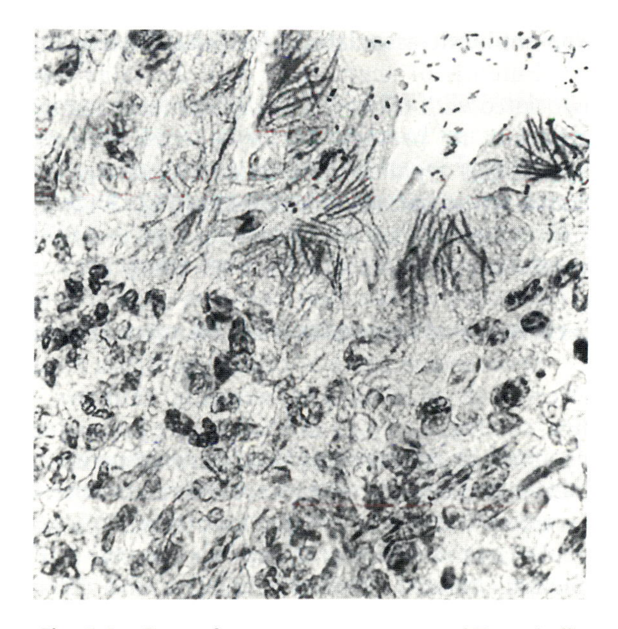

Fig. 3.9. Cecum from a spontaneous case of Tyzzer's disease in young hamster. There is an acute necrotizing typhlitis extending to the deeper layers. Bacilli are seen present in many enterocytes (Warthin-Starry stain). (Courtesy of R.J. Hampson)

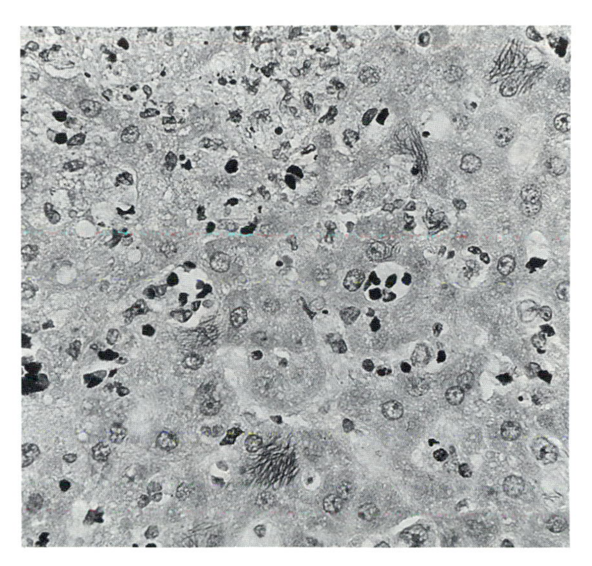

Fig. 3.10. Liver from a case of Tyzzer's disease in hamster (Warthin-Starry stain). Note the bundles of beaded bacilli in hepatocytes.

SIGNIFICANCE. Although documented cases of Tyzzer's disease are relatively rare, it does represent one cause of morbidity and mortality in hamsters. There is a possibility of interspecies transmission of the disease (Motzel and Gibson 1990).

Salmonellosis

Although recognized epizootics are relatively rare in most countries, hamsters are very susceptible to *Salmonella* infections. In one report from India, *Salmonella* was isolated from close to 50% of diseased hamsters (Ray and Mallick 1970). *S. enteritidis* serotypes *typhimurium* and *enteritidis* are the most frequent isolates in this species. Transmission is probably primarily by the ingestion of contaminated food or bedding, and interspecies transmission is likely to occur. Explosive outbreaks of salmonellosis are characterized by depression, ruffled hair coat, anorexia, dyspnea, and high mortality.

PATHOLOGY. At necropsy, there may be multifocal, pinpoint-size, pale areas in the liver, with patchy pulmonary hemorrhage and reddened hilar lymph nodes. Microscopic changes in the lung are characterized by multifocal interstitial pneumonitis, with intraalveolar hemorrhage. In the pulmonary veins and venules, there may be a septic thrombophlebitis (Innes et al. 1956), with thrombi containing leukocytes, and erosion of venous walls (Fig. 3.11). Focal splenic necrosis and focal necrotizing hepatitis, with leukocytic infiltration, and venous thrombosis are typical lesions. Embolic glomerular lesions and focal splenitis may also occur.

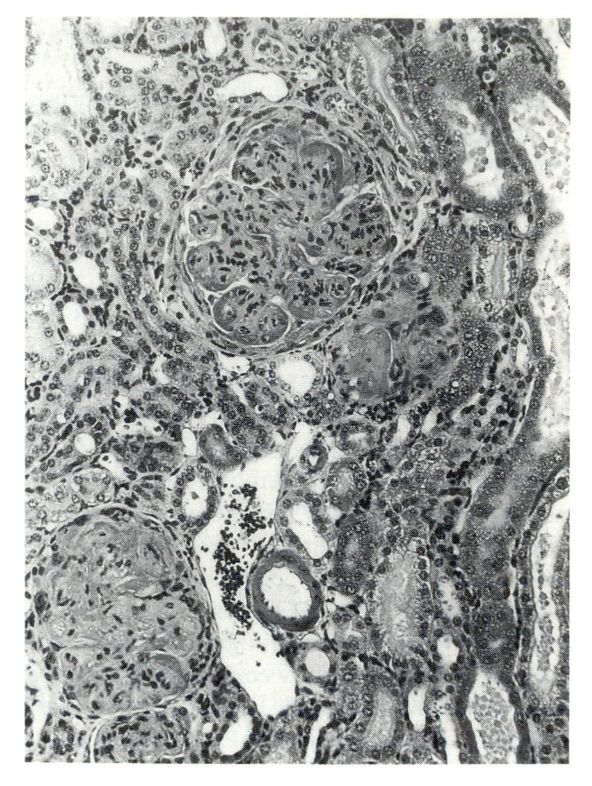

Fig. 3.19. Section of renal cortex from a spontaneous case of hamster nephropathy. Note the marked thickening of glomerular basement membranes, with obliteration of the normal architecture.

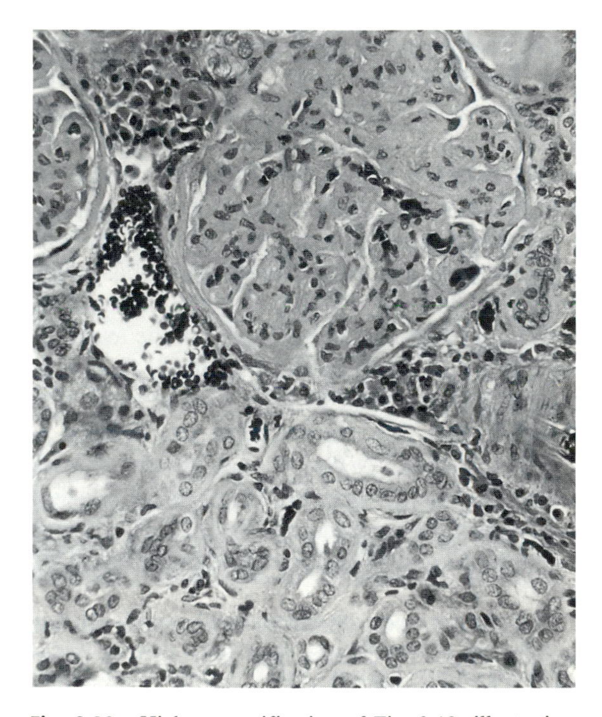

Fig. 3.20. Higher magnification of Fig. 3.19, illustrating thickening of basement membranes of glomeruli and tubules.

and there is a marked variation in the incidence, depending on the colony under study. The disease is most common in females. Amyloid deposition may be detected as early as 5 mo, but it is much more common in hamsters examined at 15 or more mo of age. There may be a drop in serum albumin and a rise in serum globulins. Amyloidosis may be produced experimentally in adult hamsters with regular injections of casein. Testosterone administration will inhibit the expression of amyloid in female hamsters (Coe and Ross 1990).

PATHOLOGY. The kidneys can be pale and irregular (Fig. 3.21), and affected livers are swollen, with a prominent lobular pattern. On microscopic examination, the liver, kidneys, and adrenal glands are most frequently involved (Fig. 3.22). Other tissues that can be affected include spleen, stomach, testis, and intestine. In the liver, deposition of eosinophilic, homogeneous material is evident around portal triads and within vessel walls, with variable involvement of the sinusoidal regions. Amyloid deposition frequently occurs initially in the glomerular tufts. The early changes may be characterized by the appearance of PAS-positive hyalinlike deposits along the glomerular basement membranes (Gleiser et al. 1971). The early deposits may have the typical amyloid fibrils evident by electron microscopy but may be negative for amyloid (paramyloid), using the usual histochemical stains (Gruys et al. 1979). In addition to deposition along glomerular basement membranes, the basement membranes of tubules are also frequently affected. In the adrenal glands, extensive cortical deposition may occur, with distortion of the normal architecture.

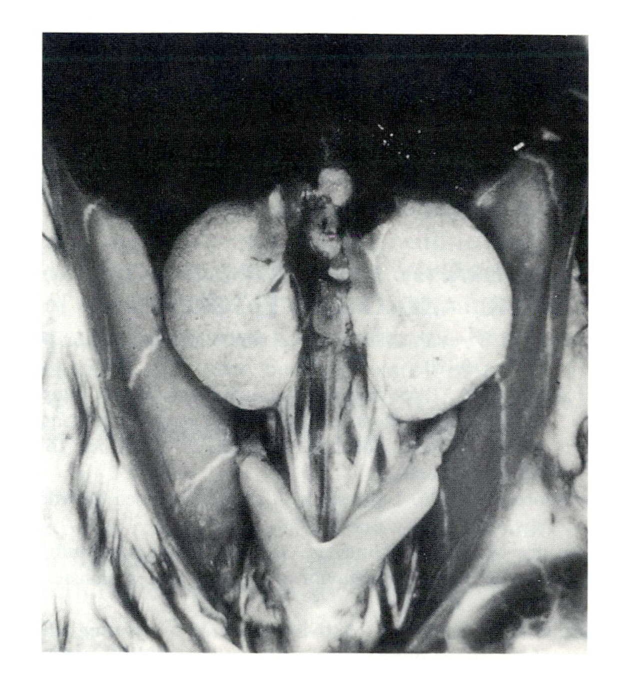

Fig. 3.21. Renal amyloidosis in aged hamster. The kidneys are pale and markedly swollen.

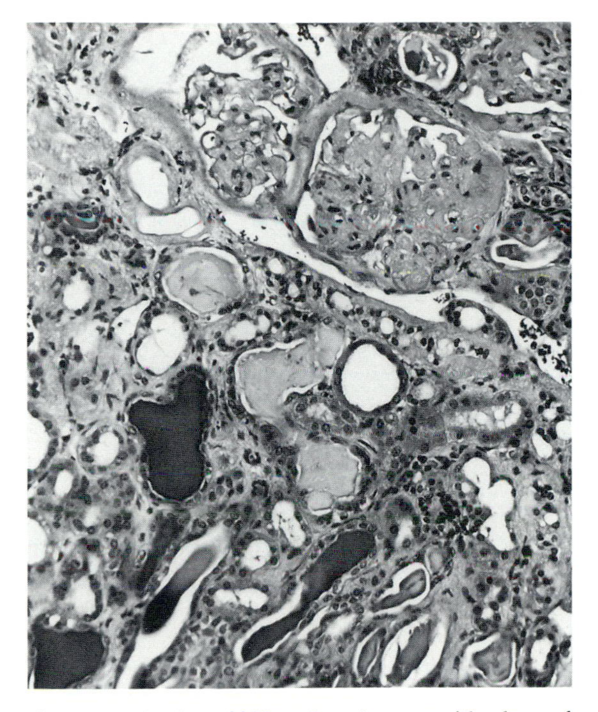

Fig. 3.22. Section of kidney from hamster with advanced renal amyloidosis. There is complete obliteration of the glomerular architecture.

DIAGNOSIS. The presence of amyloid normally can be verified using techniques, such as Congo red or thioflavin T procedures. Deposits may be negative for amyloid using the alcian blue–PAS staining method. The primary *differential diagnosis* is hamster nephrosis.

SIGNIFICANCE. Amyloidosis is an important cause of renal insufficiency and mortality in older hamsters. In advanced cases, other organs with extensive amyloid deposition, such as the adrenal glands, may have severely compromised function.

Atrial Thrombosis

EPIZOOTIOLOGY AND PATHOGENESIS. The process, which usually involves the left auricle and atrium, is a common occurrence in older hamsters in some colonies. Frequently females are affected earlier than males, and the syndrome is often associated with amyloidosis (McMartin and Dodds 1982). Changes also occur in coagulation and fibrinolytic parameters consistent with consumptive coagulopathy. Atrial thrombosis may be due in part to local blood stasis secondary to cardiac insufficiency. Frequently there is concurrent myocardial degeneration and left-sided congestive heart failure.

PATHOLOGY. Hamsters with this disorder often present with severe dyspnea, due to congestive heart failure. The thrombus is usually present in the left auricle and atrium. A moderately firm to friable, pale thrombus is adherent to the adjacent endocardium (Fig. 3.23). Bilateral ventricular hypertrophy is a common finding. Lungs may be congested and edematous. Microscopically, there may be some degree of organization of the layered thrombus (Fig. 3.24). Focal to diffuse myocardial degeneration, when present, is characterized by nuclear hypertrophy, vacuolation of sarcoplasm, fiber atrophy, and interstitial fibrosis. There may be concurrent focal medial degeneration and calcification of coronary arteries. In the valves, there may be fibrosis and myxomatous change.

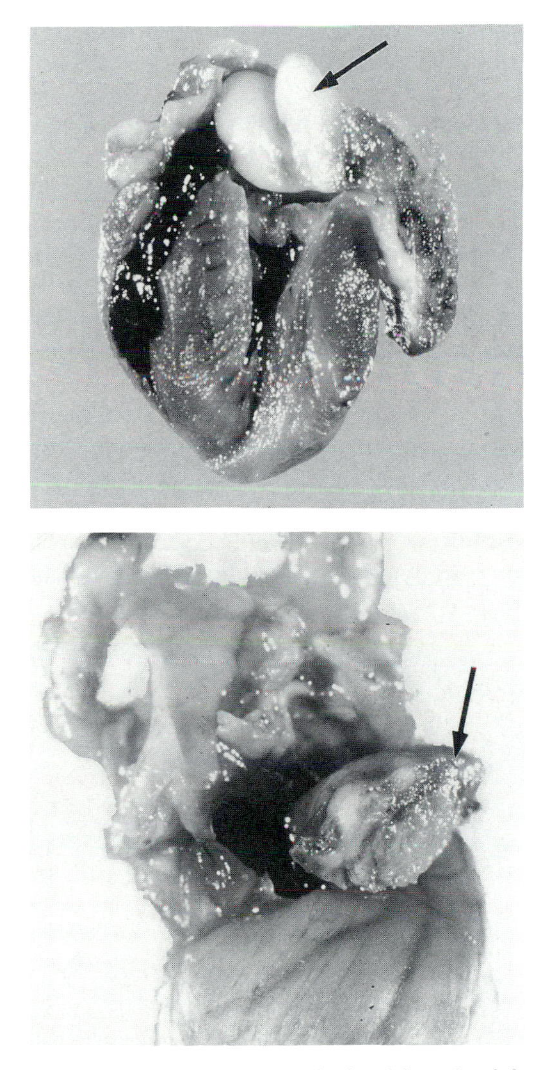

Fig. 3.23. Spontaneous thrombosis involving the left atrium and auricle (*arrows*) in two aged hamster, one important cause of spontaneous deaths in older animals.

SIGNIFICANCE. In some facilities, atrial thrombosis is a common cause of mortality in older hamsters. Necropsy procedures in this species should always include a careful examination of the chambers of the heart for evidence of thrombotic change.

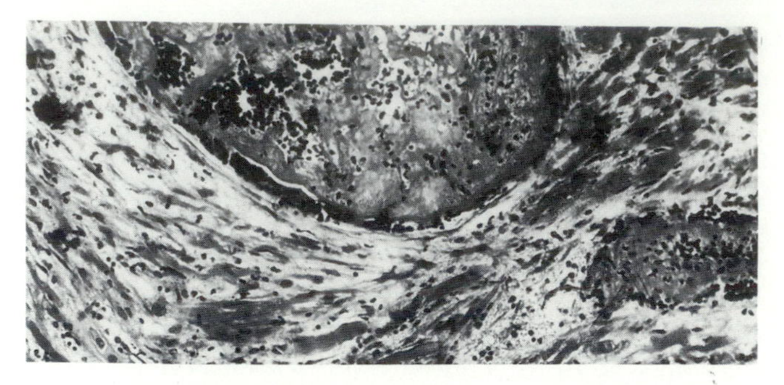

Fig. 3.24. Histological section of organizing thrombus adherent to the endocardium.

Polycystic Disease (Polycystic Liver Disease)

EPIZOOTIOLOGY AND PATHOGENESIS. Multiple hepatic cysts are occasionally seen in older hamsters at necropsy. They are considered to be of congenital origin and due to either failure of fusion of the intralobular and interlobular ducts, or failure of superfluous bile ducts to disappear. Raised, cystic areas of variable size, up to 2 cm in diameter, are present on the capsule and within the parenchyma of the liver (Fig. 3.25). True cysts may also be present in tissues such as epididymis, seminal vesicles, pancreas, and endometrium. In one report, over 75% of hamsters studied had cystic lesions at necropsy and many had lesions at multiple sites. Cysts were most common in the liver and epididymis, followed by seminal vesicles and pancreas.

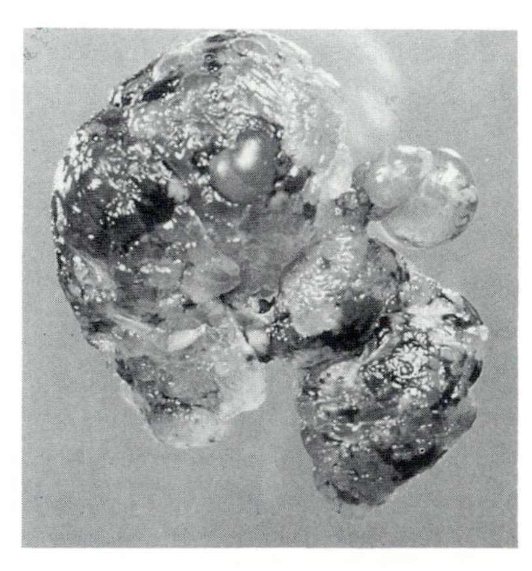

Fig. 3.25. Polycystic disease in liver from hamster necropsied at approximately 20 mo of age. There are multiple cystic areas scattered on the surface of the liver and extending into the parenchyma.

PATHOLOGY. The cysts are thin-walled, and contain clear, straw-colored fluid. On microscopic examination, there are multiple unilocular and multilocular cystic areas composed of a band of collagenous tissue and lined by flattened to cuboidal epithelial cells (Fig. 3.26). In the adjacent parenchyma of the liver, changes may include pressure atrophy of hepatic cords, hemosiderin deposition, proliferation of bile ducts, and periportal lymphocytic infiltration.

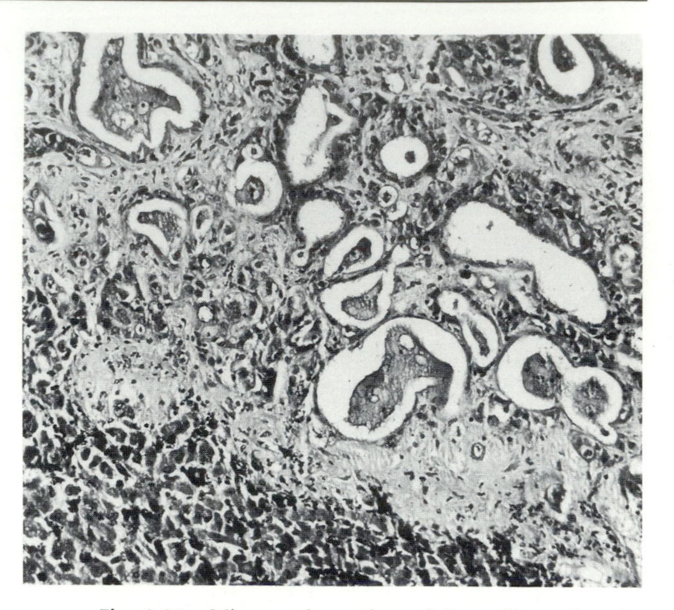

Fig. 3.26. Microscopic section of liver, illustrating the cystic areas lined by squamous to cuboidal epithelilum (polycystic disease).

SIGNIFICANCE. Hepatic cysts occasionally occur as an incidental finding at necropsy in older hamsters. They are interpreted to be of congenital origin.

Hepatic Cirrhosis

This spontaneous disorder occurs sporadically among laboratory hamsters, reaching an incidence of up to 20% in some colonies (Chesterman and Pomerance 1965). It occurs in aged animals, particularly females. There is uniform nodularity to the capsular surface grossly, with microscopic evidence of periportal fibrosis and bile duct proliferation, analagous to the liver lesion encountered in aging rats. There is also nodular hepatocellular proliferation with concurrent degeneration, necrosis, and mixed leukocyte infiltration.

Other Changes Associated with Aging

Alveolar histiocytosis, fibrinoid degeneration of arterioles, and cerebral mineralization are examples of lesions that have been observed in older animals. Focal cerebral mineralization may be seen microscopically as an incidental finding at necropsy. There are foci of mineralization in the neuropil, with displacement of adjacent structures and minimal cellular response.

For additional information on age-related changes, see Schmidt et al. (1983), Hubbard and Schmidt (1987), Pour et al. (1976), and Pour et al. (1979).

Bibliography for Diseases Associated with Aging

Chesterman, F.C., and Pomerance, A. 1965. Cirrhosis and liver tumours in a closed colony of golden hamsters. Br. J. Cancer 19:802–11.

Coe, J.E., and Ross, J.J. 1990. Amyloidosis and female protein in the Syrian hamster: Concurrent regulation by sex hormones. J. Exper. Med. 171:1257–66.

Doi, K., et al. 1987. Age-related non-neoplastic lesions in the heart and kidneys of Syrian hamsters of the APA strain. Lab. Anim. 21:241–48.

Gleiser, C.A., et al. 1971. Amyloidosis and renal paramyloid in a closed hamster colony. Lab. Anim. Sci. 21:197–202.

Gleiser, C.A., et al. 1970. A polycystic disease of hamsters in a closed colony. Lab. Anim. Care 20:923–29.

Gruys, E., et al. 1979. Deposition of amyloid in the liver of hamsters: An enzyme-histochemical and electron-microscopical study. Lab. Anim. 13:1–9.

Hubbard, G.B., and Schmidt, R.E. 1987. Noninfectious diseases. In *Laboratory Hamsters*, ed. G.L. Van Hoosier, Jr., and C.W. McPherson, pp. 169–78. New York: Academic.

McMartin, D.N., and Dodds, W.J. 1982. Atrial thrombosis in aging Syrian hamsters: An animal model of human disease. Am. J. Pathol. 107:277–79.

Newberne, P.M. 1978. Nutritional and metabolic diseases. In *Pathology of Laboratory Animals*, ed. C.K. Benirschke et al., pp. 2065–2214. New York: Academic.

Pour, P., et al. 1979. Spontaneous tumors and common diseases in three types of hamsters. J. Natl. Cancer Inst. 63:797–811.

Pour, P., et al. 1976. Spontaneous tumors and common diseases in two colonies of Syrian hamsters. 1. Incidence and sites. J. Natl. Cancer Inst. 56:931–35.

Schmidt, R.E., et al. 1983. *Pathology of Aging Syrian Hamsters.* Boca Raton, Fla.: CRC.

Slausen, D.O., et al. 1978. Arteriolar nephrosclerosis in the Syrian hamster. Vet. Pathol. 15:1–11.

Somvanshi, R., et al. 1987. Polycystic liver disease in golden hamsters. J. Comp. Pathol. 97:615–18.

Van Marck, E.A.E., et al. 1978. Spontaneous glomerular basement membrane changes in the golden hamster (*Mesocricetus auratus*): A light and electron microscopic study. Lab. Anim. 12:207–11.

NEOPLASMS

Although newborn hamsters are commonly used in vivo to screen for potentially oncogenic viruses, the incidence of spontaneous tumors in this species is relatively low. There is a marked variation in the incidence of neoplasms in different colonies. This probably reflects the influence of genetic factors, and possibly environmental conditions, on the occurrence of spontaneous tumors in this species. The majority of tumors are benign and they frequently arise from the endocrine system or alimentary tract. *Adrenocortical adenomas* are one of the most frequently recorded tumors. *Cutaneous lymphoma* resembling mycosis fungoides, the epidermotropic lymphoma seen in humans, has been observed in adult hamsters. Lethargy, anorexia, weight loss, patchy alopecia, and exfoliative erythroderma have been observed in affected animals (Saunders and Scott 1988). Microscopic changes include dense infiltrates of neoplastic lymphocytes in the dermis, with extension into the epidermis (Figs. 3.27 and 3.28). The most frequent malignant tumor of Syrian hamsters is lymphoma. These tumors arise in aged hamsters multicentrically, often involving thymus, thoracic lymph nodes, mesenteric lymph nodes, superficial lymph nodes, spleen, liver, and other sites. Cell types are variable.

For additional information on neoplasms see (Pour et al. (1976), Pour et al. (1979), Strandberg (1987), Turusov (1982), Van Hoosier and Trentin (1979), and Barthold (1992).

Fig. 3.27. Section of skin from adult hamster with spontanenous epidermotropic lymphoma. Note the lymphocytic infiltrate in the dermis and adjacent epidermis. (Courtesy of B.M. Cross)

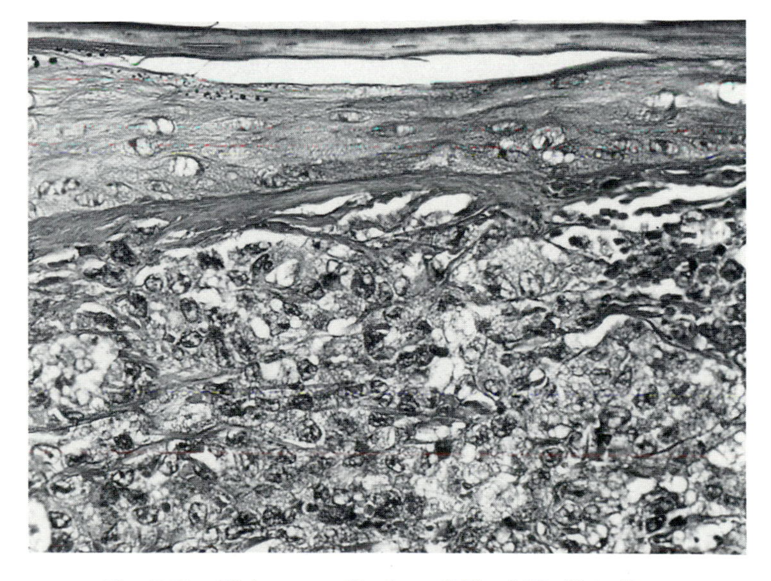

Fig. 3.28. Higher magnification of Fig. 3.27. Note the infiltrate of poorly differentiated nonnuclear cells with excavation and replacement of the overlying epidermis. (Courtesy of B.M. Cross)

Bibliography for Neoplasms

Barthold, S.W. 1992. Hemolymphatic tumors. In *The Pathology of Tumours in Laboratory Animals. III. Tumors of the Hamster*, ed. U. Mohr et al. Lyon, France: IARC Scientific Publications.

Pour, P., et al. 1979. Spontaneous tumors and common diseases in three types of hamsters. J. Natl. Cancer Inst. 63:797–811.

Pour, P., et al. 1976. Spontaneous tumors and common diseases in two colonies of Syrian hamsters. 1. Incidence and sites. J. Natl. Cancer Inst. 56:931–35.

Saunders, G.K., and Scott, D.W. 1988. Cutaneous lymphoma resembling mycosis fungoides in the Syrian hamster (*Mesocricetus auratus*). Lab. Anim. Sci. 38:616–17.

Strandberg, J.D. 1987. Neoplastic diseases. In *Laboratory Hamsters*, ed. G.L. Van Hoosier, Jr., and C.W. McPherson, pp. 157–68. New York: Academic.

Trentin, J.J. 1987. Experimental biology: Use in oncological research. In *Laboratory Hamsters*, ed. G.L. Van Hoosier, Jr., and C.W. McPherson, pp. 95–110. New York: Academic.

Turusov, V.S., et al., ed. 1982. Pathology of Tumours in Laboratory Animals. III. Tumours of the Hamster. Lyon, France: IARC Scientific Publications.

Van Hoosier, G.L., Jr., and Trentin, J.J. 1979. Naturally occurring tumors of the Syrian hamster. Prog. Exp. Tumor Res. 23:1–12.

4 GERBIL

Gerbils are members of the subfamily Gerbillinae, family Muridae, with about 14 genera and 100 species. Most gerbils that are used for research are Mongolian gerbils (*Meriones unguiculatus*), also called jirds, clawed jirds, sand rats, and antelope rats. They are desert-dwelling, burrowing rodents with a high degree of resistance to heat stress and dehydration. A few other species of *Meriones* are used for research purposes, but most of the information available on pathology of gerbils relates to *M. unguiculatus*, as does this chapter. Most commercially available gerbils are outbred, although inbred strains exist.

ANATOMIC FEATURES

Hematology

The most conspicuous peculiarity of the gerbil is a high proportion of red cells with polychromasia, basophilic stippling, and reticulocytosis (Smith et al. 1976). This is particularly obvious in young gerbils up to 20 wk of age but occurs throughout life. This may be a reflection of the short half-life of erythrocytes (approximately 10 days), compared with other species. The predominant peripheral blood leukocyte is the lymphocyte, with a 3:1 or 4:1 ratio over granulocytes (Dillon and Glomski 1975; Mays 1969). Gerbils are normally lipemic (hypercholesterolemic) on standard diets, especially adult males.

Anatomy

The gross anatomy of the gerbil is quite similar to the mouse and rat. An obvious exception is their furred tail. Gerbils are utilized in stroke research because of their susceptibility to cerebral ischemia following common carotid artery ligation. This is because gerbils often have an incomplete circle of Willis (Levine and Sohn 1969), which is of no practical significance relative to spontaneous disease. Incisor teeth grow continuously, but molar teeth are rooted. Lung lobation is similar to mice and rats. Gerbils have a prominent gland on the midline of the ventral abdomen composed of sebaceous glands and specialized hair structures (Sales 1973). It is inconspicuous in females but is prominent in sexually mature males. Gerbils do not have preputial glands. Auditory bullae are distinctively large, reflecting their highly adapted specialization for acute hearing. Microscopic adaptations in ear structure are also evident. The adrenal glands of the gerbil are quite large relative to other species of laboratory rodents. Renal function is adapted for urine concentration. The kidney has a very long papillus, and the ratio of papillus plus inner medulla to cortex is about twice that of a laboratory rat. This is a reflection of very long loops of Henle. Some Bowman's capsules in sexually mature male gerbils can be thickened due to the presence of cells that are morphologically intermediate between fibroblasts and smooth muscle cells (myofibroblasts). This lamina muscularis is unique to *Meriones* (Bucher and Kristic 1979).

Bibliography for Anatomic Features

Buchanan, J.G., and Stewart, A.D. 1974. Neurohypophysial storage of vasopressin in the normal and dehydrated gerbil (*Meriones unguiculatus*) with a note on kidney structure. J. Endocrinol. 60:381–82.

Bucher, O.M., and Kristic, R.V. 1979. Pericapsular smooth muscle cells in renal corpuscles of the Mongolian gerbil (*Meriones unguiculatus*). Cell Tissue Res. 199:75–82.

Dillon, W.G., and Glomski, C.A. 1975. The Mongolian gerbil: Qualitative and quantitative aspects of the cellular blood picture. Lab. Anim. 9:283–87.

Lay, D.M. 1972. The anatomy, physiology, functional significance and evolution of specialized hearing organs of gerbilline rodents. J. Morphol. 138:41–56.

Levine, S., and Sohn, D. 1969. Cerebral ischemia in infant and adult gerbils: Relation to incomplete circle of Willis. Arch. Pathol. 87:315–17.

Mays, A., Jr. 1969. Baseline hematological and blood biochemical parameters of the Mongolian gerbil. Lab. Anim. Care 19:838–42.

Ruhren, R. 1965. Normal values for hemoglobin concentration and cellular elements in the blood of Mongolian gerbils. Lab. Anim. Care 15:313–20.

Sales, N. 1973. The ventral gland of the male gerbil (*Meriones unguiculatus*, Gerbillidae): I. Histochemical features of the mucopolysaccharides. Ann. Histochem. 18:171–78.

Smith, R.A., et al. 1976. Erythrocyte basophilic stippling in the Mongolian gerbil. Lab. Anim. 10:379–83.

Williams, W.M. 1974. The anatomy of the Mongolian gerbil (*Meriones unguiculatus*). West Brookfield, Mass.: Tumblebrook Farms.

VIRAL INFECTIONS

There are no reported naturally occurring viral infections of gerbils, but this is probably a reflection of ignorance rather than reality. Certainly clinically significant viral infections are not currently recognized to be a problem.

BACTERIAL INFECTIONS

Tyzzer's Disease

EPIZOOTIOLOGY AND PATHOGENESIS. There have been numerous documented cases of Tyzzer's disease in this species since the first reports of *Bacillus piliformis* infection in gerbils (Carter et al. 1969). The Mongolian gerbil is very susceptible to the disease. Frequently the successful reproduction of the disease in rodents requires treatment with immunosuppressive drugs, such as cortisone. However, Tyzzer's disease has been produced readily in gerbils without benefit of immunosuppression. Young gerbils have developed the typical disease following the oral inoculation of isolates from other species (Waggie et al. 1984). Gerbils appear to be more susceptible to clinical disease following exposure to *B. piliformis* than do immunosuppressed mice. Housing sentinel gerbils on unautoclaved soiled bedding suspected to be contaminated with the organism has been used to detect carriers of the disease (Gibson et al. 1987). Thus the Mongolian gerbil is recognized to be a useful sentinel animal to detect subclinical infections or environmental contamination with *B. piliformis*.

Typical lesions associated with Tyzzer's disease in gerbils include depression, ruffled hair coat, hunched posture, anorexia, and watery diarrhea. Following oral inoculation, severely affected animals usually die within 5–7 days postinoculation. In addition to focal hepatic necrosis, bacterial antigen has been observed in ileocecal enterocytes by 3 days postexposure. Extensive lesions and bacterial antigen has been demonstrated in the jejunum, ileum, and cecum by 5–6 days postinoculation. In affected gerbils, bacterial antigen may also be present in the muscle layers of the intestine and in Peyer's patches. Ileal enterocytes and Peyer's patches may be the initial sites for bacterial growth (Yokomori et al. 1989).

PATHOLOGY. At necropsy, pinpoint, pale foci up to 2 mm in diameter are usually present in the liver. Ecchymoses on the small intestine and cecum are variable findings. The walls of the small intestine and cecum are usually edematous. Intestinal contents are fluid and sometimes contain blood. The mesenteric lymph nodes may be enlarged and edematous. On microscopic examination, liver lesions frequently are concentrated in the periportal regions. In acute cases, there are foci of coagulation to caseation necrosis, with variable leukocytic infiltration, neutrophils predominating (Fig. 4.1). Intracytoplasmic bacilli are most numerous in hepatocytes adjacent to necrotic foci (Fig. 4.2). In hepatic lesions interpreted to be sev-

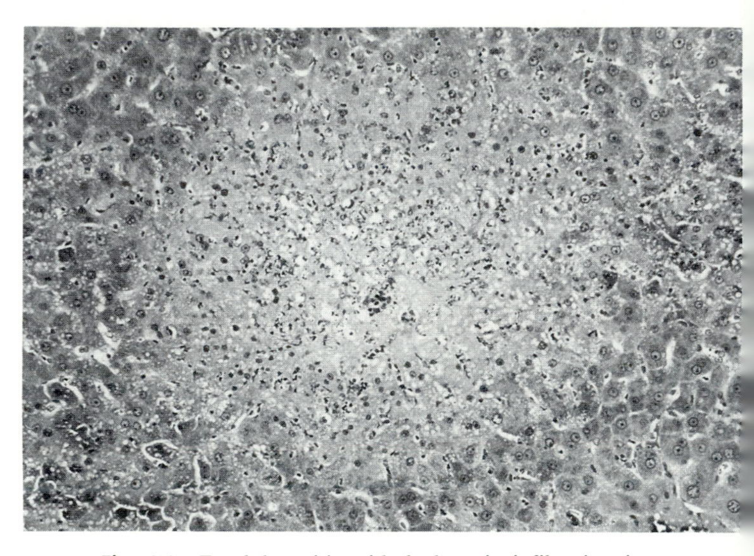

Fig. 4.1. Focal hepatitis with leukocytic infiltration in young Mongolian gerbil that died with acute Tyzzer's disease.

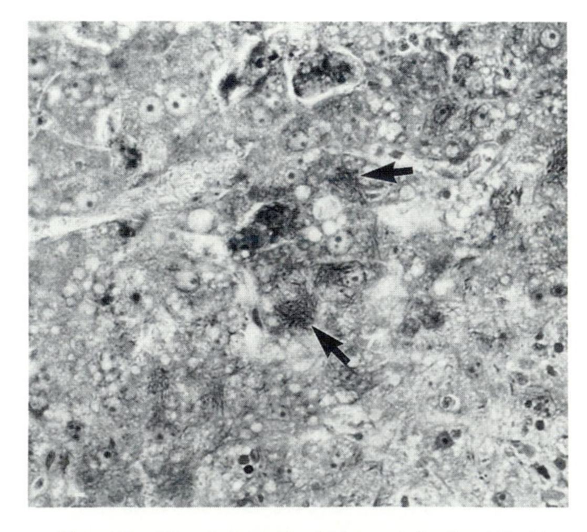

Fig. 4.2. Focal hepatic lesion stained with the Giemsa method. Note the intracytoplasmic bacilli (*arrow*) present in hepatocytes at the periphery of the lesion.

eral days in duration, there may be focal fibrosis with mineralization. Intestinal lesions are usually most extensive in the ileum and cecum. Necrosis and sloughing of enterocytes, blunting of villi in affected areas, and transmural edema occur. Leukocytic infiltrates in the lamina propria consist of neutrophils and mononuclear cells. There may be necrosis of the adjacent intestinal smooth muscle, with leukocytic infiltration. Frequently focal necrosis of Peyer's patches and mesenteric lymph nodes occurs. Intracytoplasmic bacilli are usually evident in enterocytes and sometimes in smooth muscle cells. Myocardial lesions, when present, consist of focal coagulation necrosis, with collapse of myofibers, and leukocytic infiltration. There may be mineralization of cell debris. Bundles of bacilli may be evident in myofibers bordering necrotic foci using Warthin-Starry or Giemsa stains. Diffuse suppurative encephalitis is another possible manifestation of Tyzzer's disease in this species (Veazy et al. 1992).

DIAGNOSIS. The presence of the typical microscopic lesions and the demonstration of the intracellular bacilli are sufficient to confirm the diagnosis. *Differential diagnoses* include acute bacterial infections, such as salmonellosis.

SIGNIFICANCE. The Mongolian gerbil is particularly susceptible to Tyzzer's disease. Aside from the complications due to morbidity and mortality, the gerbil is recognized to be a useful sentinel animal to detect the presence of *B. piliformis* in the research facility. The possibility of interspecies transmission is an important consideration (Motzel and Gibson 1990).

Salmonellosis

Disease and mortality have been observed in young gerbils 3–10 wk of age that were naturally infected with *Salmonella typhimurium*. Clinical signs include moderate to severe diarrhea, dehydration, weight loss, and leukocytosis with neutrophilia. The mortality rate may be over 90%. In one report, animals also had a heavy infestation with *Hymenolrepis nana* (Olson et al. 1977).

An outbreak of salmonellosis in a gerbil colony due to *Salmonella* group D has been reported. Dehydration, depression, testicular enlargement, and occasionally sudden death were observed. Focal hepatitis, splenic necrosis, suppurative orchitis, interstitial pneumonitis, and purulent to pyogranulomatous leptomeningitis were lesions observed microscopically. *Salmonella*-infected cockroaches were implicated as a possible source of the infection (Clark et al. 1992).

PATHOLOGY. At necropsy, the gastrointestinal tract is frequently distended with gas and fluid ingesta. Fibrinopurulent exudate may be present in the peritoneal cavity in some animals. Microscopic changes are characterized by multifocal hepatitis. Hepatic lesions may vary from foci of leukocytic infiltration to larger foci consisting of central caseation necrosis with variable mineralization and with epithelioid cells, lymphocytes, and neutrophils oriented around the periphery. Crypt abscesses are occasionally present in the intestine. *S. enteritidis* may be recovered from sites such as the small intestine, liver, spleen, and heart blood.

DIAGNOSIS. Isolation of the organism using the appropriate bacterial media, coupled with the typical lesions, will serve to confirm the diagnosis. *Differential diagnosis*: Tyzzer's disease is the primary one to rule out in this species.

SIGNIFICANCE. Although adult gerbils may be relatively resistant to experimental *Salmonella* infection, it is evident that younger animals may be very susceptible under certain circumstances. The dangers of interspecies transmission should be emphasized.

Staphylococcal Dermatitis

Acute, diffuse dermatitis has been associated with beta-hemolytic *Staphylococcus aureus* infection. The disease appears to affect primarily young gerbils, and there may be a relatively high morbidity and mortality. The disease has been reproduced in gerbils inoculated in the nasal region with the staphylococcal isolate (Peckham et al. 1974).

PATHOLOGY. On gross examination, there may be a diffuse moist dermatitis involving the face, nose, feet, legs, and ventral body surface. Alopecia, erythema, and moist brown exudate have been associated with the typical lesions. Microscopic changes are those of a suppurative dermatitis, with neutrophils infiltrating into the superficial and deep dermis and adnexae, with concurrent acanthosis and hyperkeratosis. Ulcerations may occur. Focal suppurative hepatitis may be present in some fatal cases of the disease.

DIAGNOSIS. Isolation and identification of the organism is necessary.

SIGNIFICANCE. The isolation of a beta-hemolytic staphylococcus from the skin must be accompanied by the typical lesions.

Bordetella bronchiseptica Infection

This bacterium is a potential problem of gerbils but has not been reported as a natural disease. Young gerbils inoculated intranasally with *B.bronchiseptica* have developed severe disease with high mortality, while older gerbils appear to be more resistant. Both the *Meriones unguiculatus* and *M. shawi* species appear to be susceptible. Because of the frequency of *Bordetella* in laboratory guinea pigs and rabbits, contact with these species should be avoided.

Leptospirosis

Leptospirosis has not been reported as a natural infection in gerbils, but they are quite susceptible to experimental infection (Lewis and Grey 1961). Acute disease is characterized by hemolytic icterus, with pale, mottled livers. Microscopically, there is degeneration of renal distal convoluted tubules and centrilobular hepatocytes with conspicuous erythrophagocytosis in the spleen. Spirochetes are present in kidney and liver in large numbers. Chronic infection occurs frequently, with chronic nonsuppurative inflammation, interstitial fibrosis, and development of progressively severe tubular degeneration and cyst formation. The infection may persist in the kidney for months to years.

Nasal Dermatitis

This is a commonly encountered problem in juvenile and adult Mongolian gerbils, appearing to be most common in postpuberal animals. Nasal dermatitis is characterized by dermatitis and alopecia around the external nares and upper labial region.

EPIZOOTIOLOGY AND PATHOGENESIS. The incidence of the disease in individual colonies may be over 15%, but an incidence of around 5% is considered to be a typical finding (Theissen and Pendergrass 1982). The disease has been associated with infections with *Staphylococcus saprophyticus* and beta-hemolytic *S. aureus*. However, staphylococci of identical type have been isolated from asymptomatic gerbils, and other factors have been implicated. Mechanical trauma may contribute to the disease in some circumstances, but porphyrin-containing lacrimal gland secretions have been shown to be an important contributing factor. Secretions from the Harderian gland normally bathe the eye and conjunctival sac and then are transported down to the external nares via the nasolacrimal duct. The secretions are mixed with saliva and spread widely over the pelage during the thermoregulatory grooming procedures (Theissen and Kittrell 1980). However, if these secretions are not removed routinely from the collection site at the external nares, chemical irritation and subsequent dermatitis may occur. Removal of the Harderian glands from affected animals has resulted in a marked improvement or recovery. Marked improvement was also seen in gerbils housed on sand. It was concluded that the failure to groom properly resulted in the accumulation of protoporphyrin-containing secretions around the external nares, which resulted in local irritation, scratching, hair loss, and dermatitis (Theissen and Pendergrass 1982). Similarly, intact gerbils fitted with Elizabethan collars, which prevented self grooming, developed nasal dermatitis, while those with bilateral Harderian gland adenectomy did not develop the disease (Farrar et al. 1988). Synergism with a pathogenic bacterium such as *S. aureus* may be necessary for the development of the moist, ulcerative form of the disease (Farrar et al. 1988).

PATHOLOGY. On gross examination, there are varying degrees of dermatitis and alopecia involving the lateral and superior nasal area and the upper and lower lip (Fig. 4.3). Lesions may progress to a severe ulcerative dermatitis, with exudation and excoriation and crusting in the upper labial region. Dermatitis and hair loss may also be present on the forepaws and periocular regions. Histopathologically, there are hyperkeratosis and epidermal hyperplasia in active cases, with increased melanin deposition in the dermis. In acute, suppurative lesions, there are spongiosis, epidermal hyperplasia and necrosis, and infiltration with neutrophils. Other changes may include ulceration and epidermal abscessation.

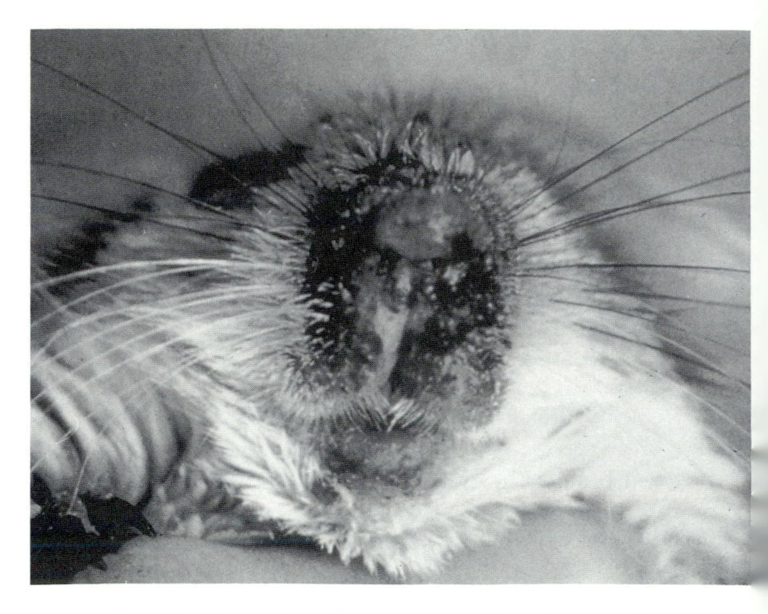

Fig. 4.3. Nasal dermatitis in mature Mongolian gerbil. There is marked reddening with encrustations around the external nares. (Courtesy of M.E. Olson)

DIAGNOSIS. The distribution and nature of the lesions are useful diagnostic criteria. Routine bacteriology should be performed, since pathogenic bacteria such as coagulase-positive *S. aureus* may play a role in the development of the lesions. *Differential diagnoses* include fighting injuries and nonspecific bacterial infections.

Bibliography for Bacterial Infections

Carter, G.R., et al. 1969. Natural Tyzzer's disease in Mongolian gerbils (*Meriones unguiculatus*). Lab. Anim. Care 19:648–51.

Clark, J.D., et al. 1992. Salmonellosis in gerbils induced by a nonrelated experimental procedure. Lab. Anim. Sci. 42:161–63.

Fujiwara, K. 1978. Tyzzer's disease. Jpn. J. Exp. Med. 48:467–77.

Gibson, S.V., et al. 1987. Diagnosis of subclinical *Bacillus piliformis* infection in a barrier-maintained mouse production colony. Lab. Anim. Sci. 37:786–91.

Harkness, J.E., and Wagner, J.E. 1984. *Biology and Medicine of Rabbits and Rodents*. Philadelphia: Lea and Febiger.

Lewis, C., and Grey, J.E. 1961. Experimental *Leptospira pomona* infection in the Mongolian gerbil (*Meriones unguiculatus*). J. Infect. Dis. 109:194–204.

Motzel, S.L., and Gibson, S.V. 1990. Tyzzer disease in hamsters and gerbils from a pet store supplier. J. Am. Vet. Med. Assoc. 197:1176–78.

Olson, G.A., et al. 1977. Salmonellosis in a gerbil colony. J. Am. Vet. Med. Assoc. 171:970–72.

Peckham, J.C., et al. 1974. Staphylococcal dermatitis in Mongolian gerbils (*Meriones unguiculatus*). Lab Anim. Sci. 24:43–47.

Van Nunen, M.C.J., et al. 1978. Prevalence of viruses in colonies of laboratory rodents. Z. Versuch. 20:201–8.

Veazy, R.S., et al. 1992. Encephalitis in gerbils due to naturally occurring infection with *Bacillus piliformis* (Tyzzer's disease). Lab. Anim. Sci. 42:516–18.

Waggie, K.S., et al. 1984. Experimentally induced Tyzzer's disease in Mongolian gerbils (*Meriones unguiculatus*). Lab. Anim. Sci. 34:53–57.

Winsser, J. 1960. A study of *Bordetella bronchiseptica*. Proc. Anim. Care Panel 10:87–104.

Yokomori, K., et al. 1989. Enterohepatitis in Mongolian gerbils (*Meriones unguiculatus*) inoculated perorally with Tyzzer's organism (*Bacillus piliformis*). Lab. Anim. Sci. 39:16–20.

Nasal Dermatitis

Bresnahan, J.F., et al. 1983. Nasal dermatitis in the Mongolian gerbil. Lab. Anim. Sci. 33:258–63.

Farrar, P.L., et al. 1988. Experimental nasal dermatitis in the Mongolian gerbil: Effect of bilateral Harderian gland adenectomy on development of facial lesions. Lab. Anim. Sci. 38:72–76.

Theissen, D.D., and Kittrell, E.M.W. 1980. The Harderian gland and thermoregulation in the gerbil (*Meriones unguiculatus*). Physiol. Behav. 24:417–24.

Theissen, D.D., and Pendergrass, M. 1982. Harderian gland involvement in facial lesions in the Mongolian gerbil. J. Am. Vet. Med. Assoc. 181:1375–77.

Vincent, A.L., et al. 1979. The pathology of the Mongolian gerbil (*Meriones unguiculatus*): A review. Lab. Anim. Sci. 29:645–51.

PARASITIC DISEASES

Ectoparasitic Infections

Mite Infections (Acariasis)

The name *Demodex meriones* has been proposed, but it may represent *D. aurati* or *criceti* (hamster mites), since mites resembling both of these species have been found on the gerbil. Copra itch mites (*Tyrophagus castellani*), probably introduced through the food, have been found incidentally on gerbils.

Demodex Infection

Demodex mites have been demonstrated in skin scrapings from a 4-year-old gerbil with diarrhea, cachexia, and rough hair coat. A lesion on the tail head was characterized by scaliness, hyperemia, and focal ulcerations. Old age and debilitation were considered to be important predisposing factors, and demodex infections are not regarded to be a problem in clinically healthy gerbils. The nature of the host-parasite relationship and the morphology of the mites are similar to the *Demodex* mite infections that occur in hamsters (Schwartzbrott et al. 1974).

Liponyssoides Infection

Liponyssoides sanguineus, an ectoparasite occasionally seen in house mice, has also been observed in Mongolian and Egyptian gerbils. Mites were also identified on laboratory mice and wild house mice on the same premises. No manifestations of disease were observed in affected animals. Mites were also present in the bedding in the cages (Levine and Lage 1984).

Endoparasitic Infections

Protozoal Infections: Giardiasis

Giardiasis has not been reported as a natural disease in gerbils, but they are highly susceptible to infection with *Giardia* cysts of human origin (Belosevic et al. 1983). Trophozoites can be found in the upper small intestine, and in heavy infections, they occur throughout the bowel.

Helminth Infections: Pinworms (Oxyuriasis)

Gerbils can become infected with several oxyurid nematodes, but none cause clinical problems. *Dentostomella translucida* has been reported in a variety of gerbils (Wightman et al. 1978a,b). It occurs in the anterior small intestine and has also been noted in the large intestine. Gerbils are susceptible to contact infection with the mouse and rat pinworms, *Syphacia obvelata* and *S. muris* (Kellogg and Wagner 1982; Ross et al. 1980).

Helminth Infections: Tapeworms

Severe infections with the "dwarf tapeworm" have been reported in pet gerbils (Lussier and Loew 1970). Dehydration and mucoid diarrhea are possible presenting signs. In another report describing an epizootic of salmonellosis in Mongolian gerbils, affected animals were also heavily parasitized with *H. nana* (Olson et al. 1977). *H. diminuta* has also been identified at necropsy in gerbils.

PATHOLOGY. At necropsy, small tapeworms are present in the small intestine. On microscopic examination of smears of intestinal mucosa, or of paraffin-embedded sections of small intestine, eggs and cysticercoids are readily identified.

SIGNIFICANCE. *H. nana* infections have been associated with debilitation and diarrhea in gerbils. In view of the direct life cycle, there is a possibility of transmission to human contacts.

Bibliography for Parasitic Diseases

Belosevic, M. 1983. *Giardia lamblia* infections in Mongolian gerbils: An animal model. J. Infect. Dis. 147:222–26.

Kellogg, H.S., and Wagner, J.E. 1982. Experimental transmission of *Syphacia obvelata* among mice, rats, hamsters and gerbils. Lab. Anim. Sci. 32:500–501.

Levine, J.F., and Lage, A.L. 1984. House mouse mites infesting laboratory rodents. Lab. Anim. Sci. 34:393–94.
Lussier, G., and Loew, F.M. 1970. Natural *Hymenolepis nana* infection in Mongolian gerbils (*Meriones unguiculatus*). Can. Vet. J. 11: 105–7.
Olson, G.A., et al. 1977. Salmonellosis in a gerbil colony. J. Am. Med. Assoc. 171: 970–72.
Ross, C.R., et al. 1980. Experimental transmission of *Syphacia muris* among rats, mice, hamsters and gerbils. Lab. Anim. Sci. 30:35–37.
Schwartzbrott, S.S., et al. 1974. Demodicidosis in the Mongolian gerbil (*Meriones unguiculatus*): A case report. Lab. Anim. Sci. 24:666–68.
Vincent, A.L., et al. 1975. Spontaneous lesions and parasites of the Mongolian gerbil, *Meriones unguiculatus*. Lab. Anim. Sci. 25:711–22.
Wightman, S.R., et al. 1978a. *Dentostomella translucida* in the Mongolian gerbil (*Meriones unguiculatus*). Lab. Anim. Sci. 28:290–96.
______. 1978b. *Syphacia obvelata* in the Mongolian gerbil (*Meriones unguiculatus*): Natural occurrence and experimental transmission. Lab. Anim. Sci. 28:51–54.

GENETIC DISORDERS

Epilepsy

Epileptiform seizures are common among Mongolian gerbils that are subjected to stress, which may include cage changing. Susceptibility begins at around 2 mo of age and can reach an incidence of 40–80% within 6–10 mo and persist throughout life. The trait is inherited as a single autosomal locus with at least one dominant allele, with variable penetrance (Loskota et al.1974). The incidence therefore varies with different populations or lines of gerbils. Seizure-sensitive and -resistant strains have been selected for experimental purposes. Clinical signs include twitching of vibrissae and pinnae, motor arrest, myoclonic jerks, clonic-tonic seizures, vestibular aberrations, and occasionally death. Histopathologic lesions have not been found.

Periodontal Disease and Dental Caries

Gerbils that are maintained on a standard laboratory pelleted diet and water may develop progressively severe periodontal disease, which is first manifested at around 6 mo of age and is readily apparent by 1 yr. Advanced disease is present in gerbils over 2 yr of age, often with tooth loss (Moskow et al. 1968). They are also prone to the development of dental caries, which can be enhanced with cariogenic diets (Fitzgerald and Fitzgerald 1965).

Malocclusion

Lack of opposing occlusal contact results in tooth overgrowth in all species of rodents, including gerbils (Loew 1967). Reported cases in gerbils are rare but have been due to loss of the upper incisors with overgrowth of the lower teeth. Molar teeth of gerbils do not grow continuously.

Behavioral Disease

Gerbils are usually relatively docile and easily handled. They are intermittently active day and night. Foot stomping is a common signal of startling, communication, and aggression. They tolerate each other very well if grouped before maturity, but mixing adult gerbils will usually provoke fighting, with death of the weaker animal.

BIBLIOGRAPHY FOR CONGENITAL/HEREDITARY DISORDERS

Afonsky, D. 1957. Dental caries in the Mongolian gerbil. N Y State Dent. J. 23:315–16.
Fitzgerald, D.B., and Fitzgerald, R.J. 1965. Induction of dental caries in gerbils. Arch. Oral Biol. 11:139–40.
Loew, F.M. 1967. A case of overgrown mandibular incisors in a Mongolian gerbil. Lab. Anim. Care 17:137–39.
Loskota, W.J., et al. 1974. The gerbil as a model for the study of the epilepsies: Seizure patterns and ontogenesis. Epilepsia 15:109–19.
Moskow, B.S., et al. 1968. Spontaneous periodontal disease in the Mongolian gerbil. J. Periodont. Res. 3:69–83.

TOXIC AND METABOLIC DISORDERS

Streptomycin Toxicity

Members of the neomycin-streptomycin group of antibiotics can cause a direct neuromuscular blocking effect at excessive doses by inhibition of acetylcholine release. Although other rodents and rabbits are susceptible to this effect, they are less likely to be treated with these drugs and are, in addition, big enough to receive the proper dose. The margin of safety for streptomycin is low, and antibiotic preparations are seldom prepared so that an appropriate dose in a small volume can be administered to a rodent. Gerbils treated with these preparations have developed acute toxicity, characterized by depression, ascending flaccid paralysis, coma, and death within minutes of administration (Wightman et al. 1980).

Lead Toxicity

Because of their urine-concentrating ability, gerbils are prone to accumulation of lead and chronic lead toxicity (Boquist 1977). They are used for this purpose experimentally (Port et al. 1974), but the potential for natural toxicity in a laboratory environment is high because of their gnawing behavior. Chronically toxic animals become emaciated. Their livers become small and pigmented, their kidneys pale and pitted. Microscopic findings include acid-fast intranuclear inclusions in proximal convoluted tubular epithelium and chronic progressive nephropathy. Occasional intranuclear inclusions may be found in liver, but the predominant finding is lipofuscin pigment granules in hepatocytes and Kupffer's cells. Gerbils also develop a microcytic, hypochromic anemia with basophilic stippling. *Differential diagnoses* should include age-related renal disease, which is both mild and relatively rare in gerbils, and erythrocytic basophilic stippling, which should be differentiated from the condition that occurs normally in the gerbil, but to a lesser degree.

Amyloidosis

Amyloidosis occurs only rarely in gerbils but has been reported in gerbils experimentally infected with a filariid worm (Vincent et al. 1975). Clinical signs include weight loss, dehydration, anorexia, and death. Amyloid infiltrates occurred in kidney, liver, heart, pancreas, and intestine.

Obesity and Diabetes

Approximately 10% of gerbils maintained on standard laboratory diet can became obese. This condition can be associated with reduced glucose tolerance, elevated insulin, and hyperplastic or degenerative changes in the endocrine pancreas (Boquist 1972).

Hyperadrenocortism/Cardiovascular Disease of Breeding Gerbils

A disease complex, attributed to hyperadrenocortism, has been described in repeatedly bred male and female, but not virgin, gerbils. Breeding females, and to a lesser extent males, develop mild to severe plaques of intimal and medial ground substance alterations with mineralization in the aorta and mesenteric, renal, and peripheral arteries. Breeders may have grossly visible plaques of the abdominal aorta, as well as aortic arch and the entire aorta in severe cases. Breeding animals have elevated serum triglycerides, enlarged pancreatic islets, fatty livers, thymic involution, adrenal hemorrhage, and adrenal lipid depletion, and some may have pheochromocytomas (Wexler et al. 1971). Male breeders have been found to have a high incidence of focal myocardial necrosis and fibrosis. This phenomenon is also linked to diabetes and obesity. Cause and effect relationships have not been firmly established, but it is clear that these lesions occur frequently in gerbil populations and seem to occur in higher prevalence among breeders.

Bibliography for Toxic and Metabolic Disorders

Boquist, L. 1975. The Mongolian gerbil as a model for chronic lead toxicity. J. Comp. Pathol. 85:119–31.

______. 1972. Obesity and pancreatic islet hyperplasia in the Mongolian gerbil. Diabetologia 8:274–82.

Nakama, K. 1977. Studies on diabetic syndrome and influences of long-term tolbutamide administration in Mongolian gerbils (*Meriones unguiculatus*). Endocrinol. Jpn. 24:421–33.

Port, C.D., et al. 1974. The Mongolian gerbil as a model for lead toxicity: I. Studies of acute poisoning. Am. J. Pathol. 76:79–94.

Vincent, A.L., et al. 1975. Spontaneous lesions and parasites of the Mongolian gerbil, *Meriones unguiculatus*. Lab. Anim. Sci. 25:711–22.

Wexler, B.C., et al. 1971. Spontaneous arteriosclerosis in male and female gerbils (*Meriones unguiculatus*). Atherosclerosis 14:107–19.

Wightman, S.R., et al. 1980. Dihydrostreptomycin toxicity in the Mongolian gerbil, *Meriones unguiculatus*. Lab. Anim. Sci. 30:71–75.

Diseases Associated with Aging

Focal Myocardial Degeneration

Focal myocardial degeneration and fibrosis are relatively common microscopic findings in older gerbils. In general, 50% or more of male breeders may be affected, and a smaller percentage of breeding females. Lesions are probably ischemic in origin, but the etiopathogenesis has not been adequately studied. On microscopic examination there are foci of degeneration of myofibers, with interstitial fibrosis (Fig. 4.4).

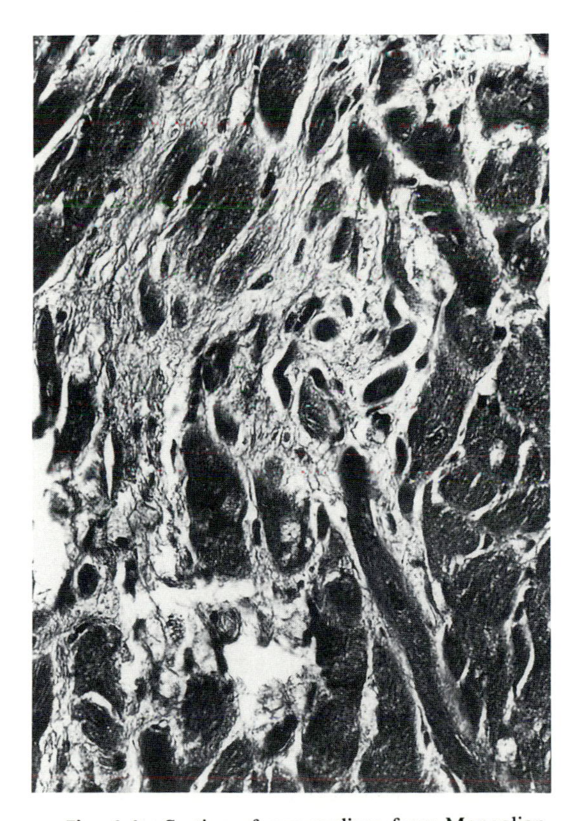

Fig. 4.4. Section of myocardium from Mongolian gerbil, 16 mo of age. There are foci of myocardial degeneration with interstitial fibrosis.

Chronic Nephropathy

Glomerular hypercellularity, tubular degeneration, and dilation and cast formation in tubules are changes seen in the kidneys of aging gerbils. Mononuclear cell infiltration consistent with chronic interstitial nephritis may be present in affected kidneys.

Aural Cholesteatoma

Spontaneous aural cholesteatomas occur in high frequency among adult gerbils, with an incidence of over 50% at 2 yr of age (Henry et al. 1983; Chole et al. 1981). These masses of keratinized epithelium arise from the outer surface of the tympanic membrane and external auditory canal. As keratin is accumulated, it displaces the tympanic membrane into the middle ear. Compression and secondary inflammation result in destruction of temporal bone and inner ear structures. Clinical signs include head tilt and accumulation of keratin plugs in the external ear canal. *Differential diagnosis* includes otitis media/interna, but this is rare in gerbils because of the vertical configuration of their eustachian tubes.

Cystic Ovaries

Female gerbils are prone to the development of ovarian cysts. Nearly 50% of gerbils over 400 days of age may be affected. Cysts range in size from 1 to 50 mm in diameter. Ovulation and corpus luteum formation continue to occur in the presence of cysts, but litter sizes are reduced and severely affected females become infertile (Norris and Adams 1972).

Ocular Proptosis

Aged gerbils can develop protrusion of the nictitating membrane and conjunctiva with bulbar proptosis. The underlying cause has not been characterized.

Bibliography for Diseases Associated with Aging

Chole, R.A., et al. 1981. Cholesteatoma: Spontaneous occurrence in the Mongolian gerbil, *Meriones unguiculatus*. Am. J. Otol. 2:204–10.

Henry, K.R., et al. 1983. Age-related increase of spontaneous aural cholesteatoma in the Mongolian gerbil. Arch. Otolaryngol. 109:19–21.

Norris, M.L., and Adams, C.E. 1972. Incidence of cystic ovaries and reproductive performance in the Mongolian gerbil, *Meriones unguiculatus*. Lab. Anim. 6:337–42.

NEOPLASMS

In general, the incidence of spontaneous tumors in this species is relatively low, with increasing incidence in gerbils over 2 yr of age. There is frequently a striking variation in the percentage of tumors that occur in different colonies of Mongolian gerbils. Ovarian, adrenocortical, and cutaneous tumors are the most commonly recognized neoplasms in this species. Of the ovarian tumors, granulosa cell tumors appear to be the most common in aged females (Meckley and Zwicker 1979; Rowe et al. 1974). Granulosa cell tumors are frequently bilateral and vary from fleshy and lobulated to cystic masses (Fig. 4.5). Granulosa cells are the predominant cell type (Fig. 4.6). Dysgerminomas, luteal cell tumors, leiomyomas, and rarely thecal cell carcinoma have been identified. Adrenal cortical adenomas and carcinomas also occur (Benitz and Kramer 1965). There is a relatively low incidence of tumors of the pituitary, mammary gland, and lung in the Mongolian gerbil. In a study involving other species of Gerbillinae, neoplasms included squamous carcinoma of the ear, thymoma, uterine adenocarcinoma, squamous cell carcinoma of the ventral marking gland (Figs. 4.7 and 4.8), adrenocortical tumors, and primary ovarian tumors (Rowe et al. 1974).

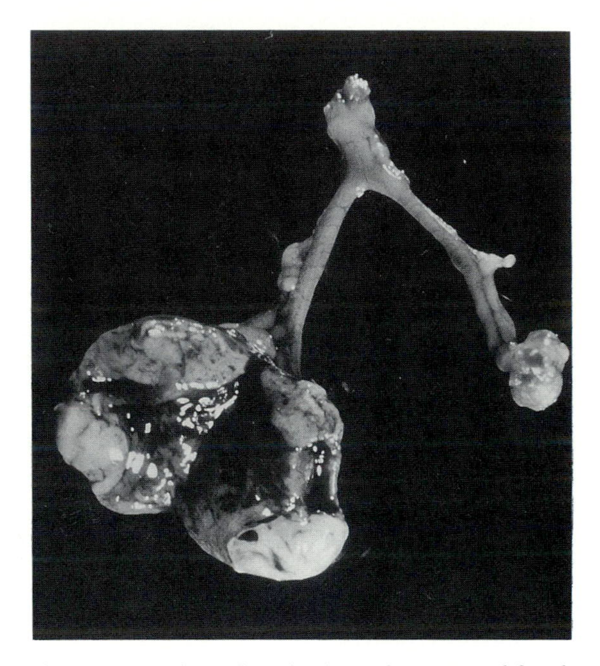

Fig. 4.5. Ovaries and uterine horns from an aged female Mongolian gerbil. The left ovary, which is markedly enlarged, fleshy, and lobulated, with variable dark red to pale tan areas, is a granulosa cell carcinoma. (Courtesy of D. Schlafer)

Bibliography for Neoplasms

Benitz, K.F., and Kramer, A.W. 1965. Spontaneous tumors in the Mongolian gerbil. Lab. Anim. Care 15:281–94.

Meckley, P.E., and Zwicker, G.M. 1979. Naturally-occurring neoplasms in the Mongolian gerbil (*Meriones unguiculatus*). Lab. Anim. 13:203–6.

Rowe, S.E., et al. 1974. Spontaneous neoplasms in aging Gerbillinae. Vet. Pathol. 11:38–51.

Vincent, A.L., et al. 1979. The pathology of the Mongolian gerbil (*Meriones unguiculatus*): A review. Lab. Anim. Sci. 29:645–61.

Vincent, A.L., et al. 1975. Spontaneous lesions and parasites of the Mongolian gerbil, *Meriones unguiculatus*. Lab. Anim. Sci. 25:711–22.

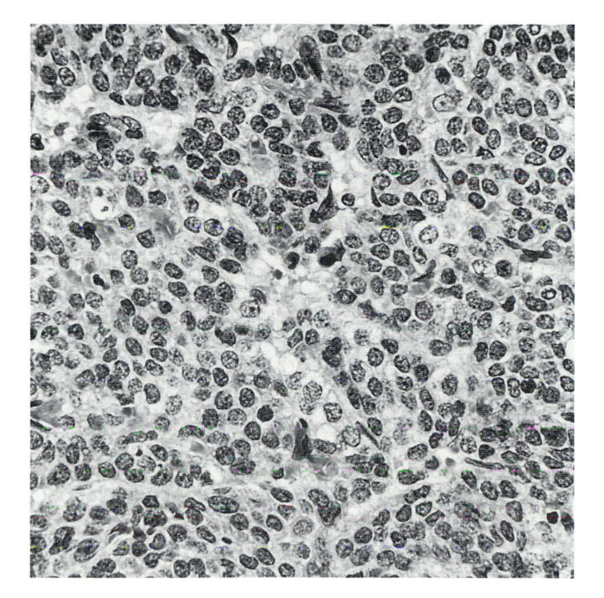

Fig. 4.6. Histological section from the granulosa cell tumor in Fig. 4.5.

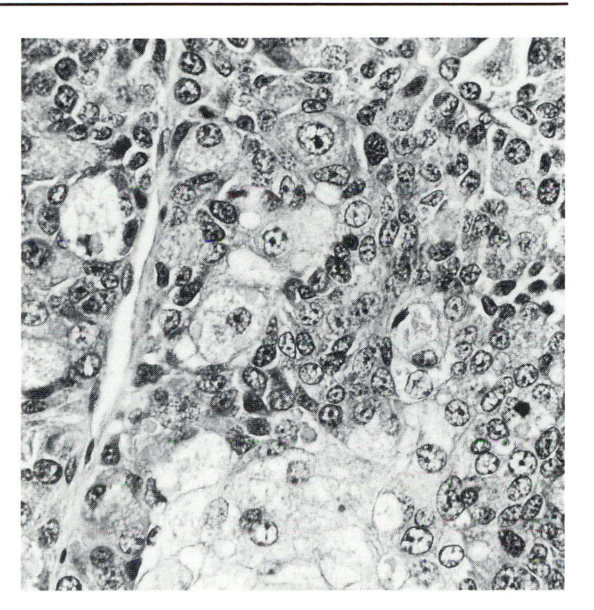

Fig. 4.8. Higher magnification of Fig. 4.7. (Courtesy of B.M. Cross)

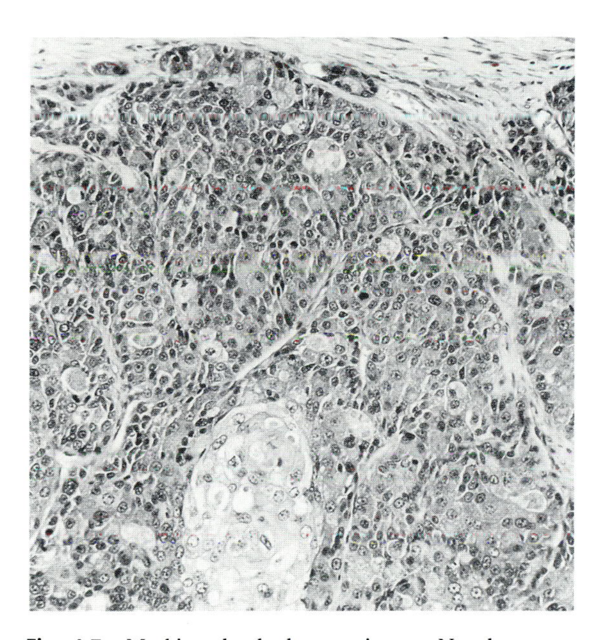

Fig. 4.7. Marking gland adenocarcinoma. Neoplasm was removed from the ventral midline (umbilical region) of a male Mongolian gerbil approximately 20 mo of age. Note the marked anisokaryosis, the variations in cytoplasmic volume and the staining properties, which vary from finely granular to vacuolated cytoplasm. (Courtesy of B.M. Cross)

5 GUINEA PIG

There are relatively few strains of domesticated guinea pigs recognized in the world today; the so-called English shorthair or Hartley strain is most commonly used in the research facility. However, the long-haired Peruvian and Abyssinian strains are popular in cavy club circles. Nervous in temperament, guinea pigs may refuse to eat or drink for some time following any significant change in location, feed, or management practices. Their long gestation period and relatively large offspring at full term may result in dystocia, particularly in sows that farrow with their first litter at 6 or more mo of age. This has been attributed to failure of the iliosacral ligaments to relax sufficiently to permit passage of the fetuses at term. Unlike some of the smaller rodents, guinea pigs have relatively few viral infections recognized to be a significant cause of disease. In general, scurvy (either clinical or subclinical), respiratory tract infections, and enteric disease are the major diagnostic problems seen in this species.

Guinea pigs live in groups with a strong male-dominance hierarchy and a loose female hierarchy. They tend to live in family units centered around an alpha male. Mature boars, particularly strangers, will fight savagely, sometimes with fatal outcome. Females will also fight on occasion. Guinea pigs are active during daylight hours. They eat frequently, are coprophagic, and do not cache their food, as many rodents do. They require a constant source of water and tend to contaminate their water bowls with ingesta as they drink. They do not lick sipper tubes without training, a consideration that can lead to dehydration and death. They are indiscriminate defecators and renowned for their tendency to sit in and soil their food bowls. Breeding activity occurs year around and infants are highly precocious at birth. At parturition both males and females assist in grooming infants and eating placentas, and lactating sows will nurse other infants. Infants do not receive much maternal attention, other than anogenital grooming, which stimulates defecation and urination. They are normally weaned within 3 wk but can be weaned as early as 3–4 days if anogenital stimulation is provided. Vocalization is well developed and complex. They respond to sudden auditory stimulation and unfamiliar surroundings by freezing in place, whereas sudden movement often elicits a random stampede, which may result in injury to young animals (Harper 1976).

ANATOMIC FEATURES

Hematology

Heterophils are the counterpart of the neutrophil in guinea pigs. These cells have distinct eosinophilic cytoplasmic granules. Lymphocytes are the predominant leukocyte in the peripheral blood, and both small and large forms are normally found (Sanderson and Phillips 1981).

Kurloff Cells (Foa-Kurloff Cells)

These unique mononuclear leukocytes are found routinely in certain tissues in guinea pigs. In nonpregnant animals, Kurloff cells are located primarily in the sinusoids of the spleen and in stromal tissues of the bone marrow and thymus. They are not normally found in lymph nodes. Kurloff cells can be readily identified in impression smears of the spleen. They are usually present in the sinusoids of the spleen in tissue sections, particularly in females. The cells contain a finely fibrillar to granular structure (Kurloff body) 1–8 μm in diameter within a cytoplasmic vacuole, with displacement of the nucleus (Fig. 5.1). Theories re-

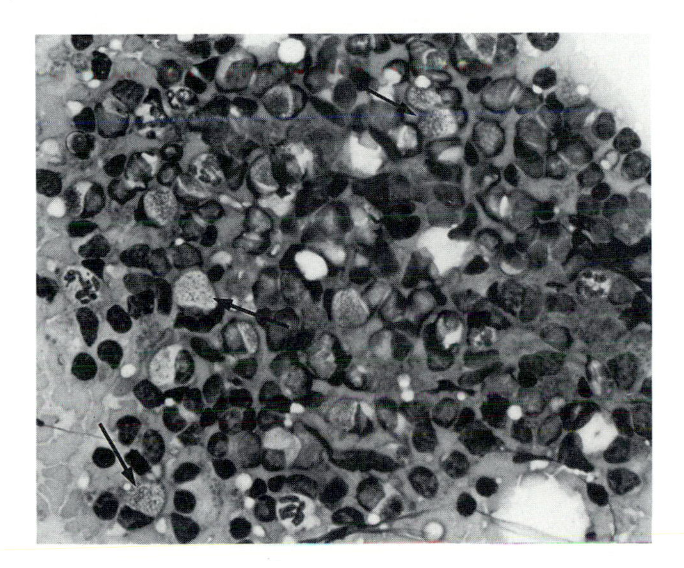

Fig. 5.1. Impression smear of spleen from adult female guinea pig, illustrating typical Kurloff cells (*arrows*). Note the large, finely granular Kurloff bodies within the cytoplasm of these mononuclear cells.

garding the origins and functions of Kurloff cells have been varied and at times highly speculative. Interpretations have included intracellular parasites, phagocytic cells, nuclear remnants, and secretory cells. It is now generally agreed that Kurloff cells are probably a member of the lymphoid series. The intracytoplasmic material is PAS-positive and stains positive for fibrinoid material with the Lendrum stain. It is interpreted to be mucopolysaccharide material secreted by the cell, and glycoprotein associated with a protein-polysaccharide material (Marshall et al. 1971). On ultrastructural examination, inclusions are membrane-bound, and cytoplasmic organelles in these cells are consistent with secretory activity (Revell et al. 1971)).

The possible function of these cells has been open to speculation. Kurloff cells, which are rarely seen in newborn guinea pigs, are present in relatively large numbers in adult female guinea pigs. Their numbers fluctuate with the stage of the estrous cycle. Treatment with estrogens produces a dramatic rise in the number of Kurloff cells in the viscera and circulation in both boars and sows. Increased numbers (e.g., more than 1–2%) are usually present in the peripheral blood during pregnancy (Ledingham 1940). In addition, large numbers of these cells aggregate in the placental labyrinth in pregnant sows (Revell et al. 1971). Kurloff cells have been shown to release the material from the inclusion into the trophoblast and fetal endothelium of the placental labyrinth. In vitro studies have demonstrated that this material has a toxic effect on macrophages. It has been suggested that Kurloff cells may therefore play a role in preventing the maternal rejection of the fetal placenta during pregnancy (Marshall et al. 1971).

SIGNIFICANCE. It is important that pathologists be aware of these cells and not misinterpret the presence of Kurloff cells as a disease process. They have on occasion been interpreted as lupus erythematosus cells when found in blood samples collected from this species.

Respiratory Tract

Pulmonary arteries and arterioles have marked medial thickening in this species (Fig. 5.2). This

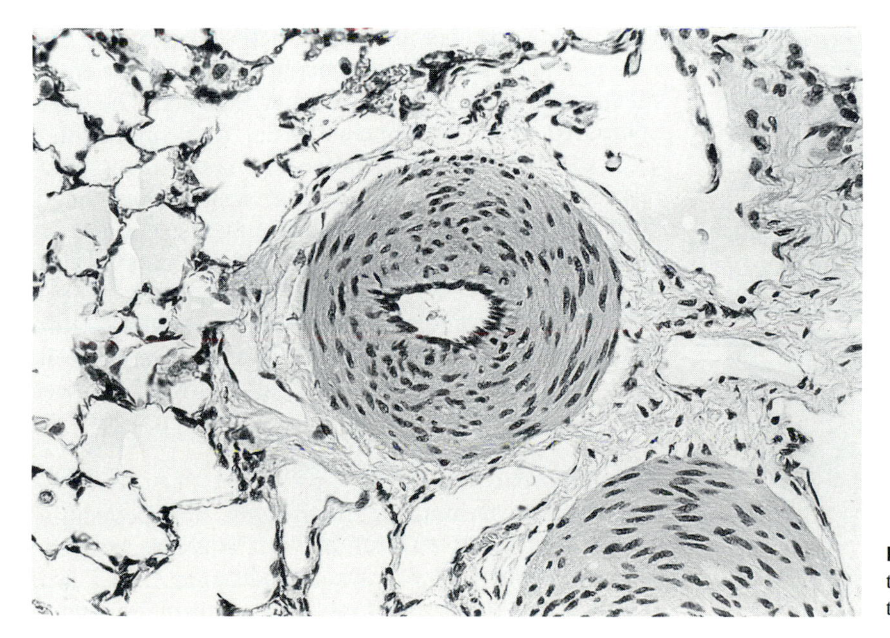

Fig. 5.2. Section of lung from guinea pig, illustrating typical medium-size pulmonary artery in this species. Note the medial thickness.

should not be interpreted to be an abnormal finding. Larger airways are surrounded by prominent concentric bands of smooth muscle. The marked contraction of the peribronchial muscle may result in marked distortion and thickening and sloughing of the respiratory epithelium that lines affected airways. On occasion, such artifacts have been interpreted to be a bronchial tumor by the unwary pathologist.

Perivascular Lymphoid Nodules. Aggregations of lymphocytes in the adventitia of pulmonary vessels are a frequent incidental finding in guinea pigs. A variety of strains may be affected. Microscopic foci have been observed in animals as young as 5 days of age, but nodules are more common in older animals (Thompson et al. 1962). The incidence and extent of the involvement varies. Frequently animals of all ages are found to be free of these changes at necropsy. The perivascular changes are normally found only in the lung. At necropsy, close examination may reveal circumscribed, pale, pinpoint subpleural foci up to 0.5 mm in diameter. On microscopic examination, concentric to eccentric aggregations of small- to medium-size lymphocytes are oriented around small arteries and veins (Figs. 5.3 and 5.4). Nodules are focal to segmental in distribution in the perivascular regions. There may be focal to diffuse infiltrates of lymphocytes in alveolar septa in some animals, but the airways and alveoli are free of exudate in the typical cases. Ultrastructural studies have revealed morphologically normal lymphocytes, and there was no evidence of viral agents associated with the cellular infiltrates (Baskerville et al. 1982). The lymphoid nodules

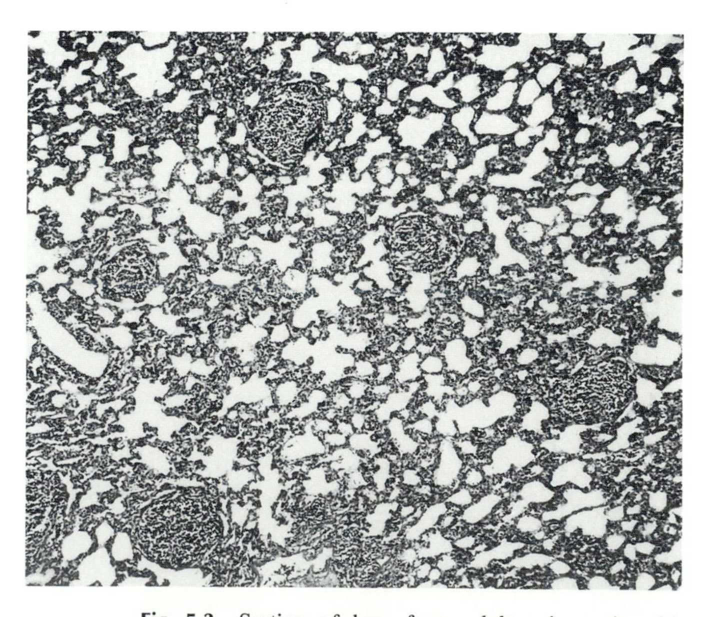

Fig. 5.3. Section of lung from adult guinea pig with prominent perivascular lymphoid nodules. These may be present as an incidental finding, particularly in adult animals.

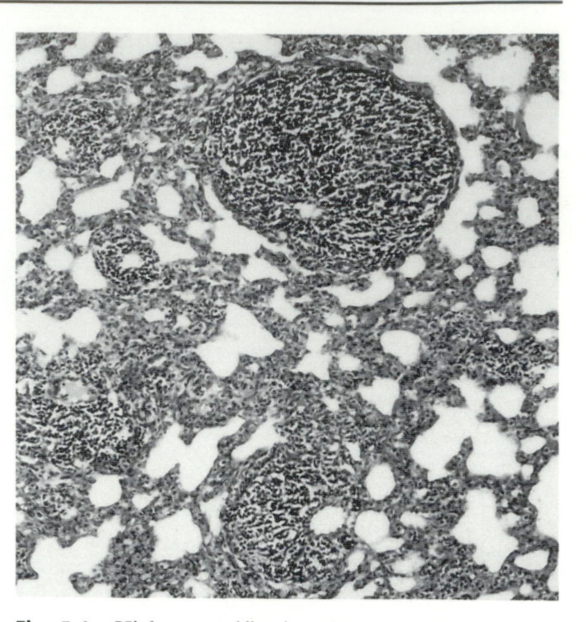

Fig. 5.4. Higher magnification of Fig. 5.3. The infiltrates are present in the adventitia of arterioles and are composed of relatively well-differentiated lymphocytes. Note the thickening of alveolar septa, with mononuclear cell infiltration.

have been ascribed to a variety of antigenic stimuli, but the pathogenesis and significance of these changes are currently not well understood.

SIGNIFICANCE. The presence of perivascular nodules, if present at necropsy, should be duly recorded. However, in the absence of other significant changes, undue importance should not be placed on this finding. They must be differentiated from the focal granulomatous pulmonary lesions seen in guinea pigs posttreatment with Freund's adjuvant (Schiefer and Stunzi 1979).

Osseous Metaplasia (Metaplastic Nodules) in the Lung. Bony spicules are occasionally observed in the lung in guinea pigs. Similar changes have been seen in other species, such as the rat and hamster. They are composed of dense, lamellar bone, with varying degrees of calcification. Usually there is no or minimal reaction in the adjacent alveolar septa. They have been interpreted to be inhaled fragments of bone of dietary origin (Innes et al. 1956), but there is evidence that in the guinea pig, they are foci of osseous metaplasia. Large numbers of metaplastic osseous foci, including well-differentiated bone marrow, have been observed in the lungs of guinea pigs following X-irradiation (Knowles 1984).

Hematopoietic System

Thymus. Degenerate thymocytes are frequently observed in close association with Hassall's corpuscles, particularly in younger animals (Fig. 5.5). This should not be interpreted to be an abnormal change,

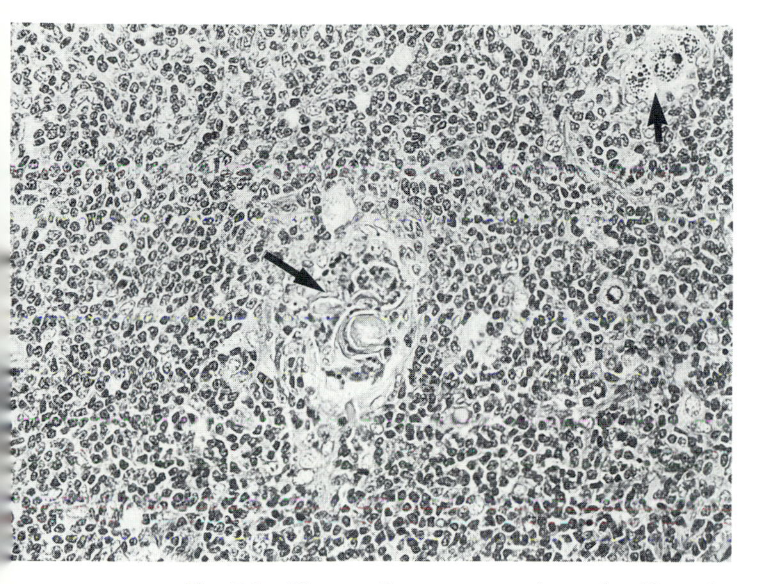

Fig. 5.5. Thymus from young guinea pig. Note the fragmentation of cells (*arrows*) associated with Hassall's corpuscles, a normal feature of the thymus in this species.

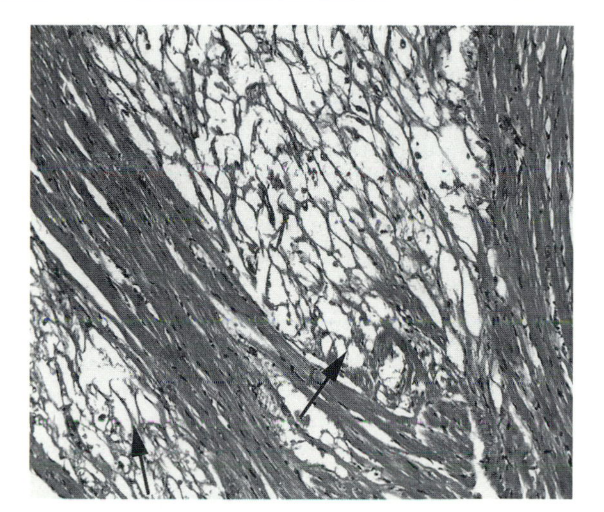

Fig. 5.6. Myocardium from adult guinea pig illustrating segmental rhabdomyomatosis, an incidental finding at necropsy. There is marked vacuolation of the sarcoplasm of affected myofibers (*arrows*).

as it is common in the guinea pig. These areas of degeneration may evolve into thymic cysts.

Heart: Rhabdomyomatosis (Nodular Glycogen Infiltration). This condition is occasionally observed as an incidental finding in guinea pigs of various ages (Vink 1969; Hueper 1941). It has been interpreted to be a degenerative condition and a congenital tissue malformation with "blastemoid" characteristics. However, the current assessment is that rhabdomyomatosis is a congenital disease related to a disorder of glycogen metabolism.

PATHOLOGY. Smaller lesions are not visible on macroscopic examination. Occasionally larger areas appear as pale pink, poorly delineated foci or streaks. Rhabdomyomatosis has been observed in various regions of the heart, including ventricle, atria, interventricular septum, and papillary muscles. Lesions are most frequently found in the left ventricle. Microscopic examination reveals a spongy network of vacuolated myofibers composed of finely fibrillar to granular, eosinophilic cytoplasm (Fig. 5.6). Vacuoles are rounded to polygonal in shape and usually fill the sarcolemmal sheath. Vacuoles contain large quantities of glycogen, which is washed out in the fixation and processing procedures. Glycogen is readily demonstrated in PAS-stained, alcohol-fixed specimens. There may be displacement and flattening of myocyte nuclei in some affected fibers. In other fibers, there may be a cytoplasmic marginal rim with a round nucleus projecting into the vacuole (Fig. 5.7). Myofibers with centrally located nuclei and radiating fibrillar processes have been called "spider cells." Interspersed within the affected myofibers, there may be poorly differentiated fibers with identifiable cross-striations.

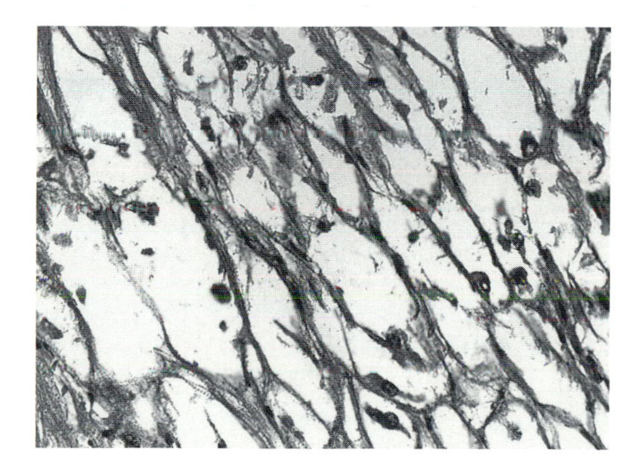

Fig. 5.7. Higher magnification of affected myocardium with rhabdomyomatosis. Note the cytoplasmic vacuolation, with fibrillar strands radiating from the nuclei.

SIGNIFICANCE. Rhabdomyomatosis is considered to be an incidental finding in the guinea pig, and normally it does not significantly compromise cardiac function.

BIBLIOGRAPHY FOR ANATOMIC FEATURES

Hematology

Sanderson and Phillips 1981 (*see* General Bibliography).

Kurloff Cells

Christensen, H.E., et al. 1970. The cytology of the Foa-Kurloff reticular cells of the guinea pig. Acta. Pathol. Microbiol. Scand. [Suppl.] 212:15–24.

Ledingham, J.C.G. 1940. Sex hormones and the Foa-Kurloff cell. J. Pathol. Bacteriol. 50:201–19.

Marshall, A.H.E., et al. 1971. Studies on the function of the Kurloff cell. Int. Arch. Allergy 40:137–52.

Revell, P.A., et al. 1971. The distribution and ultrastructure of the Kurloff cell in the guinea pig. J. Anat. 109:187–99.

Perivascular Lymphoid Nodules

Baskerville, A., et al. 1982. Ultrastructural studies of chronic pneumonia in guinea pigs. Lab. Anim. 16:351–55.

Schiefer, B., and Stunzi, H. 1979. Pulmonary lesions in guinea pigs and rats after subcutaneous injection of complete Freund's adjuvant or homologous pulmonary tissue. Zentralbl. Vetinaermed. Med. [A] 26:1–10.

Thompson, S.W., et al. 1962. Perivascular nodules of lymphoid cells in the lungs of normal guinea pigs. Am. J. Pathol. 40:507–17.

Osseous Metaplasia in the Lung

Innes, J.R.M., et al. 1956. Note on the origin of some fragments of bone in the lungs of laboratory animals. Arch. Pathol. 61:401–6.

Knowles, J.F. 1984. Bone in the irradiated lung of the guinea pig. J. Comp. Pathol. 94:529–33.

Heart

Hueper, W.C. 1941. Rhabdomyomatosis of the heart in a guinea pig. Am. J. Pathol. 17:121–26.

Vink, H. 1969. Rhabdomyomatosis of the heart in guinea pigs. J. Pathol. 97:331–34.

General Bibliography

Breazile, J.E., and Brown, E.M. 1976. Anatomy. In *The Biology of the Guinea Pig*, ed. J.E. Wagner and P.J. Manning, pp. 54–62. New York: Academic.

Harper, L.V. 1976. Behavior. In *The Biology of the Guinea Pig*, ed. J.E. Wagner and P.J. Manning, pp. 31–51. New York: Academic.

Sanderson, J.H., and Phillips, C.E. 1981. *An Atlas of Laboratory Animal Haematology*. Oxford, Engl.: Clarendon.

VIRAL INFECTIONS

DNA Viral Infections

Cytomegaloviral Infection

The cytomegalovirus (CMV) group are species-specific members of the family Herpesviridae. Natural infections occur in several mammals, including human and nonhuman primates, mice, rats, and guinea pigs. Members of the CMV group produce characteristic large, intranuclear and intracytoplasmic inclusion bodies and may persist in the host as an inapparent or latent infection for years. Under natural conditions, primary target tissues for CMV in the guinea pig are the salivary glands, kidney, and liver (Van Hoosier et al. 1985; Cook 1958).

EPIZOOTIOLOGY AND PATHOGENESIS. Most guinea pigs housed under conventional conditions may seroconvert by a few months of age. A high percentage have been shown to have salivary gland lesions in some surveys. The virus may be transmitted by exposure to infected saliva or urine or as a transplacental infection. Systemic disease, with associated lesions, has been produced in weanling guinea pigs inoculated subcutaneously with CMV. In the experimental disease, focal lesions with intranuclear inclusion bodies may be present in salivary glands, liver, spleen, lung, and kidney. Pregnant guinea pigs have been shown to develop more extensive visceral lesions when inoculated with CMV than do nonpregnant animals (Griffith et al. 1983). Lymphoproliferative disease, with mononucleosislike syndrome and lymphadenopathy, has been observed in guinea pigs postinoculation with CMV (Lucia et al. 1985). However, naturally occurring CMV infections rarely cause detectable clinical disease in the guinea pig, a pattern similar to that observed in human CMV infections (Weller 1971). There is one report of systemic CMV infection with visceral lesions in two young guinea pigs introduced into a conventional facility. Focal destructive lesions with large intranuclear and cytoplasmic inclusion bodies were observed in various tissues including spleen, liver, kidney, and lung (Van Hoosier et al. 1985)).

PATHOLOGY. Lesions are usually regarded as an incidental finding at necropsy and are confined primarily to the ductal epithelial cells of the salivary glands. Large eosinophilic inclusion bodies are associated with marked karyomegaly and margination of the nuclear chromatin in affected cells (Fig. 5.8). Intracytoplasmic inclusions are occasionally present in ductal epithelial cells. There may be a concurrent mononuclear cell infiltration around infected ducts (Fig. 5.9). In the acute systemic form of the disease, interstitial pneumonitis and multifocal areas of necrosis may be present in the lymph nodes, spleen, liver, kidney, lung, and other viscera. Intranuclear and intracytoplasmic inclusion bodies may be present in affected foci.

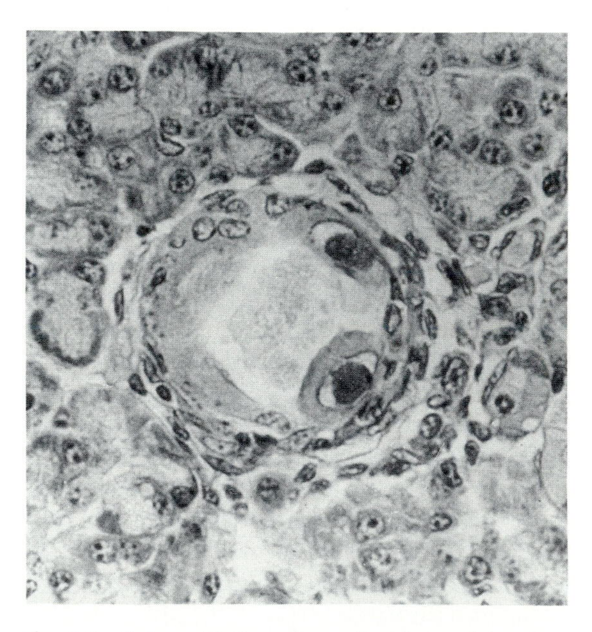

Fig. 5.8. Submandibular salivary gland from adult guinea pig with cytomegalovirus infection (an incidental finding). Note the karyomegaly of infected cells, the large intranuclear inclusion bodies, and the margination of nuclear chromatin. (Courtesy of G.D. Hsiung)

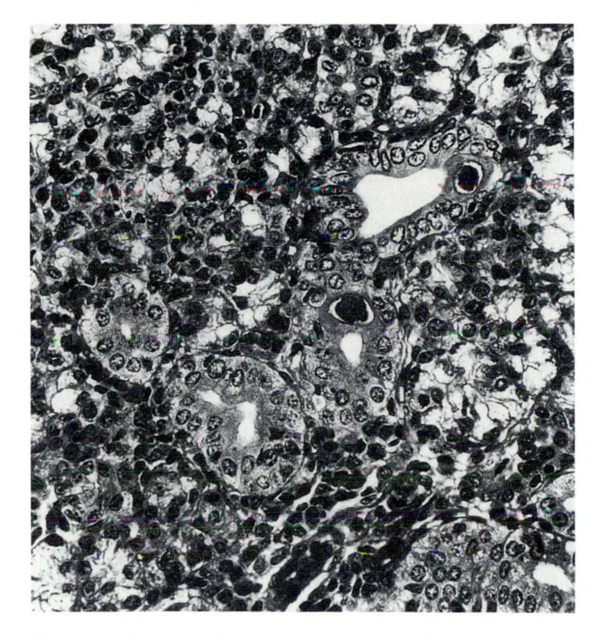

Fig. 5.9. Submandibular salivary gland from guinea pig with localized cytomegalovirus infection. Note the mononuclear inflammatory cell response. (Courtesy of G.D. Hsiung)

SIGNIFICANCE. Subclinical CMV infections are common in guinea pigs housed in conventional facilities. Spontaneous generalized cytomegalovirus infection with mortality seldom occurs in guinea pigs, and there has been speculation that these cases may be associated with a compromised immune system. Lymphoid hyperplasia, lymphadenopathy, and mononucleosis have been observed in Hartley guinea pigs inoculated subcutaneously with guinea pig CMV (Lucia et al. 1985). CMV infections in guinea pigs are considered to be a useful animal model for CMV infections in other species, including human patients.

Other Cavian Herpesviral Infections

Guinea pig "herpeslike virus" (GPHLV) has been isolated from degenerating primary kidney cell cultures prepared from guinea pigs, but to date GPHLV has not been shown to be capable of producing disease in the natural host (Bhatt et al. 1971). Guinea pig "X virus" (GPXV) is a herpesvirus originally isolated from the leukocytes of strain 2 guinea pigs. Based on serological studies and DNA analyses, GPXV is different from either GPHLV or guinea pig CMV. Following experimental inoculation of GPXV into Hartley strain guinea pigs, viremia, focal hepatic necrosis, and mortality may occur (Bia et al. 1980).

SIGNIFICANCE. Current information suggests it is unlikely that GPHLV and GPXV will prove to be an important primary pathogen in the guinea pig. However, they represent a possible complicating factor, should either occur as an inapparent infection in guinea pigs under experiment in the laboratory.

Adenoviral Pneumonitis

Outbreaks of respiratory disease attributed to an adenovirus have been recognized in Europe (Naumann 1981) and North America (Brennecke et al. 1983). The disease is characterized by low morbidity and a mortality rate in clinically affected animals of up to 100%. In those cases described to date, frequently the animals have been subjected to experimental manipulations that may have resulted in impairment of the immune response.

PATHOLOGY. Consolidation of the cranial lobes of the lung and hilus is a characteristic finding at necropsy. Microscopic changes are those of a necrotizing bronchitis and bronchiolitis, with desquamation of lining epithelial cells and leukocytic infiltration, mononuclear cells predominating. Some airways may be obliterated by cell debris, leukocytes, and fibrinous exudate. Numerous necrotic foci may be scattered throughout the lung. The nuclei of affected epithelial cells often contain round to oval basophilic inclusion bodies 7–15 μm in diameter (Fig. 5.10). The virus has not been recovered and characterized to date, but electron microscopic examination has revealed typical adenovirus particles in affected nuclei. Using homogenates of lung prepared from a spontaneous case of the

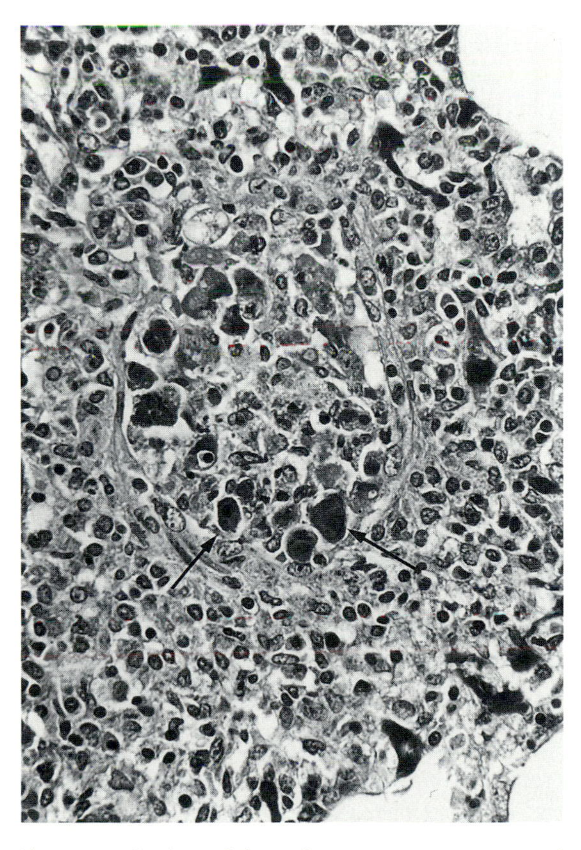

Fig. 5.10. Section of lung from spontaneous case of adenoviral bronchoalveolitis in young guinea pig. Prominent intranuclear inclusion bodies (*arrows*) are present in bronchial epithelial cells and in a few alveolar septal cells.

disease, typical lesions have been produced in intranasally inoculated newborn guinea pigs. The incubation period is 5–10 days. Older inoculated animals are relatively refractory to the disease (Kunstyr et al. 1984). Typical adenoviral particles may be found in the nuclei of pulmonary epithelial cells of the lung (Kaup et al. 1984).

DIAGNOSIS. The presence of nonsuppurative bronchitis and bronchiolitis in young guinea pigs with typical intranuclear basophilic inclusion bodies is consistent with adenoviral pneumonitis. The diagnosis may be confirmed by the demonstration of adenoviral particles in affected cells by electron microscopy. *Differential diagnoses* include cytomegalovirus infection and infections of the lower respiratory tract with bacteria, such as *Bordetella bronchiseptica*.

SIGNIFICANCE. Adenoviral infections in colonies of guinea pigs are likely more prevalent than currently recognized. Clinical disease appears to occur primarily in young animals. Other than the pulmonary lesions recognized to occur during the acute stages of the disease, the possible effects of adenoviral infections on other systems are currently unknown.

Viral Infections Associated With Cavian Leukemia

Cavian leukemia has been associated with a retroviral infection, and C-type viral particles have been visualized in transformed lymphocytes. GPHLV has also been associated with guinea pig leukemia. However, herpesviral particles interpreted to be GPHLV have been demonstrated in cells of leukemic guinea pigs that also were infected with a retrovirus (Nayak 1971). Based on current information, it seems unlikely that GPHLV plays an important role in leukemia in this species.

RNA Viral Infections

Lymphocytic Choriomeningitis (LCM) Viral Infection

EPIZOOTIOLOGY AND PATHOGENESIS. LCM is a relatively rare disease in guinea pigs, but it does represent an infection that can complicate research projects and is of public health significance. Lesions observed in guinea pigs with LCM have included lymphocytic infiltrates in the meninges, choroid plexi, and ependyma, and in liver, adrenals, and lungs. There is a wide host range, including wild mice, and exposure may occur by inhalation or ingestion, and apparently in guinea pigs through the intact skin (Hotchin 1971; Shaughnessy and Zichis 1940).

DIAGNOSIS. Confirmation requires the demonstration of viral antigen in affected tissues and/or positive serological tests.

SIGNIFICANCE. LCM virus infection has been shown to prolong the life of guinea pigs carrying a leukemia agent (Jungeblut and Kodza 1962), which emphasizes the potential for the virus to be an important complicating factor in certain types of research. Several species, including humans, are susceptible to LCM viral infection.

Coronavirus-like Infection in Young Guinea Pigs

A syndrome characterized by wasting, anorexia, and diarrhea has been observed in young guinea pigs following their arrival at a research facility. The disease was characterized by a low morbidity and mortality.

PATHOLOGY. There is an acute to subacute necrotizing enteritis involving primarily the distal ileum, and copious amounts of mucoid material may be present throughout the gastrointestinal tract. On microscopic examination, lesions are particularly prominent in the terminal small intestine. There is blunting and fusion of affected villi, with necrosis and sloughing of enterocytes from the tips of villi and frequently syncytial giant cell formation in the intestinal mucosa. Viral particles consistent with the morphology of a coronavirus were demonstrated in fecal samples examined by electron microscopy (Jaax et al. 1990).

SIGNIFICANCE. The importance of suspected coronaviral infections in this species are currently unknown. However, until additional information is available, it should be considered in the differential diagnoses in cases of enteritis and/or wasting in young guinea pigs.

Other Viral Infections

Serology studies indicate that guinea pigs are susceptible to other rodent viruses, including Sendai virus, murine poliovirus, and reovirus 3 (Van Hoosier and Robinette 1976). The significance of these findings has not yet been determined.

Bibliography for Viral Infections

Cytomegaloviral and Other Herpesviral Infections

Bhatt, P.N., et al. 1971. Isolation and characterization of a herpes-like (Hsiung-Kaplow) virus from guinea pigs. J. Infect. Dis. 123:178–89.

Bia, F.J., et al. 1980. New endogenous herpesvirus of guinea pigs: Biological and molecular characterization. J. Virol. 36:245–53.

Connor, W.S., and Johnson, K.P. 1976. Cytomegalovirus infection in weanling guinea pigs. J. Infect. Dis. 134:442–49.

Cook, J.E. 1958. Salivary-gland virus disease of guinea pigs. J. Natl. Cancer Inst. 20:905–9.

Griffith, B.P., et al. 1983. Enhancement of cytomegalovirus infection during pregnancy in guinea pigs. J. Infect. Dis. 147:990–98.

Griffith, B.P., et al. 1976. Cytomegalovirus-induced mononucleosis in guinea pigs. Infect. Immunol. 13:926–33.

Jungeblut, C.W., and Opler, S.R. 1967. On the pathogenesis of cavian leukemia. Am. J. Pathol. 51:1153–60.

Lucia, H.L., et al. 1985. Lymphadenopathy during cytomegalovirus-induced mononucleosis in guinea pigs. Arch. Pathol. Lab. Med. 109:1019–23.

Ma, B.I., et al. 1969. Detection of virus-like particles in germinal centers of normal guinea pigs. Proc. Soc. Exp. Biol. Med. 130:586–90.

Nayak, D.P. 1971. Isolation and characterization of a herpesvirus from leukemic guinea pigs. J. Virol. 8:579–88.
Opler, S.R. 1968. New oncogenic virus producing acute lymphocytic leukemia in guinea pigs. Proc. 3rd Int. Symp. Comp. Leuk. Res. 31:81–88.
Van Hoosier, G.L., Jr., et al. 1985. Disseminated cytomegalovirus in the guinea pig. Lab. Anim. Sci. 35:81–84.
Weller, T.H. 1971. The cytomegaloviruses: Ubiquitous agents with protean manifestations. New Engl. J. Med. 285:203–14, 267–72.

Adenoviral Pneumonitis

Brennecke, L.H., et al. 1983. Naturally occurring virus-associated respiratory disease in two guinea pigs. Vet. Pathol. 20:488–91.
Feldman, S.H., et al. 1990. Necrotizing viral bronchopneumonia in guinea pigs. Lab. Anim. Sci. 40:82–83.
Kaup, F.-J., et al. 1984. Experimental viral pneumonia in guinea pigs: An ultrastructural study. Vet. Pathol. 21:521-527.
Kunstyr, I., et. al. 1984. Adenovirus pneumonia in guinea pigs: An experimental reproduction of the disease. Lab. Anim. 18:55–60.
Naumann, S., et al. 1981. Lethal pneumonia in guinea pigs associated with a virus. Lab. Anim. 15:255–42.
Shaughnessy, H.J., and Zichis, J. 1940. Infection of guinea pigs by application of virus of lymphocytic choriomeningitis to their normal skins. J. Exp. Med. 72:331–43.

LCM Viral Infection

Hotchin, J. 1971. The contamination of laboratory animals with lymphocytic choriomeningitis virus. Am. J. Pathol. 64:747–69.
Jungeblut, C. W., and Kodza, H. 1962. Interference between lymphocytic choriomeningitis virus and the leukemia transmitting agent of leukemia L2C in guinea pigs. Arch. Gesamte Virusforsch. 12:522–60.

Coronavirus-like Infection

Jaax, G.P., et al. 1990. Coronavirus-like virions associated with a wasting syndrome in guinea pigs. Lab. Anim. Sci. 40:375–78.

General Bibliography

Van Hoosier, G.L., Jr., and Robinette, L.R. 1976. Viral and chlamydial diseases. In *The Biology of the Guinea Pig*, ed. J.E. Wagner and P.J. Manning, pp. 137–52. New York: Academic.

BACTERIAL INFECTIONS

Bordetella bronchiseptica Infection

Bordetella bronchiseptica is a small, gram-negative rod that is an important cause of respiratory disease in the guinea pig (Baskerville et al. 1982).

EPIZOOTIOLOGY AND PATHOGENESIS. The organism is harbored in the upper respiratory tract in several other species, including dogs, cats, and rabbits, and it is likely that interspecies transmission can occur. To date, there is no evidence that there are variations in virulence of the isolates when tested in guinea pigs. Guinea pigs of all ages are susceptible and develop respiratory tract lesions following intranasal inoculation. However, disease and mortality occur most often in young guinea pigs, particularly during winter months. In some outbreaks, there may be other identifiable manipulations or environmental factors that could precipitate disease (Ganaway 1976). Guinea pigs may harbor the organism in the upper respiratory tract and trachea as an inapparent infection. In enzootically infected colonies, the incidence of nasal shedders may be relatively high. Infection rates are usually highest in the winter months. Most animals appear to develop solid immunity and eventually eliminate the organism, but a small percentage may remain carriers. The organism is readily transmitted as an airborne infection (Traham 1987). It has an affinity for ciliated respiratory epithelium and has been shown to cause ciliostasis in other species (Bemis and Wilson 1985). During epizootics of bordetellosis, pregnant sows may die, abort, or produce stillborn offspring. *Bordetella* has also been isolated from a case of *pyosalpinx* in the guinea pig (Sinka and Sleight 1968).

PATHOLOGY. At necropsy the external nares, nasal passages, and trachea frequently contain mucopurulent or catarrhal exudate. Consolidated areas vary from dark red to gray, are anteroventral in distribution, and may involve entire lobes or individual lobules (Fig. 5.11). Mucopurulent exudate is present in affected airways, pleuritis occasionally occurs, and purulent exudate may be present in the tympanic bullae.

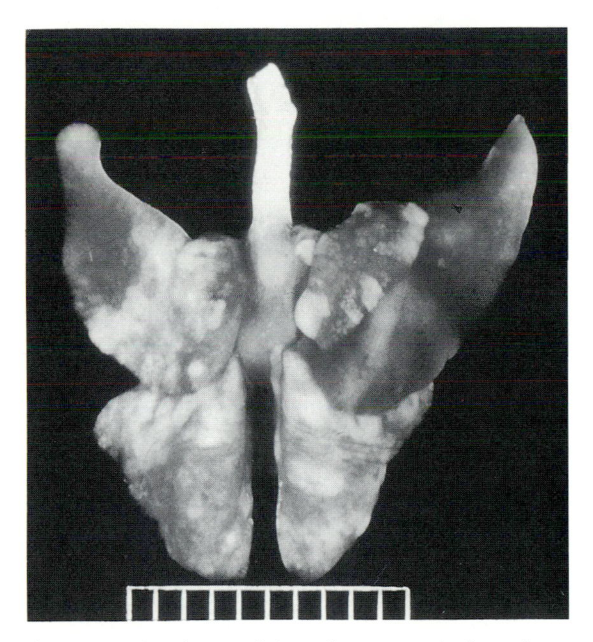

Fig. 5.11. Cranioventral bronchopneumonia from juvenile guinea pig with acute *Bordetella bronchiseptica* infection.

Histologically, there is an acute to chronic suppurative bronchopneumonia, with marked infiltration by heterophils in airways and affected alveoli and obliteration of the normal architecture (Figs. 5.12 and 5.13). In some acute cases, there may be fibrinous exudation into terminal airways and alveoli.

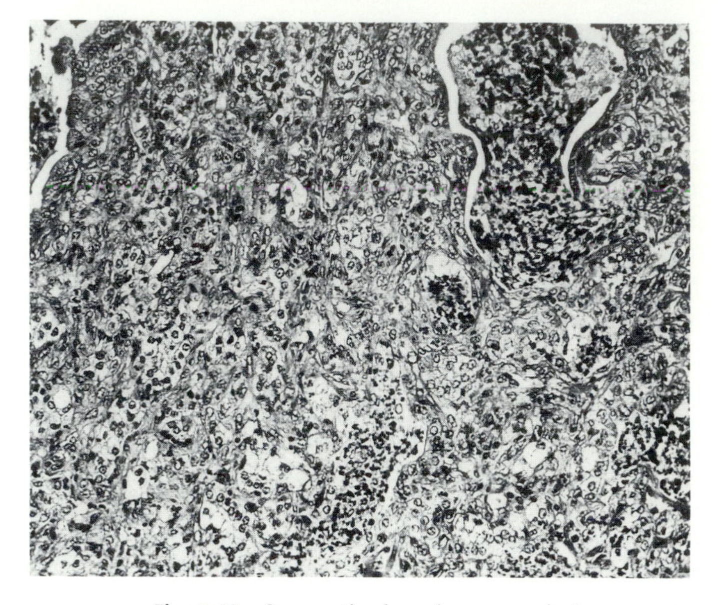

Fig. 5.12. Suppurative bronchopneumonia from spontaneous case of *Bordetella* infection. There is marked infiltration with heterophils and mononuclear cells, with obliteration of the normal architecture.

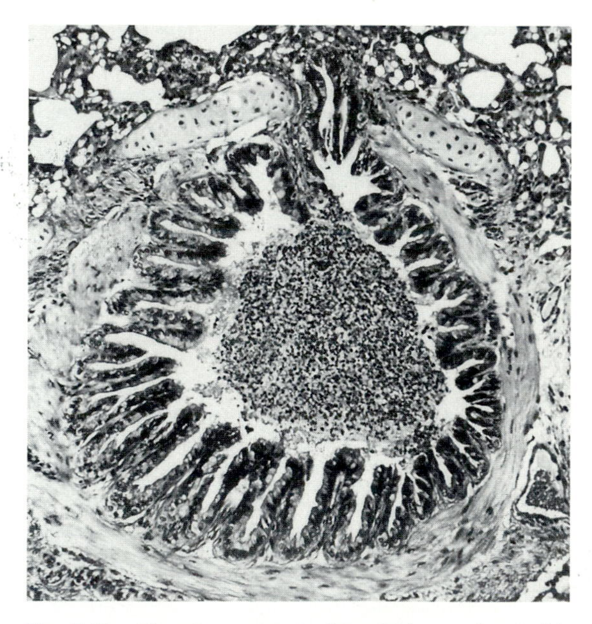

Fig. 5.13. Chronic suppurative bronchitis associated with *Bordetella* infection in mature guinea pig.

DIAGNOSIS. The organism can usually be readily recovered on blood agar cultures from the respiratory tract, affected tympanic bullae, and in cases of metritis, the uterus. If the results of bacterial cultures are unrewarding, or not available, *differential diagnoses* include acute pneumococcal, *Klebsiella*, or staphylococcal infections, and a systemic form of *Streptococcus zooepidemicus* infection. If there is a history of prior administration of Freund's adjuvant, chronic interstitial fibrosis and granulomatous pulmonary lesions may occur. These must be differentiated from resolving lesions due to other causes, including *B. bronchiseptica* infections.

SIGNIFICANCE. *B. bronchiseptica* infections are a major cause of respiratory disease in this species. Exacerbations of subclinical infections may occur. Attempts have been made to correlate cases of interstitial pneumonitis and perivascular lymphoid nodules with previous *Bordetella* infections, but without success (Baskerville et al. 1982). The dangers of interspecies transmission should be emphasized. Commercial and autogenous bacterins have been used to reduce the incidence of disease (Ganaway et al. 1965). However, immunization does not necessarily eliminate the carrier state.

Streptococcal Infections

Streptococcal Lymphadenitis/Septicemia. *Streptococcus zooepidemicus* of Lancefield's group C is a gram-positive encapsulated coccus that produces beta hemolysis on blood agar plates. It is associated with suppurative lymphadenitis in this species.

EPIZOOTIOLOGY AND PATHOGENESIS. The organism may be carried in the nasopharynx and conjunctiva as an inapparent infection. Females have been shown to be more susceptible to the disease than are males, and there is a strain-related variation in susceptibility. Steroid treatment does not appear to increase susceptibility to the disease (Olson et al. 1976). The usual route of invasion appears to be via abrasions in the oral mucosa (Mayora et al. 1978), but inhalation, skin abrasions, and invasion of the genital tract at farrowing are other possible portals of entry. The disease has also been produced in young guinea pigs by inoculation of the intact nasal and conjunctival mucous membranes (Murphy et al. 1991). Lymphadenitis has been produced consistently in guinea pigs inoculated sublingually with *S. zooepidemicus* (Olson et al. 1976). Lesions have been readily produced in guinea pigs by oral inoculation of the organism following abrasion of the oral mucosa, but not in unabraided animals. Following penetration of the oral mucosa and invasion of the underlying tissue, the organism is likely transported to the draining cervical lymph nodes via the lymphatics. The pyogenic organism then proliferates, producing a chronic suppurative inflammatory process.

PATHOLOGY. Affected adults usually have lesions confined to the regional lymph nodes. In the localized form of the disease, there is bilateral enlargement of the cervical lymph nodes. The nodes are freely movable, firm to soft, and frequently nonfluctuant and they contain thick purulent exudate (Fig. 5.14). Localized abscessation involving other sites such as mesenteric lymph nodes is an infrequent finding. Retroorbital abscessation is another possible manifestation of the disease. Otitis media may also occur. Occasionally

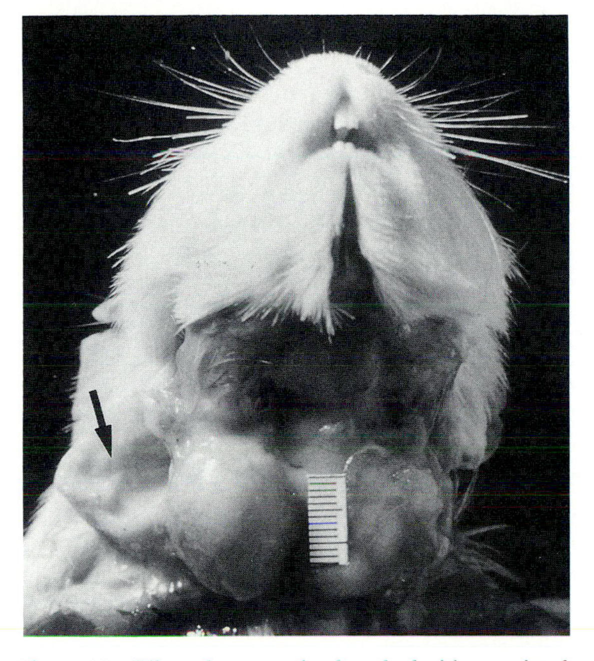

Fig. 5.14. Bilateral suppurative lymphadenitis associated with *Streptococcus zooepidemicus* infection. Note the purulent exudate (*arrow*) in the incised cervical lymph node.

there is an acute systemic form of the disease, particularly in younger animals. In this case, fibrinopurulent bronchopneumonia, pleuritis, and pericarditis may be present at necropsy. On rare occasions, arthritis and abortions have been attributed to *S. zooepidemicus* infections. Microscopically, changes present in the cervical lymph nodes are those of a chronic suppurative lymphadenitis with central necrosis, peripheral fibrosis, and marked infiltration with heterophils. In the acute systemic form, fibrinopurulent pericarditis, focal myocardial degeneration, focal hepatitis, and acute lymphadenitis may be evident on histological examination. In one report, acute bronchopneumonia, hemopericardium, and hemothorax were associated with acute *S. pyogenes* infections (Okewole et al. 1991).

DIAGNOSIS. The typical beta-hemolytic streptococci can usually be readily recovered from affected tissues, except in some cases of chronic lymphadenitis of some duration. *Differential diagnoses* in the acute systemic form of the disease include acute pneumococcal septicemia and acute *Bordetella* infection.

SIGNIFICANCE. Enzootic infections with *S. zooepidemicus* represent a potential complication in research. Culling the affected animals and thorough disinfection are required. Vaccination using a less virulent strain of the organism has provided a significant degree of protection against the disease (Mayora et al. 1978).

Diplococcal (Pneumococcal) Pneumonia. The organism, *Streptococcus pneumoniae*, is a lancet-shaped, gram-positive encapsulated coccus that occurs in pairs and short chains. Capsular polysaccharide type 19 is most frequently isolated from guinea pigs (Parker et al. 1977; Keyhani and Naghshineh 1974). Type 4 has also been identified.

EPIZOOTIOLOGY AND PATHOGENESIS. Pneumococcal infections have been recognized to occur in guinea pigs for decades, but the disease seldom occurs in well-managed facilities today. The organism may be carried as an inapparent infection in the upper respiratory tract. In affected colonies, up to 50% of the animals may be carriers of *S. pneumoniae*. Transmission is probably primarily by aerosols. Epizootics occur most often during winter months, and younger animals and pregnant sows are considered to be particularly at risk. Other predisposing factors include changes in environmental temperature, poor husbandry and experimental procedures, and inadequate nutrition. During epizootics, high mortality, abortions, and stillbirths may occur. The organisms do not produce toxins but are protected from phagocytosis primarily through their abundant polysaccharide capsules. Many pneumococci can activate the alternate complement pathway; thus complement activation may be the important stimulus for the early tissue changes (Yoneda and Coonrod 1980).

PATHOLOGY. At necropsy, typical lesions include fibrinopurulent pleuritis, pericarditis, peritonitis, and marked consolidation of affected lobes of lung. Microscopic changes are those of an acute bronchopneumonia with fibrinous exudation and polymorphonuclear cell infiltration. Thrombosis of pulmonary vessels may occur in acute cases (Fig. 5.15). Infiltrat-

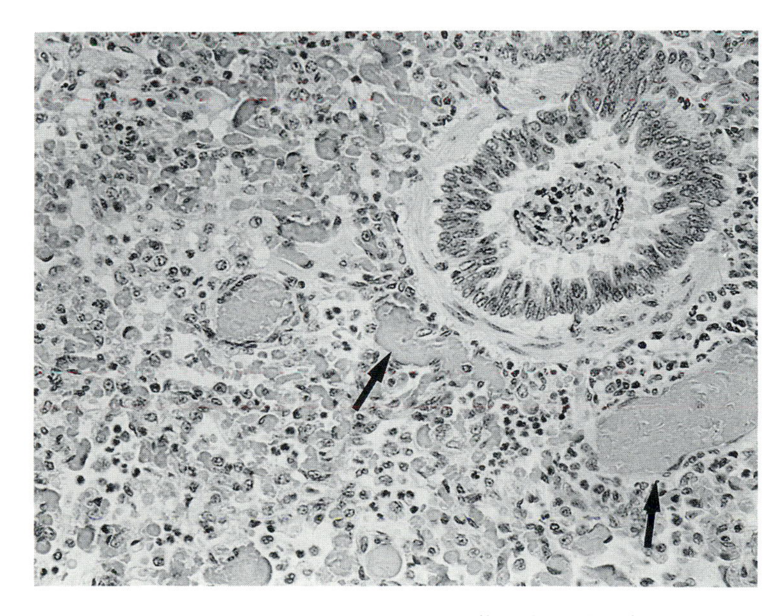

Fig. 5.15. Section of lung from juvenile guinea pig with peracute infection with *Streptococcus pneumoniae.* There is alveolar flooding, and thrombi (*arrows*) are present in small pulmonary vessels.

ing cells may be elongated and fusiform, forming pallisading patterns within affected airways and alveoli (Fig. 5.16). Fibrinopurulent pleuritis, pericarditis, and epicarditis frequently occur. Splenitis, fibrinopurulent meningitis, metritis, focal hepatic necrosis, lymphadenitis, and ovarian abscessation have also been observed (Parker et al. 1977). *S. pneumoniae*–associated suppurative arthritis and osteomyelitis have been reported to occur in guinea pigs with borderline vitamin C deficiency (Witt et al. 1988).

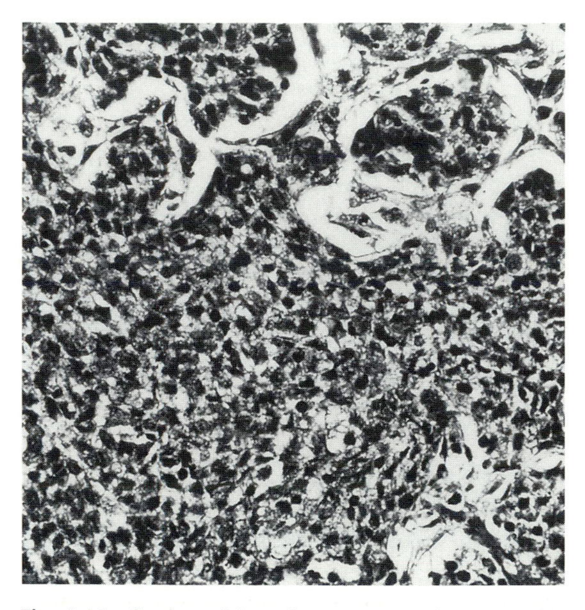

Fig. 5.16. Section of lung from young guinea pig with spontaneous diploccoal pneumonia (*Streptococcus pneumoniae*). There is necrotizing alveolitis with leukocytic infiltration and obliteration of the normal architecture.

DIAGNOSIS. Direct smears of Gram-stained inflammatory exudate should reveal the typical gram-positive diplococci. Using blood agar or enrichment media, the organism should be recoverable from affected tissues. (*S. pneumoniae* is more fastidious in growth requirements than are most other streptococci.) *Differential diagnoses* include acute septicemia due to *S. zooepidemicus* and acute *Bordetella* infections.

SIGNIFICANCE. Pneumococcal infections may be an important cause of disease and mortality in enzootically infected colonies and represent a potential complicating factor in research. Serotypes isolated from guinea pigs are identical to human isolates. The possibility of interspecies transmission is feasible, but not proven to date.

Staphylococcal Infections

Ulcerative pododermatitis. Bumblefoot is frequently associated with coagulase-positive staphylococcal infections (Taylor et al. 1971). Predisposing factors include trauma due to defective or rusty cage wire and poor sanitation. The volar surface of the forefeet typically are swollen, painful, and encrusted with necrotic tissue and clotted blood. In some advanced cases, amyloid deposition has been observed in the spleen, liver, adrenals, and islets. Staphylococcal infections have also been associated with isolated cases of suppurative pneumonia, purulent mastitis, and conjunctivitis.

Acute Staphylococcal Dermatitis (Exfoliative Dermatitis). This condition was reported to occur most frequently in strain 13 guinea pigs (Ishihara 1980). In clinically affected animals, there was an age-related variation in mortality that was negligible in adults and relatively high in young animals, particularly those born to affected dams. Clinically, the disease was characterized by alopecia and erythema in the ventral abdominal region, with exfoliation of the epidermis. In survivors, skin lesions usually regressed within 2 wk, with subsequent new hair growth.

PATHOLOGY. At necropsy, there was erythema and hair loss, with dull red scabs and cracks in the epidermis, particularly along the ventral abdomen and the medial aspect of the extremities. On microscopic examination, there was marked epidermal cleavage, with parakeratotic hyperkeratosis and minimal inflammatory response. *Staphylococcus aureus* was isolated from the lesions in the majority of affected animals, and the disease was reproduced in young guinea pigs inoculated with a coagulase-positive strain of *S. aureus* isolated from an affected animal. The organism was also isolated from the upper respiratory tract and pharynx of many of the affected animals and from a few clinically normal guinea pigs (Ishihara 1980). Abrasions of the skin may have been an important predisposing factor, resulting in colonization with pathogenic staphylococci and invasion of the epidermis.

SIGNIFICANCE. Staphylococcal infections may occur as an inapparent infection in colonies of guinea pigs and may be recoverable from a relatively high percentage of clinically normal animals. The infection may persist in a colony for years. The possibility of interspecies transmission is an important consideration (Blackmore and Francis 1970; Markham and Markham 1966).

Salmonellosis

Salmonella infections were a common occurrence in the first half of this century. However, with the current standards of husbandry and hygiene, improved wild rodent control, and the feeding of good-quality prepared feeds, the disease seldom occurs in well-managed facilities. *S. typhimurium* and *S. enteritidis* are the most common isolates from guinea pigs

(Olfert et al. 1976; Fish et al. 1968). *S. dublin* has also been isolated from fatal cases of salmonellosis in this species (John et al. 1988).

EPIZOOTIOLOGY AND PATHOGENESIS. Salmonellosis affects guinea pigs of all ages and strains, but young weanlings and sows around farrowing time are considered to be particularly at risk. Inapparent carriers may occur. Recovered animals may shed the organism intermittently in the feces, representing a possible source of reinfection. Ingestion of contaminated feces or feed is considered to be the usual source of the organism. Clinical signs observed include depression, conjunctivitis, and abortions. Diarrhea is a variable manifestation of the disease. The mortality rate may be 50% or higher.

PATHOLOGY. Gross lesions are similar to those observed in salmonellosis in other mammals. Pinpoint-size, pale foci may be present on the liver and spleen. Massive splenomegaly frequently occurs. Necrotic miliary foci may also be present in other viscera, including lymph nodes. Lesions may be absent in peracute cases. Frequently there is a multifocal granulomatous hepatitis, splenitis, and lymphadenitis, with infiltration by histiocytic cells and polymorphs. Focal suppurative lesions may also occur in lymphoid tissues of the intestinal tract.

DIAGNOSIS. Recovery of the organism from heart blood, spleen, and feces is best accomplished with media selective for *Salmonella*. In the absence of bacteriology, the characteristic paratyphoid nodules seen in organs such as liver and spleen are useful morphologic criteria. *Differential diagnoses* include clostridial enterotoxemia, yersiniosis, Tyzzer's disease, and pneumococcal septicemia.

SIGNIFICANCE. Aside from the dangers of interspecies spread in the animal facility, the zoonotic potential must be emphasized. Identical phage types have been recovered from guinea pigs and human contacts in epizootics of the disease. Slaughter is recommended. In one reported epizootic, shedders were identified by mass fecal sampling and culture, then removed. Strict hygienic measures and the culling of all contact animals have been used to eliminate the organism from an infected colony (Olfert et al. 1976).

Klebsiella Infection

Epizootics of acute infections due to *Klebsiella pneumoniae* have been reported on rare occasions. Patterns of disease vary from acute septicemia to acute necrotizing bronchopneumonia, with pleuritis, pericarditis, peritonitis, and splenic hyperplasia (Ganaway 1976).

Citrobacter freundii Infection

An epizootic of *citrobacter* septicemia with high mortality was reported in guinea pigs. Pneumonia, pleuritis, and enteritis were observed, and *C. freundii* was isolated from lung, liver, spleen, and intestine at necropsy. No predisposing factors or possible sources of the infection were identified (Ocholi et al. 1988).

Tyzzer's Disease

Spontaneous cases of Tyzzer's disease have been recognized in this species. In some reported cases in young guinea pigs, lesions were confined to the intestinal tract. Typical *Bacillus piliformis* organisms were identified in enterocytes. Large numbers of spirochetes were associated with lesions in the gut. The typical disease has been produced in young guinea pigs inoculated orally with *B. piliformis*. Lesions were observed in the ileum, large intestine, and liver by 4 days postinoculation. Typical bacilli were demonstrated in the gut lesions at 4–10 days and in the liver only at 8–10 days postinoculation (Waggie et al. 1987). Vertical transmission has been reported to occur in a hysterectomy-derived, gnotobiotically reared guinea pig (Boot and Walvoort 1984).

PATHOLOGY. Necrotizing ileitis and typhlitis, frequently with transmural involvement, are typical findings in cavian Tyzzer's disease. Hepatic lesions, when present, are characterized by focal coagulative necrosis in periportal regions, with variable polymorphonuclear cell infiltration. Intracellular bacilli are best demonstrated by the Warthin-Starry or Giemsa stains. The significance of the spirochetes seen in association with *B. piliformis* infections in guinea pigs in some outbreaks has not been determined (Zwicker et al. 1978; McLeod et al. 1977).

Adenomatous Intestinal Hyperplasia Associated with *Campylobacter*-like Organisms

Segmental epithelial hyperplasia of the duodenum attributed to *Campylobacter*-like organisms was observed in guinea pigs on steroid treatment (Elwell et al. 1981). A spontaneous outbreak of diarrhea, with weight loss and mortality, was reported from Asia. Adenomatous changes were observed in the jejunum and ileum, and intracellular organisms interpreted to be *Campylobacter* were observed in immature crypt epithelial cells by electron microscopy. The changes observed were very similar to those seen in *Campylobacter*-associated hamster ileitis (Muto et al. 1983).

Yersiniosis

Spontaneous outbreaks of disease and mortality due to *Yersinia pseudotuberculosis* are relatively rare. Inapparent carriers may also occur. Yersiniosis may be produced experimentally in guinea pigs inoculated orally or parenterally with *Y. pseudotuberculosis* (Obwolo 1977).

PATHOLOGY. In the acute form of the disease, miliary, cream-colored nodules are present in the intes-

PARASITIC DISEASES

ECTOPARASITIC INFECTIONS

Mite Infections (Acariasis)

***Trixacarus caviae* Infection.** Sarcoptic mange is associated with *T. caviae* infection. This pathogenic sarcoptid mite appears to be widespread in some conventional colonies of guinea pigs. Lesions are usually distributed over the neck, shoulders, inner thighs, and abdomen. Changes in the skin seen grossly are keratosis with scaling and crusting and alopecia. Marked pruritis may occur, and in severe cases, animals become thin and lethargic. Hematological changes include heterophilia, monocytosis, eosinophilia, and basophilia (Rothwell et al. 1991). Flaccid paralysis and convulsions have been observed (Kummel et al. 1980). Vigorous scratching may precipitate convulsive seizures. Untreated animals with extensive lesions may die. On microscopic examination of the typical lesions, there is epidermal hyperplasia, with orthokeratotic and parakeratotic hyperkeratosis. Irregular burrows in the stratum corneum contain mites and eggs. There may be spongiosis, with leukocytic infiltration in the underlying dermis. Hair follicles are normally not invaded by the parasite.

DIAGNOSIS. Skin scrapings of hair and scale cleared with 10% KOH and examined microscopically should reveal the typical mites and eggs. The parasites can also be demonstrated in paraffin-embedded sections of affected skin. *Differential diagnoses* include pediculosis, dermatophytosis, trauma, and idiopathic alopecia, a condition that is seen occasionally in guinea pigs.

SIGNIFICANCE. Urticaria may occur in human contacts (Dorrestein and Van Bronswijk 1979). *T. caviae* infestations are an important cause of dermatitis in this species.

***Chirodiscoides caviae* Infection.** These mites have been identified in guinea pigs from commercial suppliers, in laboratory facilities, and in pet animals (Wagner et al. 1972). The parasite tends to be concentrated in the lumbar region and lateral aspect of the hindquarters. Even parasite loads of up to 200/cm^2 appear to evoke minimal or no clinical evidence of pruritus or damage to the skin. Other predisposing factors, including concurrent disease, may have a significant influence on the incidence of acariasis in this species. Microscopic examination of the adult mite is necessary for positive identification.

***Demodex caviae* Infection.** Although *D. caviae* has been recovered from guinea pigs in the absence of clinical signs, the incidence and significance of these infections in the laboratory guinea pigs is currently unknown. Infestations with ectoparasites such as *Myocoptes musculinus* and *Notoedres muris* are rare and may be due to interspecies infections (Ronald and Wagner 1976).

Louse Infection (Pediculosis)

Gliricola porcelli and *Gyropus ovalis* are large biting lice that are associated with pediculosis in guinea pigs. Frequently moderate infections are not accompanied by clinical signs. Pruritus, rough hair coat, and alopecia are seen in heavy infections (Figs. 5.19 and 5.20).

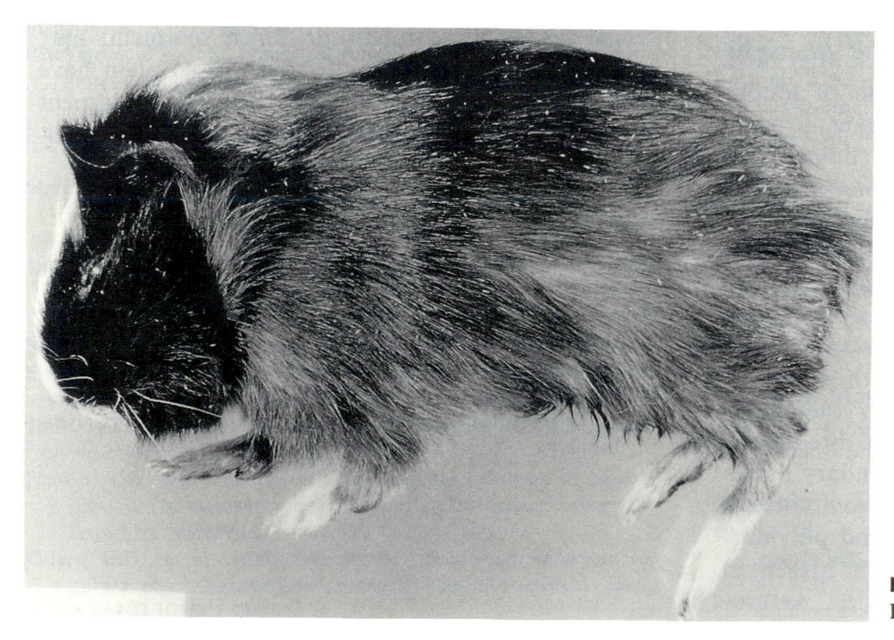

Fig. 5.19. Pediculosis due to *Gliricola porcelli*. Lice are attached to many hair shafts.

Fig. 5.20. Closer view of pellage in Fig. 5.19.

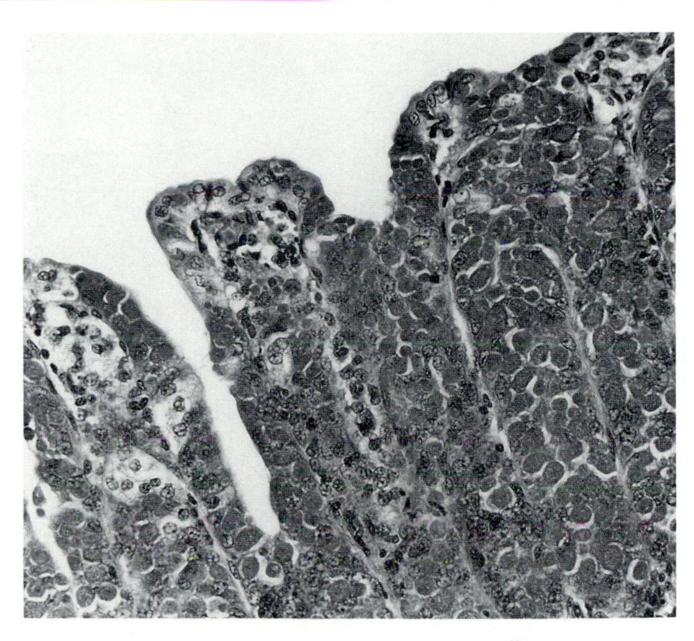

Fig. 5.21. Section of large intestine from case of spontaneous coccidiosis due to *Eimeria caviae*. Note the large numbers of macro- and microgametocytes.

Endoparasitic Infections

Protozoal Infections

Coccidiosis. In the guinea pig, coccidiosis is associated with *Eimeria caviae* infections. Following ingestion of the sporulated oocysts, sporozoites penetrate the intestinal mucosa, and schizogony is detectable by 7–8 days postinfection. Endogenous stages occur primarily in the cryptal cells of the anterior colon, although the cecum may also be involved. Diarrhea usually occurs at 10–13 days postexposure. The prepatent period is around 11 days, but severely affected animals may succumb with profuse diarrhea before oocysts are evident on fecal flotation. The time required for sporulation of oocysts to occur is from 2 to 3 days to up to 10 days. Clinical outbreaks of diarrhea occur predominantly in weanling animals. Seasonal fluctuations may occur. Mortality rates are variable but usually are relatively low (Muto et al. 1985a,b).

PATHOLOGY. At necropsy, the large intestine usually contains fluid, fetid material, sometimes with brown flecks of blood. The mucosa is congested and edematous, with variable petechial hemorrhages. Microscopic changes are characterized by colonic hyperplasia and progress to sloughing of enterocytes, edema of the lamina propria, and infiltration with polymorphonuclear and mononuclear cells. Micro- and macrogametocytes are usually present in large numbers in the cecal and colonic mucosa (Fig. 5.21).

DIAGNOSIS. The demonstration of the organisms by mucosal scrapings, histopathology, and fecal flotation will confirm the diagnosis. Deaths may occur before oocysts are evident on fecal flotation. *Differential diagnoses* include cryptosporidiosis, clostridial enteropathies, and antibiotic-induced intestinal dysbacteriosis.

SIGNIFICANCE. *E. caviae* infections are relatively common in breeding colonies of guinea pigs. The organism is regarded as moderately pathogenic, and clinical disease usually indicates a heavy infection. Improved sanitation and husbandry are essential steps in the control of the disease.

Cryptosporidiosis. *Cryptosporidium wrairi* is a recognized protozoal pathogen in the guinea pig. Clinical infections occur most frequently in juvenile animals. Infection rates of 30–40% are considered to be typical in conventional colonies. Clinical signs include diarrhea, weight loss, and emaciation. In outbreaks of the disease, morbidity and mortality rates in young animals range from negligible to up to 50% (Gibson and Wagner 1986).

PATHOLOGY. At necropsy, animals may be thin and pot-bellied, with fecal staining of the perineum. The small and large intestine usually contain watery material. Microscopically, acute lesions are usually concentrated in the jejunum, ileum, and cecum. There is hyperplasia of the crypt epithelium, edema of the lamina propria, and leukocytic infiltration. Necrosis and sloughing of enterocytes occur at the tips of the villi. In chronic lesions, villus atrophy and flattening of enterocytes commonly occur. Cryptosporidia are most numerous in acute cases. They are present within the brush border along the apices of enterocytes (Figs. 5.22 and 5.23). *Escherichia coli* has been associated with clinical cases of cryptosporidiosis.

DIAGNOSIS. Identification of the parasite by mucosal scrapings and examination by phase contrast mi-

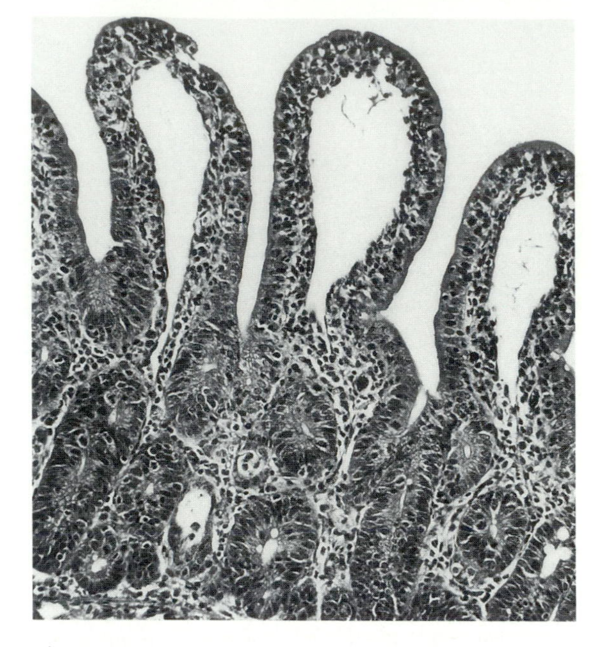

Fig. 5.22. Ileum from young guinea pig with cyrptosporidiosis. There is marked dilation of lacteals and flattening of enterocytes.

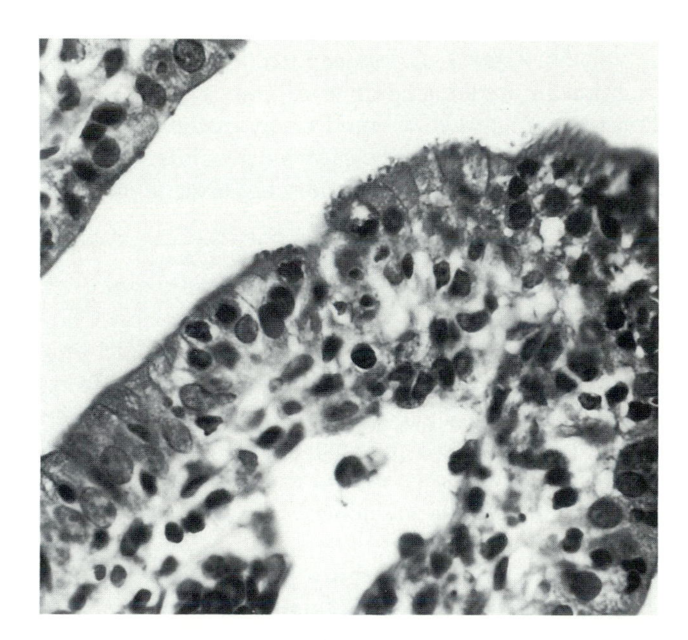

Fig. 5.23. Higher magnification of Fig. 5.22, illustrating the organisms on the mucosal surface, with mononuclear cell infiltration in the lamina propria.

croscopy is recommended. The organism may also be demonstrated in embedded sections of affected gut prepared for light or electron microscopy.

SIGNIFICANCE. Cryptosporidiosis is one recognized cause of enteritis in this species. Improved sanitation, and if necessary, sulfonamide treatment are recommended. The significance of the concurrent *E. coli* infections seen in some outbreaks (Gibson and Wagner 1986) has not been determined.

Encephalitozoonosis. Spontaneous cases of *Encephalitozoon cuniculi* infection have been recognized in the guinea pig. Multifocal granulomatous encephalitis and interstitial nephritis occur. Lesions are similar to those seen in encephalitozoonosis in other species. The presence of lesions seen histologically as an incidental finding may represent an important complication when infected animals are used in certain types of research. For additional information, see Vetterling (1976).

Toxoplasmosis. Naturally occurring infections have been reported in this species, but they rarely occur, particularly under current housing practices (Henry and Beverly 1976). Infections are frequently asyptomatic, although multifocal hepatitis and pneumonitis are possible manifestations in active infections. Cysts may be present in tissues such as myocardium and central nervous system in asymptomatic chronic infections. Animals may become infected through the ingestion of material contaminated with oocysts from *Felidae* or via the accidental injection of contaminated biological material.

Helminth Infection: *Paraspidodera uncinata*

Paraspidodera uncinata are small cecal worms up to approximately 25 mm in length that are located in the cecal and colonic mucosa. The life cycle is direct and is complete in around 65 days. No migration beyond the intestinal mucosa occurs, and infections are normally asymptomatic.

BIBLIOGRAPHY FOR PARASITIC DISEASES

Ectoparasitic Infections

Dorrestein, G.M., and Van Bronswijk, J.E.M.H. 1979. *Trixacarus caviae* as a cause of mange in guinea pigs and papular urticaria in man. Vet. Parasitol. 5:389–98.

Flynn, R.J. 1973. *Parasites of Laboratory Animals*. Ames: Iowa State University Press.

Henderson, J.D. 1973. Treatment of cutaneous acariasis in the guinea pig. J. Am. Vet. Med. Assoc. 163:591–92.

Kummel, B.A., et al. 1980. *Trixacarus caviae* infestation of guinea pigs. J. Am. Vet. Med. Assoc. 177:903–8.

Ronald, N.C., and Wagner, J.E. 1976. The arthropod parasites of the genus *Caviae*. In *The Biology of the Guinea Pig*, ed. J.E. Wagner and P.J. Manning, pp. 201–25. New York: Academic.

Rothwell, T.L.W., et al. 1991. Haematological and pathological responses to experimental *Trixacarus caviae* infection in guinea pigs. J. Comp. Pathol. 104:179–85.

Wagner, J.E., et al. 1972. *Chirodiscoides caviae* infestation in guinea pigs. Lab. Anim. Sci. 22:750–52.

Protozoal Infections

Gibson, S.V., and Wagner, J.E. 1986. Cryptosporidiosis in guinea pigs: A retrospective study. J. Am. Vet. Med. Assoc. 189:1033–34.

Moffat, R.E., and Schiefer, B. 1973. Microsporidiosis (encephalitozoonosis) in the guinea pig. Lab. Anim. Sci. 23:282–83.

Muto, T., et al. 1985a. Studies on coccidiosis in guinea pigs. 1. Clinico-pathological observation. Exp. Anim. 34:23–30.

Muto, Y., et al. 1985b. Studies on coccidiosis in guinea pigs. 2. Epizootiological survey. Exp. Anim. 34:31–39.

Vetterling, J.M. 1976. Protozoan parasites. In *The Biology of the Guinea Pig*, ed. J.E. Wagner and P.J. Manning, pp. 163–96. New York: Academic.

Toxoplasmosis

Henry, L., and Beverly, J.K.A. 1976. Toxoplasmosis in rats and guinea pigs. J. Comp. Pathol. 87:97–102.

Markham, F.S. 1937. Spontaneous toxoplasma encephalitis in the guinea pig. Am. J. Hyg. 26:193–96.

NUTRITIONAL, METABOLIC, AND OTHER DISORDERS

Scurvy (Hypovitaminosis C)

One of the major scourges of explorers for centuries, scurvy also occurred in other populations, particularly in children up to the late 19th century. Frequently scurvy was complicated by concurrent ricketts in growing children.

PATHOGENESIS. Most species synthesize ascorbic acid by the glucuronic pathway. Ascorbic acid–dependent species are genetically deficient in the enzyme L-gulonolactone oxidase, which is involved in the conversion of L-gulonolactone to L-ascorbic acid. The lack of L-gulonolactone oxidase is believed to be a genetic defect in species that normally have access to ascorbic acid in their natural diet. This biosynthetic activity occurs in the liver in mammals, but the synthesis of vitamin C occurs in the kidney in amphibians and reptiles. In addition to the inability of simian and human primates and guinea pigs to synthesize endogenous vitamin C, certain bats (e.g., Indian fruit bat), some birds (e.g., red-vented bulbul bird, northern shrike), some fish (e.g., channel catfish), and cetaceans also require supplemental vitamin C (Gillespie et al. 1980). Ascorbic acid is essential in the hydroxylase reactions necessary for the formation of hydroxyproline and hydroxylysine in the collagen molecule. Thus connective tissue cells are unable to synthesize collagen at a normal rate, resulting in deficient and defective production of interstitial osseous matrix. Vitamin C is also necessary for the catabolism of cholesterol to bile acids. In scurvy, cartilage produced in the epiphysis persists, while bone formation is suppressed. The cartilaginous lattice persists and lengthens, but it is not replaced by bone. This calcified cartilage scaffolding is relatively susceptible to mechanical forces; thus multiple microfractures occur in the epiphyseal region. Immobilization of the limb in a plaster cast will prevent the occurrence of microfractures, emphasizing the effect of the normal stresses and strains of limb movement on the development of the typical lesions (Follis 1943). There is also increased capillary fragility. There is widening of intercellular spaces between endothelial cells, vacuolar degeneration of endothelium, and depletion of subendothelial collagenous tissue (Kim 1977; Gore et al. 1965). There is also increased prothrombin time in animals with scurvy. The increased susceptibility of scorbutic guinea pigs to bacterial infections such as *Streptococcus pneumoniae* is probably due, at least in part, to impaired macrophage migration and depressed phagocytic activity of heterophils (Ganguly et al. 1976; Nungester and Ames 1948).

PATHOLOGY. At necropsy, there may be enlargement of the costochondral junctions, with hemorrhages into the regional soft tissues. Hemorrhages are present in the periarticular regions, particularly the hindlimbs (Figs. 5.24 and 5.25). Animals may be thin

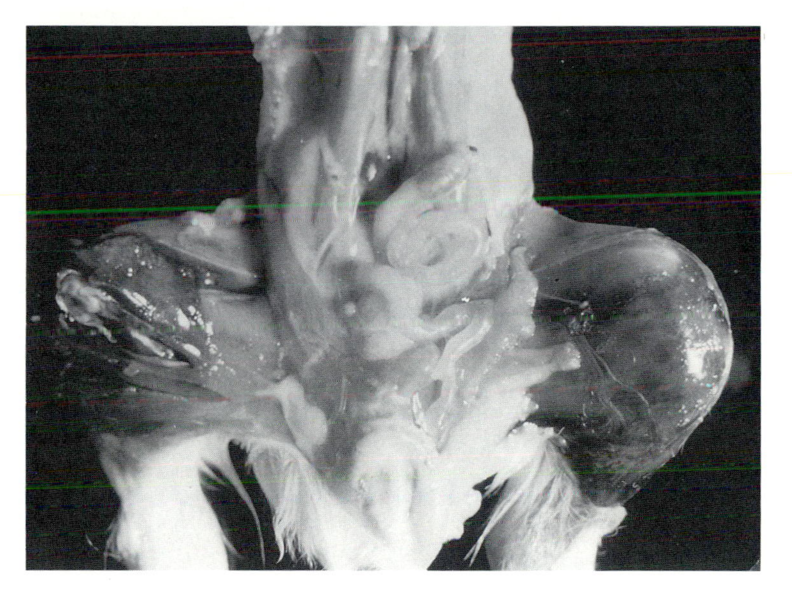

Fig. 5.24. Spontaneous scurvy in young guinea pig. Note the extensive periarticular hemorrhages.

and appear unkempt. Evidence of diarrhea is a variable finding. Occasionally blood-tinged gut contents are observed, and there may by ecchymoses in the urinary bladder. Adrenal glands are frequently markedly enlarged. Microscopically, persistence and irregularities of the epiphyseal cartilage is evident in young growing animals. Microfractures of the cartilaginous spicules and hemorrhage are common findings. There is marked proliferation of poorly differentiated fusiform mesenchymal cells in the periosteal regions and medullary cavity, with displacement of normal hematopoietic cells. Frequently there are aggregations of eosinophilic material interspersed between the mesenchymal cells (Figs. 5.26, 5.27, and 5.28). Dental abnormalities also occur. Fibrosis of the pulp and derangement of odontoblasts have been observed during the early stages of the disease. In cases of subclinical scurvy, hemosiderin-laden macrophages may be present in the lamina propria of the intestine (Clarke et al. 1980). For a more complete description of microscopic changes, see Woodard (1978).

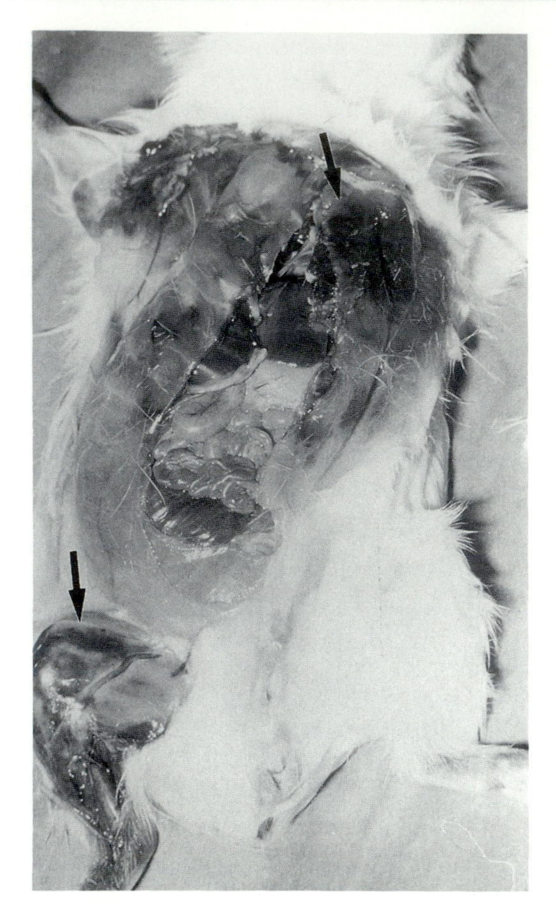

Fig. 5.25. Carcass of guinea pig with scurvy. There are prominent hemorrhages, particularly at the costochondral junctions and periarticular regions (*arrows*).

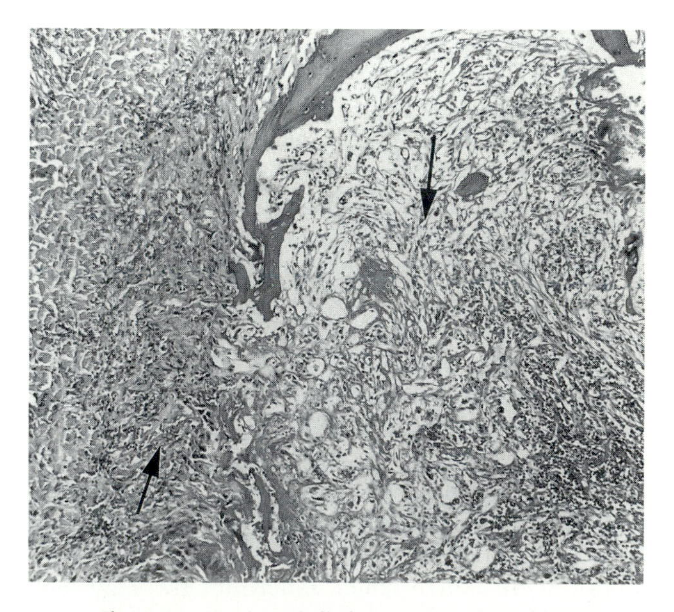

Fig. 5.26. Section of rib from a case of scurvy in guinea pig. There is marked proliferation of fusiform mesenchymal cells in the periosteal and medullary regions (*arrows*).

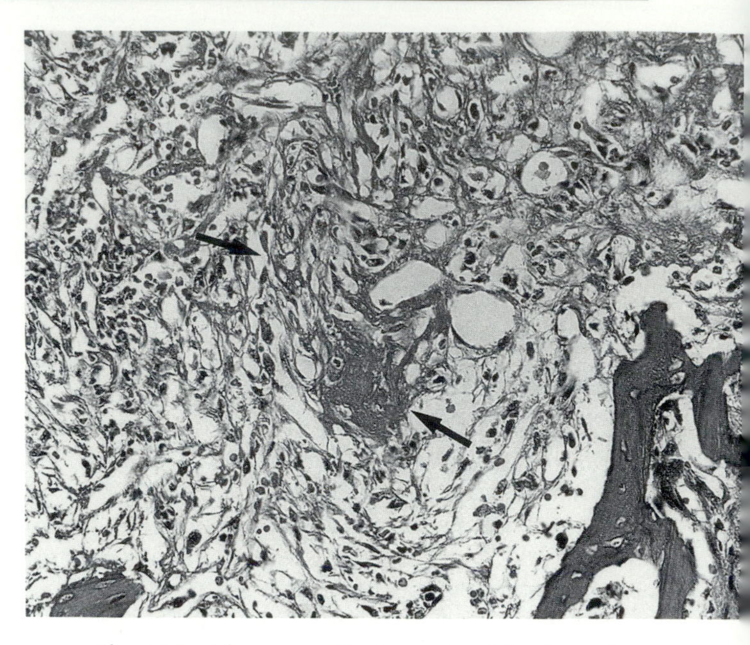

Fig. 5.27. Higher magnification of Fig. 5.26, illustrating the amorphous eosinophilic material and the typical fusiform mesenchymal cells (*arrows*).

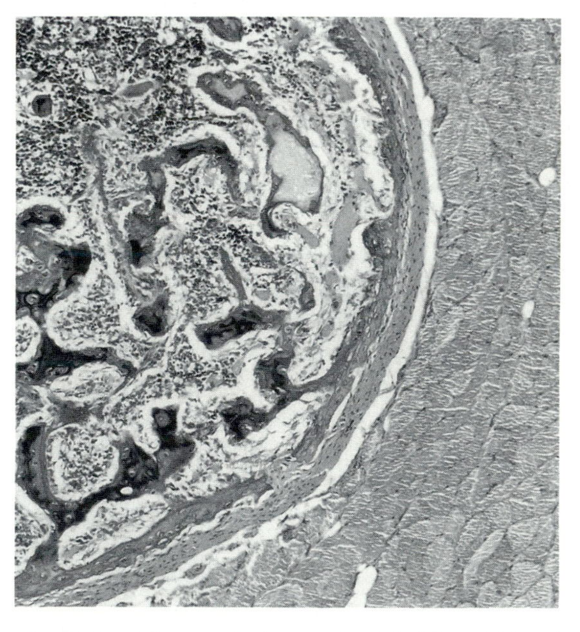

Fig. 5.28. Adjacent rib from animal in Figs. 5.26 and 5.27. The cortical architecture and adjacent soft tissue is relatively normal, emphasizing the marked variation in the degree of involvement of bony structures in one animal.

SIGNIFICANCE. Aside from the severe skeletal abnormalities and locomotor problems seen in scurvy, the disease may have significant effects on other processes, including cholesterol metabolism and resistance to bacterial infections. Prolonged clotting times and aberrations in amino acid metabolism may occur. Subclinical scurvy is identified as one important cause of diarrhea in guinea pigs. Recent pet store acquisitions may be particularly at risk. Regular dietary intake of vitamin C is essential, since animals have a limited ability to store Vitamin C. Requirements are particularly high in young growing guinea pigs and in sows during pregnancy. Congenital scurvy may also occur. Commercial ration prepared for guinea pigs should be properly stored and fed within 3 mo of the milling date to ensure that vitamin C levels are adequate, because vitamin C levels drop at a relatively rapid rate in prepared rations (Eva et al. 1976).

Myopathies

Myopathy/Myositis. Necrotizing myopathy, with necrosis of myofibers and leukocytic infiltration, has been reported in guinea pigs. Loss of cross-striations, multinucleated muscle bud formation, and variable mononuclear cell infiltration were features of the disease (Webb 1970; Saunders 1958). The etiology was not determined, although a viral infection was suggested as a likely possibility (Webb 1970). *Differential diagnoses* include nutritional muscular dystrophy and spontaneous muscular mineralization with degeneration.

Nutritional Muscular Dystrophy. Myopathy has been reported to be associated with vitamin E/selenium deficiency. Depression and conjunctivitis may be present on clinical examination. In one report, spontaneous hindlimb weakness was a prominent clinical feature of the disease. There may be marked reduction in reproductive performance in affected sows. Severely affected animals may die within 1 wk of the onset of clinical signs. Elevated serum creatine phosphokinase (CPK) is a feature of the disease. Animals may respond to alpha tocopherol therapy (Howell and Buxton 1975).

PATHOLOGY. At necropsy, there is a marked pallor of the affected muscles. Microscopic changes are characterized by coagulative necrosis and hyalinization of myofibers, fragmentation of sarcoplasm, increased basophilia of the sarcoplasm, and rowing of nuclei in regenerating myofibers. Multinucleated muscle fibers may be present in regenerating myofibers. Mineralization of myofibers was apparently not an important feature of the disease. Testicular degeneration was a later development seen in vitamin E–deficient guinea pigs (Pappenheimer and Schogoleff 1944).

Myocardial and Skeletal Muscle Degeneration with Mineralization. This is a poorly understood syndrome, and the contributing factors have not been clearly identified. Multifocal mineralization of individual muscle fibers may be seen as an incidental finding, particularly in the major muscles of the hindlimbs. Affected animals frequently are asymptomatic. On microscopic examination, there may be multifocal mineralization of skeletal muscle fibers, frequently with minimal cellular response. Myocardial degeneration with mineralization occasionally occurs. Changes are characterized by degeneration of myofibers, with variable mineralization and minimal mononuclear cell infiltration (Fig. 5.29). In chronic lesions of longer duration, there may be concurrent mineralization with fibrosis (Fig. 5.30). In one report, myocardial lesions were observed in crossbred Abyssinian/Hartley guinea pigs. Vitamin E and selenium levels were within normal limits, and genetic factors were implicated in the disease (Griffith and Lang 1987).

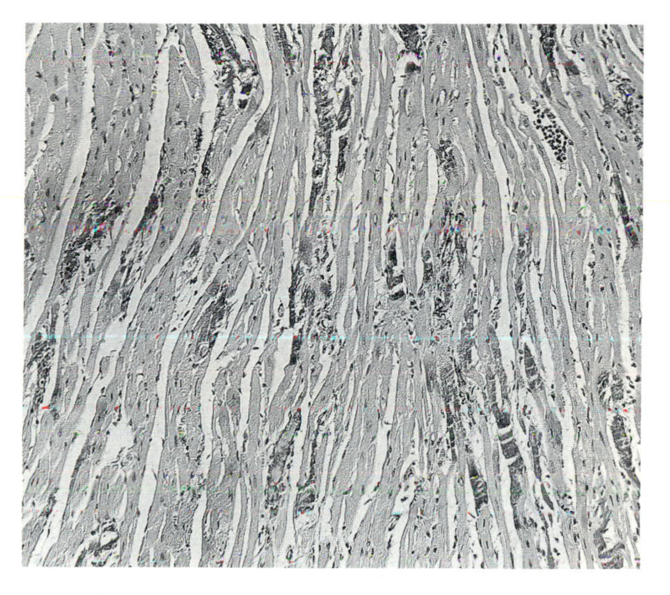

Fig. 5.29. Section of myocardium from adult guinea pig with spontaneous focal muscular degeneration and mineralization.

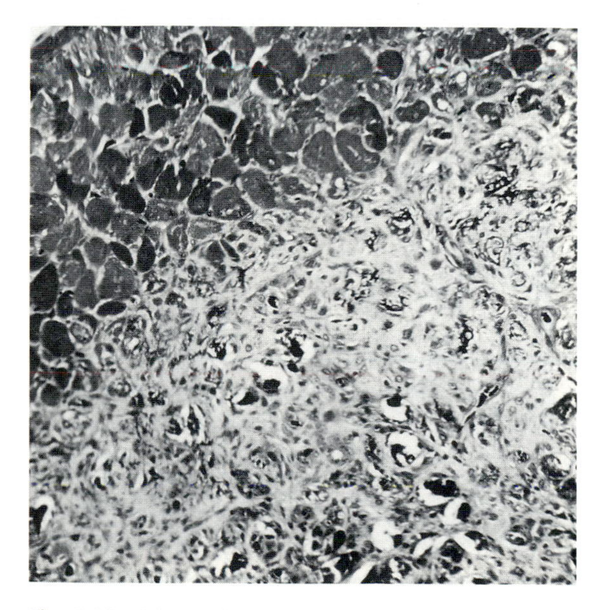

Fig. 5.30. Myocardial degeneration with extensive fibrosis and mineralization in aged guinea pig.

fluorosis, lesions were characterized by impairment of dentin and enamel formation and to excessive wear. Abnormalities of this type are not evident in typical cases of malocclusion.

Alopecia

Bilateral alopecia commonly occurs in sows in advanced pregnancy and during lactation, particularly in older animals. Nutritional and genetic factors may be involved. In pregnant animals, hair loss may be due to reduced anabolism of maternal skin during fetal growth (Wagner 1976). The hair loss frequently occurs over the back and rump, and the pellage will return to normal in due course in the typical case. *Differential diagnoses* include barbering, pediculosis, and dermatophyte infections.

Gastric Dilatation

Multiple cases of acute gastric dilatation associated with gastric volvulus have been recognized in one colony of guinea pigs (Lee et al. 1977), and it occurs sporadically in other facilities. Frequently affected animals were found dead, with no previous indication of disease. Typical cases have a 180-degree rotation along the mesenteric axis, and stomachs are distended with fluid and gas. Death has been attributed to respiratory impairment and possibly vascular shock. Contributing and/or predisposing factors have not been identified.

Cecal Torsion

Deaths due to cecal torsion are occasionally observed in this species. At necropsy, the displaced cecum is distended with fluid and gas and is edematous and hemorrhagic (Fig. 5.35).

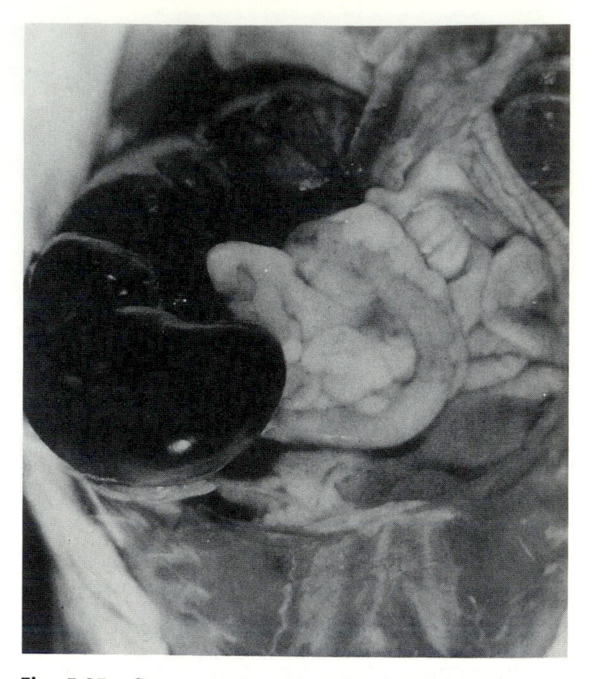

Fig. 5.35. Spontaenous cecal torsion in adult guinea pig. Note the hemorrhagic appearance of cecum associated with impaired circulation and infarction.

Focal Hepatic Necrosis

Multifocal coagulation necrosis of the liver is occasionally seen at necropsy. Affected areas are frequently subcapsular in distribution, with minimal or no inflammatory response. They are frequently interpreted to be a terminal event and may be due to hypoxic change secondary to impaired blood flow in the region. *Differential diagnoses* include bacterial hepatitis (e.g., Tyzzer's disease) and toxic change.

Chronic Idiopathic Hepatopathy

Periportal fibrosis is occasionally seen in adult guinea pigs as an enzootic problem in individual facilities. Lesions, which are usually concentrated around portal triads, are characterized by hepatocyte degeneration, proliferation of cholangioles, and interstitial fibrosis (Fig. 5.36). The changes are suggestive of a toxin-induced change, but to date the etiopathogenesis has not been resolved.

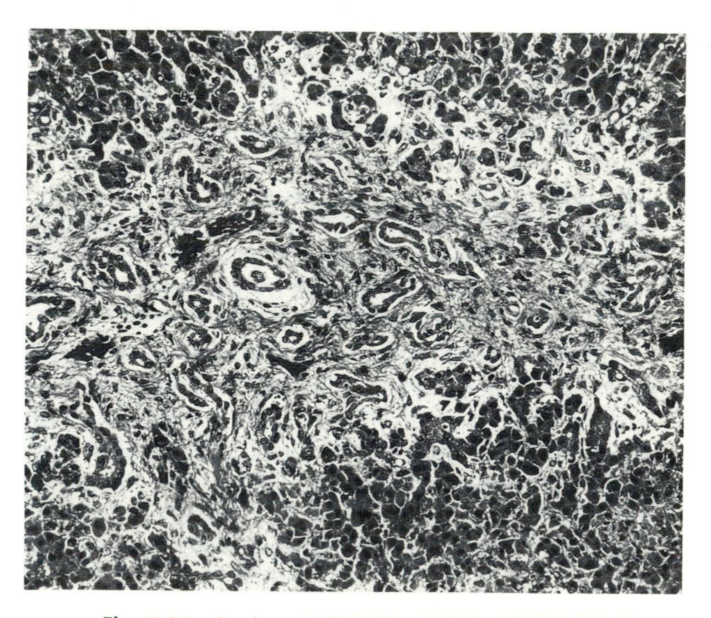

Fig. 5.36. Section of liver from adult guinea pig with spontaneous case of biliary cirrhosis of unknown etiopathogenesis. There is marked proliferation of bile ducts, with perioportal fibrosis.

Liver Contusions

Fractures of the capsule of the liver, with hemorrhage into the peritoneal cavity are occasionally observed at necropsy. Traumatic lesions of this type may be caused by events such as mishandling or falls.

Foreign Body Pneumonitis (Pneumoconiosis)

Focal pulmonary lesions associated with aspirated food or bedding occur as an incidental finding, particularly in young guinea pigs. This has been observed in guinea pigs on various bedding materials, including wood products and rice straw (Muto 1984).

PATHOLOGY. At necropsy, there may be foci of atelectasis or circumscribed nodules in the parenchyma of the lung, but frequently lesions are not detected on gross examination. On microscopic examination, in lesions of recent onset, plant fibers are lodged within small airways, with polymorphonuclear and mononuclear cell infiltration. In lesions of some duration frequently there is a focal granulomatous bronchiolitis and/or interstitial alveolitis, with mononuclear cell infiltration and foreign body multinucleated giant cell formation. Plant fibers may be identified in these areas. *Differential diagnoses* include osseous metaplasia, lesions of primary bacterial or viral origin, focal mycotic lesions (e.g., *Aspergillus* spp.), and granulomatous pulmonary lesions associated with the subcutaneous administration of Freund's adjuvant (Schiefer and Stunzi 1979).

SIGNIFICANCE. Pneumoconiosis is usually regarded as an incidental finding, particularly if lesions are circumscribed and solitary. Aspirated plant fibers can serve as a nidus for opportunistic bacterial invaders. Pneumoconiosis could complicate research data, particularly if assessment of pulmonary changes is a component of the protocol.

Adjuvant-associated Pulmonary Granulomas

Pulmonary granulomas may occur in guinea pigs and rats following subcutaneous injection with complete Freund's adjuvant (Schiefer and Stunzi 1979). Microscopic changes are characterized by multifocal granulomatous inflammatory response (Fig. 5.37). *Differential diagnoses* include perivascular lymphoid nodules, pneumoconiosis, and focal pneumonitis due to infectious agents.

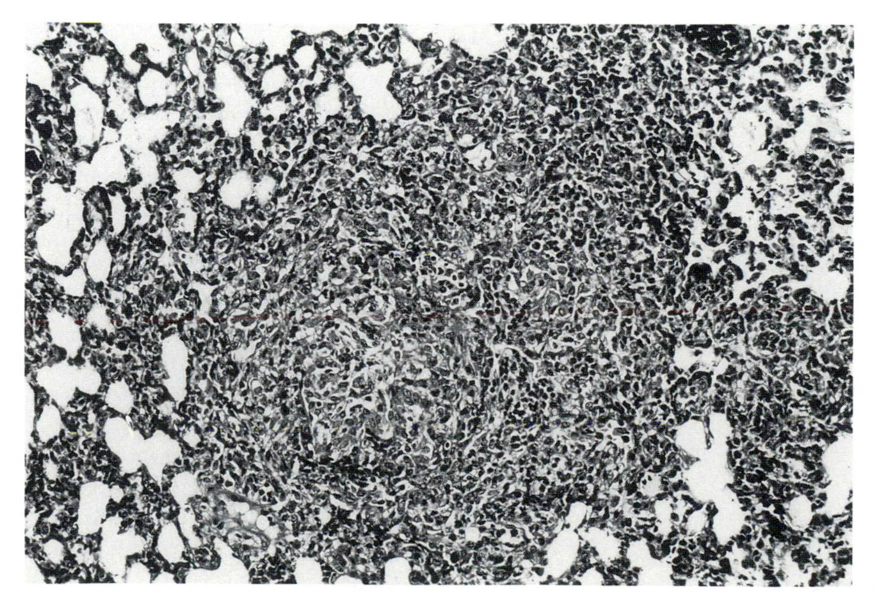

Fig. 5.37. Lung from guinea pig illustrating focal granulomas that may occur in animals treated with complete Freund's adjuvant. Note the granulomatous inflammatory response, with fibrosis and mononuclear cell infiltration.

Behavioral Diseases

Hair pulling is a common behavior pattern among guinea pigs in a group and can become an excessive activity once it is in vogue. An extension of this behavior is ear chewing, which can result in severe trauma and amputation of the ear pinnae. Frequently sexually mature boars fight when placed together, and severe lacerations or death may be the end result. Very young guinea pigs in communal housing may be trampled by older animals in group stampedes.

Bibliography for Nutritional, Metabolic, and Other Disorders

Scurvy

Clarke, G.L., et al. 1980. Subclinical scurvy in the guinea pig. Vet. Pathol. 17:40–44.

Eva, J.K., et al. 1976. Decomposition of supplementary vitamin C in diets compounded for laboratory animals. Lab. Anim. 10:157–59.

Follis, R.H. 1943. Effect of mechanical force on the skeletal lesions in acute scurvy in guinea pigs. Arch. Pathol. 35:579–82.

Gangulay, R., et al. 1976. Macrophage function in vitamin C–deficient guinea pigs. Am. J. Clin. Nutr. 29:762–65.

Gillespie, D.S. 1980. An overview of species needing vitamin C. J. Zoo. Anim. Med. 11:88–91.

Gore, I., et al. 1965. Endothelial changes produced by ascorbic acid deficiency in guinea pigs. Arch. Pathol. 80:371–76.

Kim, J.C.S. 1977. Ultrastructural studies of vascular and muscular changes in ascorbic acid–deficient guinea pigs. Lab. Anim. 11:113–17.

Nungester, W.J., and Ames, A.M. 1948. The relationship between ascorbic acid and phagocytic activity. J. Infect. Dis. 83:50–54.

Woodard, J.C. 1978. Bones. In *Pathology of Laboratory Animals*, ed. K. Benirschke et al., pp. 664–820. New York: Springer-Verlag.

Myopathies

Griffith, J.W., and Lang, C.M. 1987. Vitamin E and selenium status of guinea pigs with myocardial necrosis. Lab. Anim. Sci. 37:776–79.

Howell, J.M., and Buxton, P.H. 1975. Alpha tocopherol responsive muscular dystrophy in guinea pigs. Neuropathol. App. Neurobiol. 1:49–58.

Navia, J.M., and Hunt, C.E. 1976. Nutrition, nutritional diseases and nutrition research applications. In *The Biology of the Guinea Pig*, ed. J.E. Wagner and P.J. Manning, pp. 235–67. New York: Academic.

Pappenheimer, A.M., and Schogoleff, C. 1944. The testis in vitamin E deficiency in guinea pigs. Am. J. Pathol. 20:239–44.

Saunders, L.Z. 1958. Myositis in guinea pigs. J. Natl. Cancer Inst. 20:899–903.

Ward, G.S., et al. 1977. Myopathy in guinea pigs. J. Am. Vet. Med. Assoc. 171:837–38.

Webb, J.N. 1970. Naturally occurring myopathy in guinea pigs. J. Pathol. 100:155–62.

Metastatic Calcification

Galloway, J.H., et al. 1964. Relationship to diet and age to metastatic calcification in guinea pigs. Lab. Anim. Care 14:6–12.

Sparschu, G.L., and Christie, R.J. 1968. Metastatic calcification in a guinea pig colony: A pathological survey. Lab. Anim. Care 18:520–26.

Pregnancy Toxemia

Bergman, E.N., and Sellers, E.F. 1960. Comparison of fasting ketosis in pregnant and nonpregnant guinea pigs. Am. J. Physiol. 198:1083–86.

Ganaway, J.R., and Allen, A.M. 1971. Obesity predisposes to pregnancy toxemia (ketosis) of guinea pigs. Lab. Anim. Sci. 21:40–44.

Golden, J.G., et al. 1980. Experimental toxemia in the pregnant guinea pig (*Cavia porcellus*). Lab. Anim. Sci. 30:174–79.

Seidl, D.C., et al. 1979. True pregnancy toxemia (preeclampsia) in the guinea pig (*Cavia porcellus*). Lab. Anim. Sci. 29:472–78.

Diabetes Mellitus

Lang, C.M., et al. 1977. The guinea pig as a model of diabetes mellitus. Lab. Anim. Sci. 27:789–805.

Other Disorders

Franks, L.M., and Chesterman, F.C. 1962. The pathology of tumours and other lesions of the guinea pig lung. Br. J. Cancer 16:696–700.

Hard, G.C., and Atkinson, F.F.V. 1967. "Slobbers" in laboratory guinea pigs as a form of chronic fluorosis. J. Pathol. Bacteriol. 94:95–104.

Lee, K.J., et al. 1977. Acute gastric dilatation associated with gastric volvulus in the guinea pig. Lab. Anim. Sci. 27:685–86.

Muto, T. 1984. Spontaneous organic dust pneumoconiosis in guinea pigs. Jpn. J. Vet. Sci. 46:925–27.

O'Dell, B.L., et al. 1957. Diet composition and mineral imbalance in guinea pigs. J. Nutr. 63:65–67.

Rest, J.R., et al. 1982. Malocclusion in inbred strain-2 weanling guinea pigs. Lab. Anim. 16:84–87.

Schiefer, B., and Stunzi, H. 1979. Pulmonary lesions in guinea pigs and rats after subcutaneous injection of complete Freund's adjuvant or homologous pulmonary tissue. Zentrabl. Veterinaermed. [A] 26:1–10.

Wagner, J.E. 1976. Miscellaneous disease conditions of guinea pigs. In *The Biology of the Guinea Pig*, ed. J.E. Wagner and P.J. Manning, pp. 227–34. New York: Academic.

DISEASES ASSOCIATED WITH AGING

Segmental Nephrosclerosis

Irregularly pitted, granular renal cortices are a common finding at necropsy, particularly in guinea pigs that are at least 1 yr of age. It is frequently considered to be an incidental finding, but lesions may be extensive enough to result in renal insufficiency.

PATHOGENESIS. The pathogenesis of the segmental interstitial renal scarring has not been resolved (Bauch and Stefkovic 1986), but theories proposed have included autoimmune disease, infectious agents, and vascular disease. In one study, treatment with immunosuppressants failed to prevent the development of the disease, and gamma globulin was not demonstrated in the kidney (Takeda and Grollman 1970). The kidney lesions have been interpreted to be the result of a general vascular disturbance, resulting in focal areas of ischemia and fibrosis. In another study, using immunohistochemical techniques, glomerular changes were evaluated in guinea pigs collected from several sources. Spontaneous deposits of IgG and complement (C3) were demonstrated along the mesangial and peripheral glomerular basement membranes. It was suggested that the antigen-antibody complexes might be due to an infectious agent or endogenous tissue antigen (Steblay and Rudofsky 1971). In a search for possible infectious agents, cultures for bacteria are usually unrewarding. In one study, a herpesvirus was isolated from affected kidneys. However, the virus was also recovered from normal kidneys, and the renal lesions in inoculated guinea pigs were not reproduced (Bhatt et al. 1971). An accelerated disease process in guinea pigs fed an unusually high protein diet has been observed. Elevated blood pressures consistent with a mild degree of hypertension have been recorded in animals with renal lesions (Takeda and Grollman 1970).

PATHOLOGY. At necropsy, there are multiple, granular, pitted areas on the surface of the kidney, resulting in irregular contours in severely affected animals (Fig. 5.38). On cut surface, pale linear streaks extend down into the cortex, with some involvement of the medulla in advanced lesions. On microscopic examination, there is segmental to diffuse interstitial fibrosis, with distortion and obliteration of the normal architecture (Figs. 5.39 and 5.40). Tubular lesions are concentrated primarily in the regions of the convoluted tubules and Henle's loop. Scattered tubules are dilated and lined by poorly differentiated, cuboidal to squamous epithelial cells. Some nephrons, interpreted to be nonfunctional, consist of tubular remnants lined by poorly differentiated cuboidal epithelium with lightly eosinophilic to amphophilic cytoplasm. Tubules are occasionally dilated and contain proteinaceous material and cellular debris. In nephrons interpreted

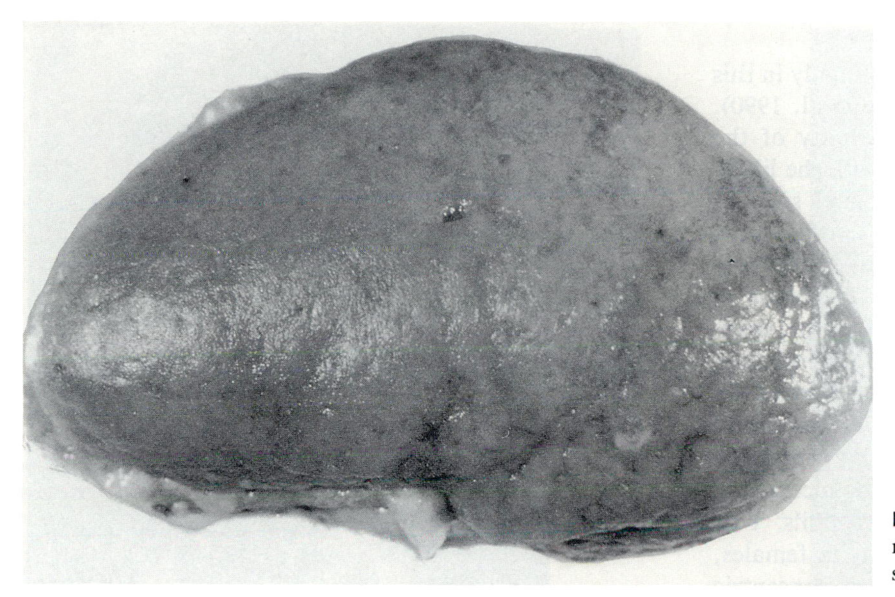

Fig. 5.38. Kidney from aged guinea pig with nephrosclerosis. A finely granular, pitted cortical surface is evident macroscopically.

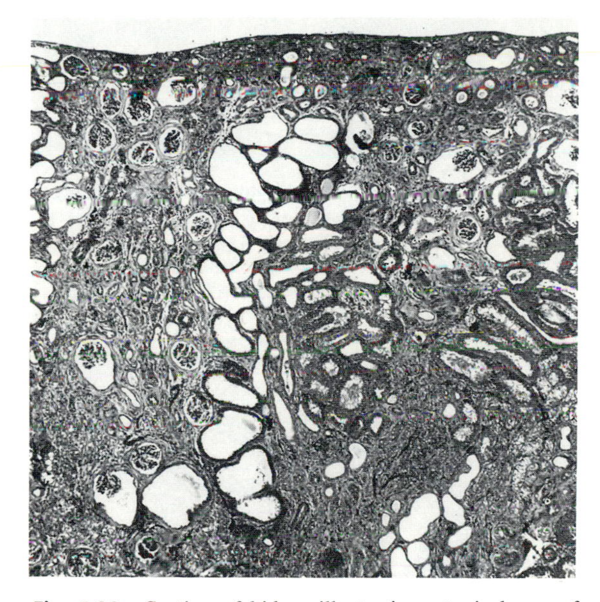

Fig. 5.39. Section of kidney illustrating a typical case of segmental nephrosclerosis. Note the interstitial fibrosis and the convoluted tubules lined by poorly differentiated epithelial cells.

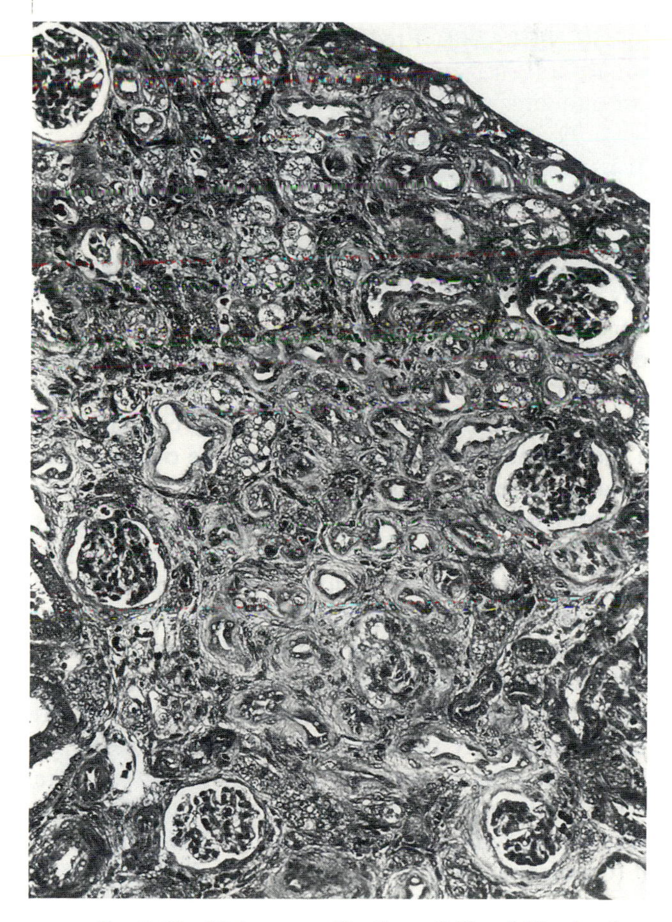

Fig. 5.40. Higher magnification of Fig. 5.39, showing thickening of basement membranes and marked interstitial fibrosis.

to be fully functional, convoluted tubules are lined by hypertrophied epithelial cells with abundant, eosinophilic cytoplasm. Most glomeruli are essentially normal histologically. Occasionally there is atrophy of individual glomeruli, with regional fibrosis. In advanced lesions, there is diffuse to segmental infiltration with fibroblasts and collagenous tissue formation. There are minimal focal aggregations of mononuclear cells consisting mainly of lymphocytes. Arterioles and arteries may have moderate medial hypertrophy, sometimes with prominent endothelial lining cells.

SIGNIFICANCE. Advanced cases of nephrosclerosis can cause disability in older guinea pigs. High BUN and serum creatinine values, nonregenerative anemia, and low urinary-specific gravity are clinical findings recorded in animals with advanced nephrosclerosis. The pathogenesis of the renal changes has not been resolved.

BIBLIOGRAPHY FOR DISEASES ASSOCIATED WITH AGING

Bauch, L., and Stefkovic, G. 1986. Searching the records for clues about kidney disease in guinea pigs. Vet. Med. 81:1127–30.
Bhatt, P.N., et al. 1971. Isolation and characterization of a herpes-like(Hsiung-Kaplow) virus from guinea pigs. J. Infect. Dis. 123:178–89.
Keller, L.S.F., and Lang, C.M. 1987. Reproductive failure associated with cystic rete ovarii in guinea pigs. Vet. Pathol. 24:335–39.
Peng, X., et al. 1990. Cystitis, urolithiasis and cystic calculi in aging guinea pigs. Lab. Anim. 24:159–63.
Quattropani, S.L. 1977. Serous cysts in aging guinea pig ovary: Light microscopy and origin. Anat. Rec. 188:351–60.
Steblay, R.W., and Rudofsky, U. 1971. Spontaneous renal lesions and glomerular deposits of IgG and complement in guinea pigs. J. Immunol. 107:1192–96.
Takeda, T., and Grollman, A. 1970. Spontaneously occurring renal disease in the guinea pig. Am. J. Pathol. 40:103–17.

NEOPLASMS

Spontaneous tumors are rare in guinea pigs under 3 yr of age and uncommon even in older animals. There appear to be variations in genetic susceptibility to spontaneous neoplasia. A serum factor (probably asparaginase) has been demonstrated in the sera of normal guinea pigs that has antitumor activity (Wriston and Yellin 1973). In addition, splenic preparations containing large numbers of Foa-Kurloff cells have been shown to inhibit transformed human epithelial cells in vitro (Bimes et al. 1981).

Cavian Leukemia

On rare occasions, leukemia occurs as a spontaneous disease in various inbred and noninbred strains of guinea pigs. Cases are most frequently seen in young adult animals. Leukocyte counts in the peripheral blood are relatively high, varying from 50,000 to over 200,000/mm. Leukemia has been produced experimentally with transplanted cells and cell-free filtrates. Leukocyte counts of up to 350,000 mm have been observed (Opler 1968). Although retrovirus appears to play an important role in this disease, C-type virus particles have also been observed in the germinal centers of lymph nodes from normal guinea pigs.

PATHOLOGY. Leukocytosis (up to 180,000 or greater) with a preponderance of lymphoblastic cells is the typical picture seen in blood samples. At necropsy, lymph nodes, such as cervical, axillary, and inguinal, are enlarged and firm, homogeneous, and tan on cut surface. There is marked splenomegaly and hepatomegaly. Microscopically, there is usually moderate to marked infiltration of lymphoblastic cells in the spleen, liver, bone marrow, interstitium of the lung, thymus, alimentary tract lymphoid tissue, heart, eyes, and adrenals. (Figs. 5.47 and 5.48).

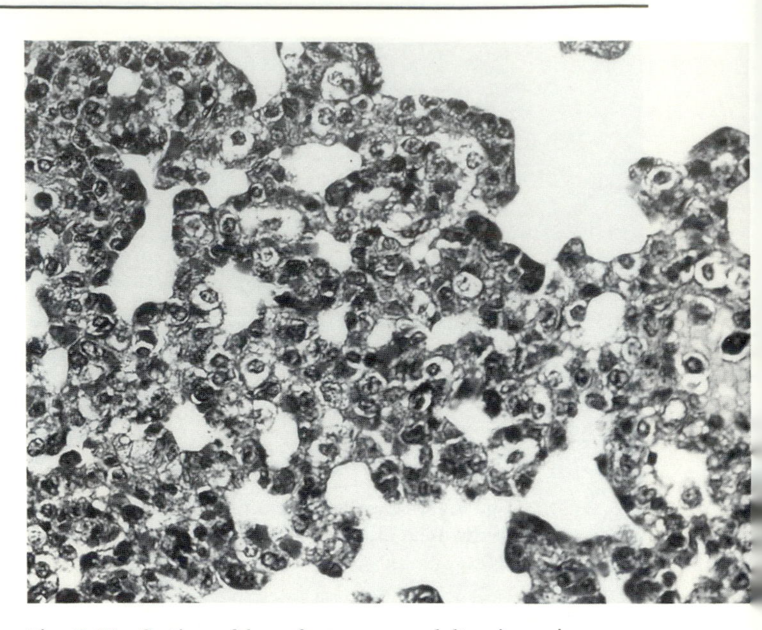

Fig. 5.47. Section of lung from young adult guinea pig with cavian leukemia. Note the marked hypercellularity of the alveolar septa.

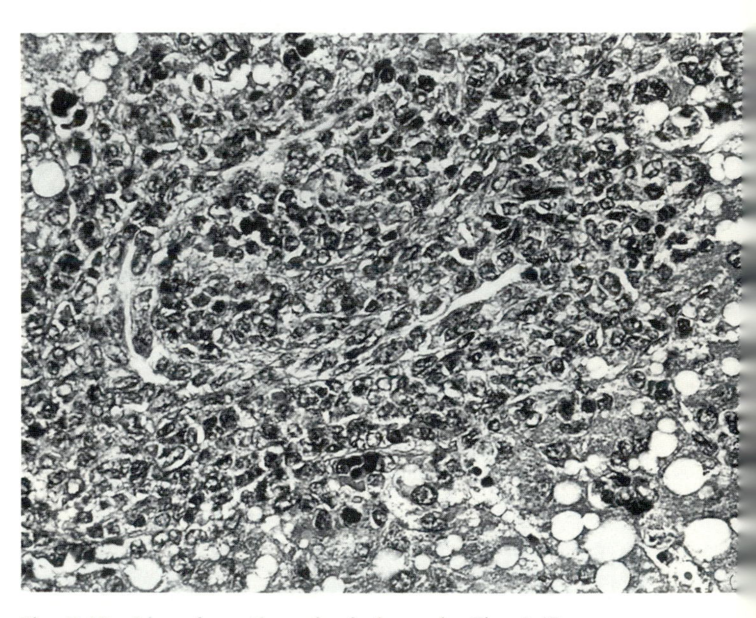

Fig. 5.48. Liver from the animal shown in Fig. 5.47. There is a marked periportal infiltrate consisting of poorly differentiated lymphopoietic cells.

SIGNIFICANCE. Guinea pig leukemia is another example of a retrovirus-associated malignancy of the lymphopoietic system.

Tumors of the Reproductive Tract

Tumors of the reproductive tract represent approximately 25% of spontaneous tumors in this species. Of the ovarian tumors, the majority are teratomas. A variety of tissue types may be evident in these tumors, including ciliated and mucous epithelial cells, striated muscle, and cells of ectodermal origin. These tumors should not be confused with cystic rete tubules seen commonly in older sows. Uterine tumors are primarily benign, and of mesenchymal origin. Most are fibromas, or leiomyomas (see Figs. 5.44 and 5.45). Rarely, malignant uterine tumors such as myxosarcomas or leiomyosarcomas have been described. Primary malignant uterine tumors may consist of poorly differentiated mesenchymal cells, with extension into the peritoneal cavity. Mammary adenocarcinomas occur in both male and female guinea pigs. The majority are interpreted to be of ductal origin (Fig. 5.49). Metastases may occur to regional lymph nodes. Some are of low-grade malignancy and remain localized to the original site.

Pulmonary Tumors

Pulmonary tumors represented approximately 35% of reported tumors in one survey. The majority were benign papillary adenomas, and most were interpreted to be of bronchogenic origin. The changes were similar to those produced by certain infectious agents, and it was suggested that there may be hyperplastic and adenomatous changes in airways and alveoli in response to various stimuli, not bona fide primary pulmonary tumors (Manning 1976). Small, white, circumscribed nodules, visible macroscopically, on microscopic examination consist of papillary structures lined by a single layer of hyperchromatic cuboidal epithelium (Figs. 5.50 and 5.51). Primary malignant tumors of the lung are rare in guinea pigs.

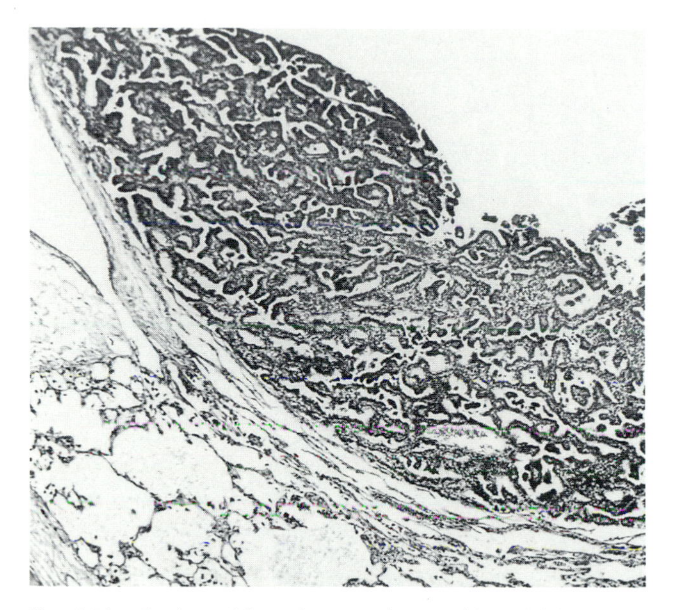

Fig. 5.50. Section of lung from aged sow with multiple bronchia adenomas. Affected airways are lined by poorly differentiated respiratory epithelial cells.

Tumors of the Skin

Skin tumors represent a small percentage of neoplasms reported in this species. The majority are benign. Trichoepitheliomas are the most common tumors of the skin (Figs. 5.52 and 5.53). Sebaceous gland adenomas, penile papillomas, lipomas, fibrosarcomas, fibromas, and carcinomas have also been described.

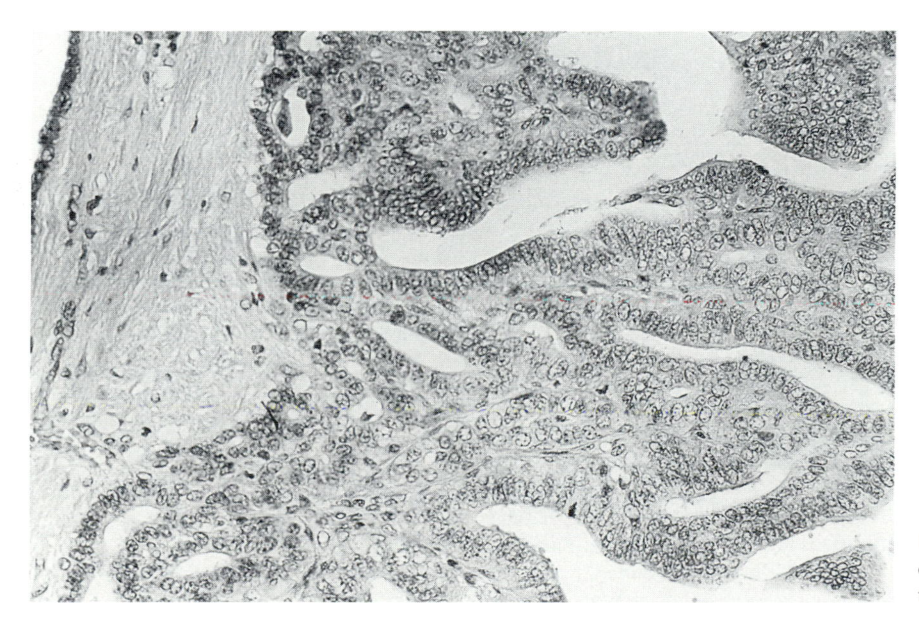

Fig. 5.49. Section from mammary adenocarcinoma in older sow. Mammary tumors of this type also occur occasionally in males.

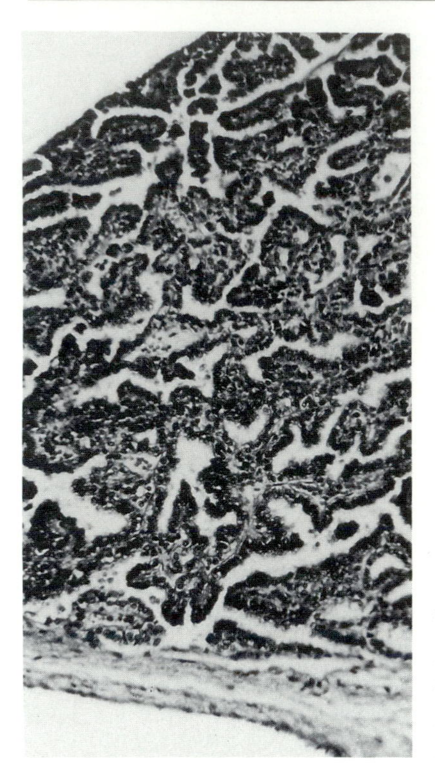

Fig. 5.51. Higher magnification of Fig. 5.50, illustrating the papillary pattern and the cuboidal epithelial cells lining the underlying stroma.

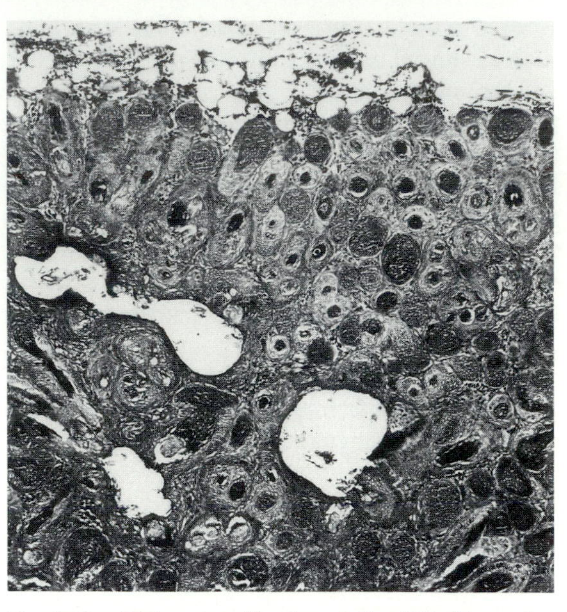

Fig. 5.53. Higher magnification of Fig. 5.52, illustrating the numerous follicular structures within the mass.

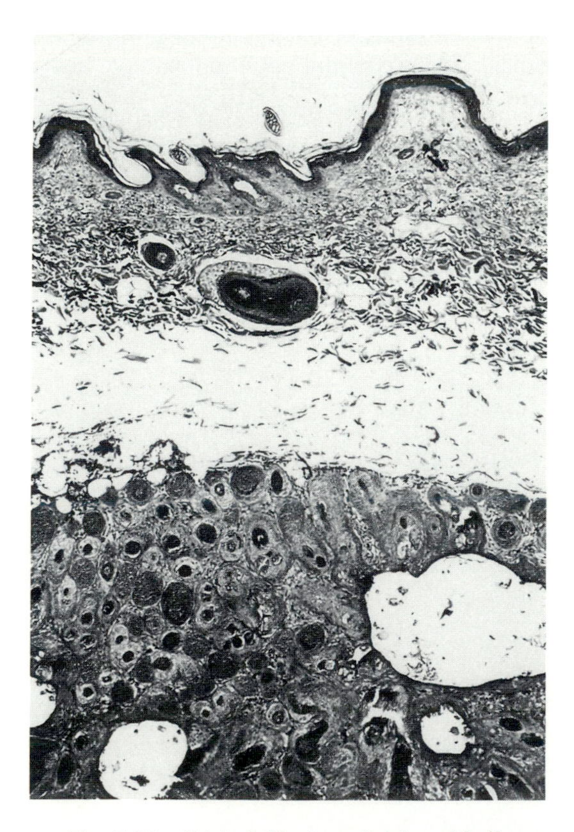

Fig. 5.52. Trichofolliculoma/trichoepithelioma from adult guinea pig. Note the aggregations of follicular structures in the subcutis and the intact epidermis overlying the tumor.

Tumors of the Endocrine System

Tumors of the endocrine system occur in guinea pigs; these include benign adrenocortical tumors. Benign mixed tumors (myxomas) are the most commonly reported tumors of the cardiovascular system. They may include well-differentiated mesenchymal components, such as cartilage, bone, and fat. Primary myocardial tumors should not be confused with rhabdomyomatosis, a congenital condition characterized by vacuolation of myofibers and glycogen deposition (see Rhabdomyomatosis, discussed in Anatomic Features). For an excellent summary of primary cavian tumors, see Manning (1976).

Bibliography for Neoplasms

Bimes, C., et al. 1981. Demonstration of a principle ensuring destruction of human cancer cells in culture. C.R. Hebd. Seances Acad. Sci. 292:293–98.

Field, K.J., et al. 1989. Spontaneous reproductive tract leiomyomas in aged guinea pigs. J. Comp. Pathol. 101:287–94.

Hong, C.C., et al. 1980. Naturally occurring lymphoblastic leukemia in guinea pigs. Lab. Anim. Sci. 30:222–26.

Jungeblut, C.W., and Opler, S.R. 1967. On the pathogenesis of cavian leukemia. Am. J. Pathol. 51:1153–60.

Manning, P.J. 1976. Neoplastic diseases. In *The Biology of the Guinea Pig*, ed. J.E. Wagner and P.J. Manning, pp. 211–25. New York: Academic.

Opler, S.R. 1967. Pathology of cavian leukemia. Am. J. Pathol. 51:1135–47.

Prattis, S.M., et al. April 1989. Cutaneous penile mass in a guinea pig. Lab. Anim. (USA) 18:15–17.

Wriston, J.C., and Yellin, T.O. 1973. L-asparaginase: A review. Adv. Enzymol. 39:185–248.

Zwart, P., et al. 1981. Cutaneous tumors in the guinea pig. Lab. Anim. 15:375–77.

6 RABBIT

Rabbits are classified as members of the order Lagomorpha. An additional pair of incisor teeth are present directly behind the large incisors on the upper jaw, thus placing them in a different category than the order Rodentia. Domestic rabbits used in the research laboratory and for meat production in the commercial rabbitry are descendants of the European wild rabbit, *Oryctolagus cuniculus*. Their size and readily accessible vessels of the ear make them ideal candidates for antibody production and blood collection. Under natural conditions, the blood flow to their large ears also represents the primary means of temperature control in the domestic rabbit. There are over 100 breeds of rabbits recognized on a world basis. However, the majority of rabbits seen in the research facility are the New Zealand White breed. Does are induced ovulators, and following parturition ("kindling"), usually nurse the kits only once daily. The practice of cecotrophy, or reingestion of the mucus-coated "night feces," occurs daily and appears to be one method of recycling these cecotrophs, which are relatively high in protein and B complex vitamins (Cheeke 1987). Domestic rabbits in the laboratory animal facility or the commercial rabbitry are frequently nervous and easily frightened. Careful, firm handling, including support of the hindlegs, is essential in order to avoid unrestrained kicking, which may result in accidental fracture of the vertebral column.

ANATOMIC FEATURES

Hematology

Erythrocytes are approximately 6.5–7.5 μm in diameter, and polychromasia may be evident in a few erythrocytes. The normal value for reticulocytes in adult rabbits is 2–5%, and the estimated mean life span for erythrocytes in circulation is approximately 50 days. Heterophils are the counterpart of neutrophils that are present in most mammals. They are 9–15 μm in diameter and have distinct, acidophilic granules present in the cytoplasm. Eosinophils are 12–16 μm in diameter, with large cytoplasmic granules that stain a dull pink-orange with conventional hematology stains. Basophils may be relatively numerous and occasionally represent up to 30% of circulating leukocytes (Sanderson and Phillips 1981). Lymphocytes are usually the predominant leukocyte in the peripheral blood of domestic rabbits. Small lymphocytes are approximately 7–10 μm in diameter; large lymphocytes may vary from 10–15 μm in diameter and sometimes have a few azurophilic cytoplasmic granules. Hematological values have been compared in healthy rabbits and in animals with various disease states (Toth and Krueger 1989; Hinton et al. 1982).

Alimentary Tract

As with other herbivores, rabbits have a large and relatively complex digestive system. They are hind gut fermenters. Fine particulate materials are selectively channeled into the cecum during passage through the large intestine, while larger particulate material is usually directed into the colon and passed as fecal pellets. The sacculus rotundus is a spherical, thick-walled enlargement at the ileocecal junction (Fig. 6.1). The increased thickness of the wall is due to the aggregations of lymphoid tissue and macrophages in the lamina propria and submucosa. The appendix constitutes the tip of the cecum. It is thick-walled and has abundant lymphoid tissue, which accounts for the marked thickness evident on gross examination. The gut-associated lymphoid tissue (GALT) represents over 50% of the total mass of lymphoid tissue in the body, which accounts for the relatively small spleen seen macroscopically in this species.

taneous myxoid mass, usually within 3–4 days. Within 6–8 days, mucopurulent conjunctivitis, subcutaneous edema, and multiple subcutaneous skin tumors are usually observed. In rabbits that die with a peracute form of the disease, the animal may be found dead, and other than redness of the conjunctiva, there may be no other evidence of disease.

PATHOLOGY. Microscopically, in the subcutaneous masses there is proliferation of large, stellate mesenchymal cells ("myxoma cells") interspersed within a homogeneous matrix of mucinous material, with a sprinkling of inflammatory cells (Fig. 6.3). Hypertrophy and proliferation of endothelial cells occur, and changes in the epithelium overlying the lesions may vary from hyperplasia to degeneration. Intracytoplasmic inclusions may be present in the affected epidermis (Fig. 6.4) and in epithelial cells of the conjunctiva. Proliferation of alveolar epithelium, hypertrophy and hyperplasia of reticulum cells in lymph nodes and spleen, focal necrosis, hemorrhage, and proliferative vasculitis have also been described. Lymphoid depletion of the spleen is a common finding.

DIAGNOSIS. The presence of characteristic gross and microscopic lesions in rabbits of the genus *Oryctolagus*, particularly in areas where the disease is known to occur, are useful diagnostic criteria. Demonstration of the virus by intracutaneous inoculation of young susceptible rabbits with suspect material, or recovery and identification of the virus from lesions, following inoculation onto chorioallantoic membrane or cell culture, are procedures used to confirm the diagnosis (Mare 1974).

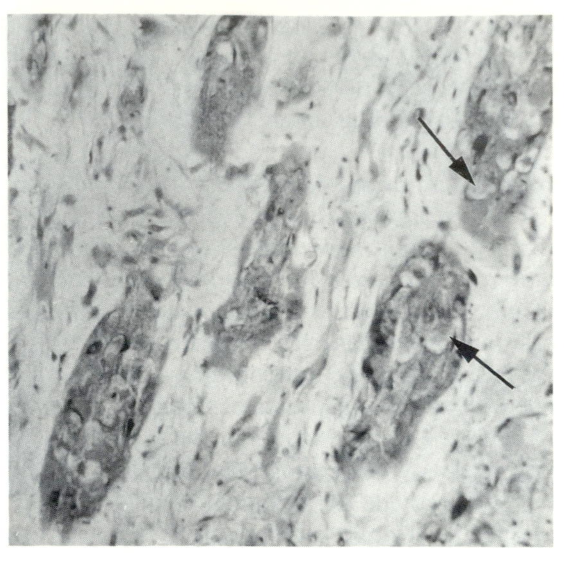

Fig. 6.4. Higher magnification of skin from myxomatosis lesion. Intracytoplasmic inclusions (*arrows*) are present in many cells of follicular epithelium. (Courtesy of J.A. Yager)

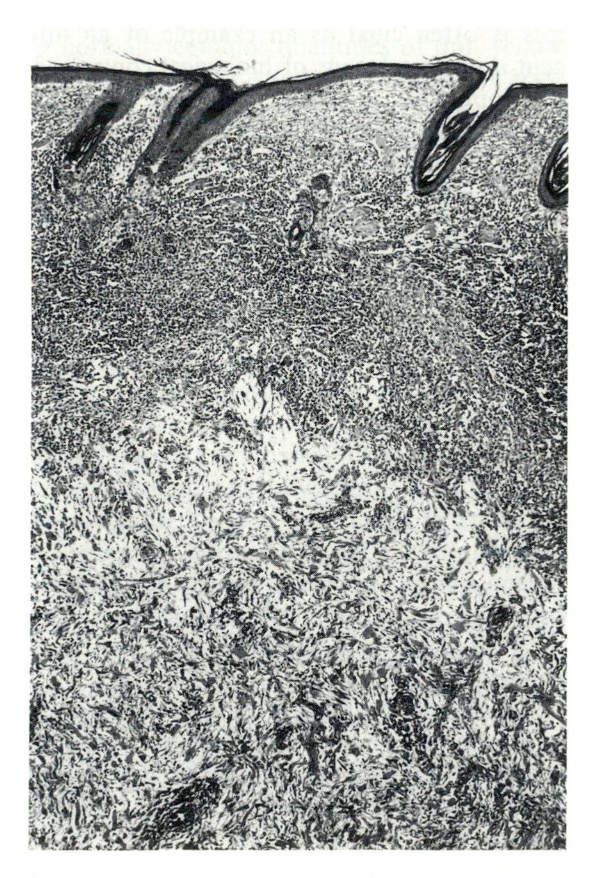

Fig. 6.3. Subcutaneous myxoid lesion in spontaneous case of myxomatosis in New Zealand White rabbit. (Courtesy of N.I. Patton).

SIGNIFICANCE. In confirmed cases of myxomatosis, identification of the source of the virus, including possible wildlife reservoirs, and the elimination and exclusion of insect vectors are primary considerations.

Rabbit (Shope) Fibromatosis. The tumor-producing, transmissible agent was first isolated from a cottontail rabbit (*Sylvilagus floridanus*) in the United States in 1932. The virus is transmissible to European rabbits (*Oryctolagus cuniculus*) and cottontails, producing localized fibromas. Shope fibroma virus infections are relatively widespread in wild cottontail rabbits in the United States and Canada. It is regarded as a benign, self-limiting disease in the wildlife population. The virus may persist for several months within lesions, and mechanical transmission by arthropod vectors appears to be the primary means of spread. Shope fibromatosis has been diagnosed in an outdoor rabbitry in Texas. Wild cottontail rabbits in the area were considered to be the most likely reservoir host, with spread by insect vectors (Joiner et al. 1971).

ETIOLOGY. Rabbit fibroma virus is a poxvirus closely related antigenically to myxomatosis virus and to the hare and squirrel fibroma viruses.

PATHOLOGY. In naturally infected cottontail and European rabbits, firm, flattened tumors occur on the legs and feet, sometimes with involvement of the muzzle, periorbital, and perineal regions. These subcutaneous tumors may be up to 7 cm in diameter, are usually freely movable, and may persist for several months. In young rabbits, metastases to abdominal viscera and bone marrow may also occur.

MICROSCOPIC FEATURES. There is localized fibroblast proliferation, with mononuclear and polymorphonuclear cell infiltration. Fibroblasts characteristically are fusiform to polygonal. In European domestic rabbits (*O. cuniculus*), subcutaneous masses may vary from myxoid in type to typical fibromas. Intracytoplasmic, eosinophilic inclusion bodies may be present in reactive mesenchymal cells (Fig. 6.5) and in epidermal cells overlying the tumors. *Differential diagnoses*: The typical gross and histological appearance of the circumscribed fibrous masses should enable the differentiation of Shope fibroma from myxomatosis and from the raised, horny, epidermal growths seen in papillomatosis. However, the myxoid forms sometimes seen histologically can be confused with myxomatosis. Considerable variation in microscopic appearance may occur.

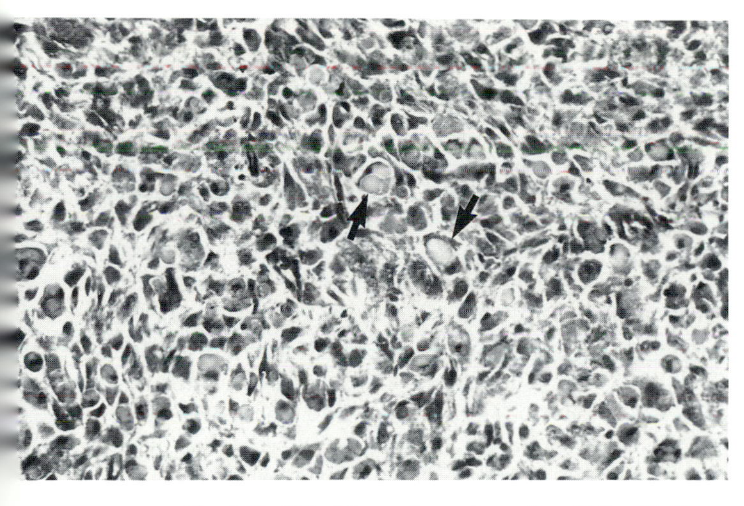

Fig. 6.5. Skin biopsy from domestic rabbit with Shope fibromatosis. There is a dense network of fusiform to polyhedral fibroblasts in the dermis, with prominent intracytoplasmic inclusion bodies (*arrows*).

Rabbit Pox. The disease is relatively rare and is characterized by the presence of localized to confluent papules on the skin, sometimes accompanied by necrosis and hemorrhage. Papular lesions may also occur in the oropharynx, respiratory tract, spleen, and liver (Mare 1974). In the "pockless" form, a few pocks were present in the oral cavity, and focal hepatic necrosis, pleuritis, and splenomegaly were also observed. Histologic changes described have included focal necrosis with leukocytic infiltration in the skin and affected viscera, and necrosis of lymphoid tissue. The diagnosis has been confirmed by virus isolation and fluorescent antibody procedures. In a comparison of the pathogenesis of the Utrecht and the Rockefeller strains of rabbit pox, it was concluded that the primary site for viral replication in the naturally occurring disease is the respiratory tract. There is a subsequent viremia, with replication in lymphoid tissues and skin (Bedson and Duckworth 1963).

Papillomaviral Infections

Rabbit (Shope) Papillomatosis

EPIZOOTIOLOGY. Primarily a benign disease of cottontail rabbits, this disease has occurred in spontaneous outbreaks in domestic rabbits. Insect vectors are probably the usual means of mechanical spread of the causative agent, a papillomavirus, from cottontail rabbits to domestic rabbits.

PATHOLOGY. The papillomas, which occur most frequently on the eyelids and ears, consist of a pedunculated, cornified surface overlying a fleshy central area. The histologic findings are consistent with a squamous papilloma (Fig. 6.6).

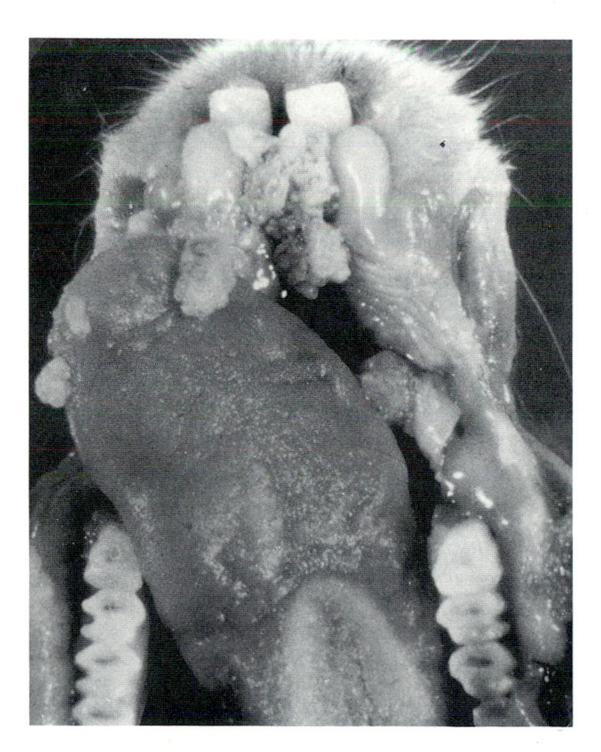

Fig. 6.6. Multiple oral papillomas in a juvenile New Zealand White domestic rabbit. There are multiple papillomas involving both the tongue and buccal cavity.

Oral Papillomatosis

EPIZOOTIOLOGY. The causative agent, rabbit oral papillomavirus, is a member of the family Papovaviridae, and is the only member of this group with the domestic rabbit as the natural host.

PATHOLOGY. The virus is probably spread by direct contact, and oral abrasions may permit the initial entry. Lesions most frequently occur in rabbits between 2 and 18 mo of age. Pedunculated lesions most commonly occur along the ventral aspect of the tongue, and usually they regress spontaneously within a few weeks (Mare 1974). They are typical squamous papillomas on microscopic examination. Basophilic intranuclear inclusions and viral antigen may be present in the stratum spinosum. Fibromas have been produced in hamsters inoculated with homogenates of oral papilloma material collected from rabbits (Sundberg et al. 1985).

Adenoviral Enteritis

Adenoviral infections have been identified in commercial rabbits in Hungary. Peak losses occurred at 6–8 wk of age. Profuse diarrhea was observed in severely affected animals, with low mortality. There was a dramatic increase in the numbers of *Escherichia coli* in the small intestine and cecum in rabbits that succumbed to the disease (Bondon and Prohaska 1980).

PATHOLOGY. Severely affected animals are dehydrated, with fluid contents in the cecum.

DIAGNOSIS. The adenovirus was isolated from the intestinal wall and gut contents, spleen, kidney, and lung when inoculated onto rabbit kidney cell cultures. A significant rise in adenoviral antibody levels was detected in convalescent sera.

SIGNIFICANCE. To date, confirmed cases of adenoviral enteritis in rabbits appear to be confined to Europe. In the outbreak described, a marked increase in coliforms was observed, indicating that *E. coli* may have played a significant role in the disease process.

Parvoviral Infection

A parvovirus has been isolated from clinically normal rabbits in Japan. In a serological survey of commercial rabbits, approximately 60% of animals evaluated had antibody to lapine parvovirus (Matsunaga et al. 1977). Oral or intravenous inoculation of young rabbits with lapine parvovirus may result in transient depression and anorexia, with no mortality. The virus may be isolated from a variety of organs for up to 2 wk postinoculation and from the small intestine on day 30 postinoculation. On microscopic examination, a mild to moderate enteritis was present in the small intestine, with exfoliation of enterocytes. In a survey of laboratory rabbits from commercial and private sources in the United States, the majority had relatively high antibody titers to lapine parvovirus. In addition, rabbit parvovirus was isolated from the kidneys of neonatal rabbits (Metcalf et al. 1989). Thus the inadvertent use of parvovirus-contaminated rabbits or primary cell cultures represents a potential complicating factor when used in research.

SIGNIFICANCE. The role, and importance (if any), of lapine parvovirus in the enteritis complex currently is unknown.

Herpesviral Infections

***Herpesvirus sylvilagus* Infection.** *H. sylvilagus* was first isolated from primary kidney cell cultures harvested from weanling cottontail rabbits (Hinze 1971b). Inoculation of young cottontail rabbits with the virus by the parenteral route produces a chronic infection with persistent viremia. Within 6–8 wk following inoculation, changes consistent with a lymphoproliferative disease are observed grossly and microscopically. Histological examination reveals alterations, which may vary from lymphoid hyperplasia to lymphosarcoma. In the malignant form of the disease, diffuse infiltration of various tissues with immature lymphocytes commonly occurs. *H. sylvilagus* infection has been proposed as a model to compare the changes that occur in Epstein-Barr virus infections in humans. Although *H. sylvilagus* replicates in kidney cells prepared from the domestic rabbit, attempts to infect New Zealand White rabbits with the virus have been unsuccessful (Hinze 1971a,b).

Herpesvirus-like Viral Infections. Systemic infections with a herpesvirus-like agent have been observed in several commercial rabbitries in Canada. The disease, which affects animals of various ages, is characterized by sudden onset. Frequently animals are found dead, with no previous evidence of disease. At necropsy, multiple hemorrhages are present in the skin, abdominal viscera, and hydropericardium. Microscopically, there are multiple foci of necrosis in the spleen, dermis, lungs, and adrenal glands. Intranuclear eosinophilic to amphophilic inclusions are present in affected areas in the lung, spleen, and skin. Typical herpesviral particles have been observed in affected tissues and in inoculated rabbit kidney cell culture (Swan et al. 1991).

RNA Viral Infections

Rotaviral Enteritis

ETIOLOGY. Rotaviruses (from the Latin *rota*, "wheel") are double-stranded, nonenveloped RNA viruses of the family Reoviridae. Cultivatable rabbit strains of rotavirus isolated to date have been identified as serotype 3.

EPIZOOTIOLOGY AND PATHOGENESIS. Rotaviral infections are recognized to be a major cause of enteritis in human and veterinary medicine. In the domestic rabbit, outbreaks of rotaviral enteritis are usually confined to suckling and weanling animals (DiGiacomo and Thouless 1986; Schoeb et al. 1986). Serological surveys have indicated that rotaviral infections are enzootic in many commercial rabbitries. A high percentage of animals, including preweaning animals, may be seropositive. Epidemiology studies have indicated that protection may be afforded by transplacentally derived maternal antibodies, which subsequently decline to low levels by 1 mo of age. However, on exposure to the virus at 30–45 days of age, there may be sufficient residual antibody to protect them from overt disease. Viral shedding in conventional diarrheic rabbits is frequently seen at 35–42 days of age. Based on serological surveys, clinically healthy rabbits may have a subclinical infection at around 4 wk, with subsequent rise in antibodies to rotavirus (DiGiacomo and Thouless 1986). In one epizootic of rotaviral enteritis in a specific-pathogen-free rabbitry, sucklings 1–3 wk of age were affected. The disease was characterized by rapid spread, with high morbidity and mortality. This was interpreted to be consistent with the introduction of a new infectious agent into a colony not previously exposed, thus with no maternal immunity to afford protection during the neonatal period (Schoeb et al. 1986). Following ingestion, rotaviruses replicate in the relatively mature enterocytes lining villi, particularly in the jejunum and ileum. Due to damage to cells that synthesize disaccharidases, lactose and other disaccharides remain in the gut lumen, causing an osmotic drain and attracting fluid into the lumen (Peeters et al. 1984). Only the monosaccharides may be absorbed. Viral infections of enterocytes may also facilitate bacterial adhesion to damaged cells.

PATHOLOGY. On gross examination, animals may be dehydrated, and fluid contents are present in the cecum. Other organs are usually grossly normal. In the small intestine, there may be moderate to severe villous atrophy, fusion of adjacent villi, and vacuolation to flattening of apical enterocytes in the jejunum and ileum, sometimes with denuding (Fig. 6.7). There may be focal areas of desquamation in the cecum, and basophilic debris may be present in the cytoplasm of affected enterocytes. Lesions may be similar in *coronaviral enteritides*. Depending on the availability of diagnostic laboratory facilities, demonstration of the virus (usually by examination of intestinal contents for viral particles by electron microscopy or by the demonstration of viral antigen by ELISA techniques), virus isolation, and serology are procedures that should enable one to identify the causative viral agent. In cases of *bacterial enteritis*, villous atrophy may occur with attaching strains of *Escherichia coli* in this species. Microscopic examination of the small intestine for evidence of bacterial attachment to enterocytes may be of assistance, but isolation and serotyping are essential steps if a primary or concomitant infection with *E. coli* is suspected. Fecal flotation and/or microscopic examination of the small and large intestine are essential steps to determine whether coccidia are playing a role in the disease.

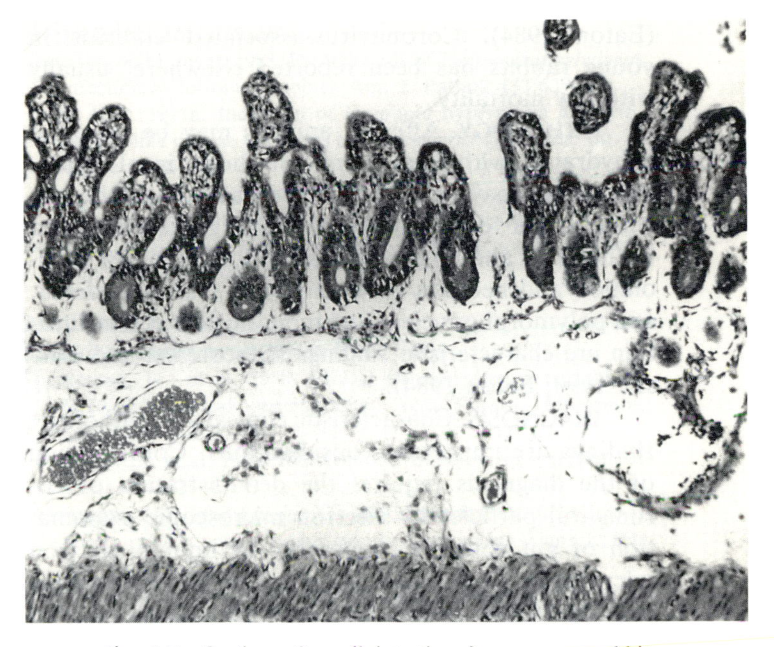

Fig. 6.7. Section of small intestine from young rabbit with a confirmed epizootic of rotaviral enteritis. There is blunting and fusion of villi. Enterocytes lining villi are cuboidal and poorly differentiated. (Courtesy of T.R. Schoeb)

DIAGNOSIS. Diagnosis is based on the history, age, and characteristic gross and microscopic features and on the demonstration of rotavirus by direct electron microscopy of fecal samples, by the ELISA technique or virus isolation, or by a rise in antibody titer to the virus. *Differential diagnoses* include coronaviral enteritis, coliform enteritis, coccidiosis, and clostridial enteropathies. ELISA kits are available from commercial suppliers for testing fecal samples for viral antigen.

SIGNIFICANCE. Demonstration of rotaviral particles in the feces does not constitute a definitive diagnosis but should be confirmed by microscopic examination of the gastrointestinal tract for characteristic lesions. In one study, rotavirus was detected in 25% of diarrheic feces and 10% of normal feces (DiGiacomo and Thouless 1986). Frequently there are other factors, including infectious agents, contributing to the disease. For example, infections with copathogens such as attaching *E. coli* and/or coccidia frequently play a significant role in this disease. The disease in rabbits has been studied as an animal model for rotaviral infections in other species (Conner et al. 1988).

Coronaviral Infections

Coronaviral Enteritis

EPIZOOTIOLOGY. Enteritis with mortality has been associated with a coronaviral infection in a barrier-maintained breeding colony in Germany. The epizootic occurred in young rabbits 3–8 wk of age, with peak mortality at 6 wk of age. Typical coronaviral particles were demonstrated by electron microscopy

chronic infections, there may be prominent peribronchial cuffing with lymphocytes. The organism can be recovered in large numbers from lesions in the respiratory tract.

SIGNIFICANCE. *B. bronchiseptica* on occasion may be considered to be a significant opportunistic infection or to act as a copathogen in this species. Occasionally outbreaks of respiratory disease occur in young rabbits which are attributed to *Bordetella* infections. Until proven otherwise, *B. bronchiseptica* should be regarded as a potential pathogen, particularly in rabbits 4–12 wk of age. Chronic lesions in the lower respiratory tract are a relatively common finding in clinically healthy conventional rabbits. Chronic interstitial pneumonitis, chronic bronchiolitis, and perivascular and peribronchial lymphocytic infiltrations are typical findings. These changes have been associated with chronic infections with *B. bronchiseptica*, but lesions of this type have also been observed with chronic *Pasteurella* infections and in the absence of bacterial isolates from the lung.

Respiratory Infection due to Cilia-associated Respiratory (CAR) Bacillus

Colonization of the apices of the respiratory cells lining the larynx, trachea, and bronchi have been observed in laboratory rabbits. The bacilli were demonstrated both in silver-stained preparations and by electron microscopy (Kurisu et al. 1990). The animals were asymptomatic, suggesting that rabbits may not develop clinical disease following exposure as readily as do rats and mice. Rabbits have been shown to seroconvert postexposure. Based on some serological studies, the organism is relatively common in commercial rabbitries. The significance of the CAR bacillus as a respiratory tract pathogen in this species is currently not known.

Staphylococcal Infections

Outbreaks of staphylococcosis occur sporadically in commercial rabbitries and in laboratory animal facilities. Manifestations of the disease vary from localized abscessation (Okerman et al. 1984) to an acute septicemic, frequently fatal form of the disease (Snyder et al. 1976; Hagen 1963).

EPIZOOTIOLOGY AND PATHOGENESIS. Hemolytic, coagulase-positive strains of *Staphylococcus aureus*, particularly type C strains, have been associated with the disease. On rare occasions, staphylococcal infections have been associated with "snuffles" and lower respiratory tract disease (Renquist and Soave 1969). However, virulent strains of the organism may be harbored as an inapparent infection in the upper respiratory tract. Respiratory tract infections may spread by direct contact, but the organism has also been recovered from the air in contaminated facilities. In cases of abscessation, genital tract infection, or mastitis associated with pathogenic strains of *S. aureus*, possible routes of invasion may include local inoculation or hematogenous spread. In the acute, systemic form of staphylococcosis, possible sites of entry include umbilical vessels, skin abrasions, or possibly the aerogenous route. Neonatal staphylococcal infections are a recognized cause of neonatal mortality in domestic rabbits in Europe, the United States, and Canada. In France, it is considered to be an important cause of mortality in suckling rabbits in commercial rabbitries (Okerman et al. 1984). The original source of the organism usually cannot be determined. Disseminated staphylococcal infections have been observed in wild rabbits (Osebald and Gray 1960). In one reported outbreak, an identical strain of *S. aureus* was isolated from the nasal passages of the attending animal technician and from the diseased rabbits (Renquist and Soave 1969).

PATHOLOGY. *Chronic suppurative lesions* may occur in the subcutis, mammary gland, genital tract, conjunctiva, or other areas, such as the upper and lower respiratory tract. Embolic suppurative lesions may occur in tissues such as kidney, brain, heart, and lung. The thick purulent exudate is similar to that seen in *Pasteurella* infections; thus bacterial culture is essential in order to confirm the diagnosis. In staphylococcal mastitis, affected glands may vary in appearance from swollen, red areas with induration of the overlying skin to chronic abscessation.

PODODERMATITIS. Pododermatitis ("sore hocks") has been associated with staphylococcal infections. It may occur in association with staph abscesses or mastitis. The cellulitis and dermatitis present in pododermatitis is usually concentrated along the ventral hock region. *Respiratory tract lesions*, when present, may be characterized by mucopurulent rhinitis, occasionally with localized bronchopneumonia or abscessation of the lung. The *acute septicemic form* of the disease usually occurs only in suckling kits during the first week of life, frequently with high mortality in affected litters. At necropsy, multifocal suppurative lesions may be present in the subcutaneous tissue and in viscera, including lung, kidney, spleen, heart, and liver (Fig. 6.14). Microscopically, focal suppurative necrotizing lesions are present in affected organs. Bacterial colonies are usually associated with the lesions (Figs. 6.15 and 6.16).

DIAGNOSIS. Recovery of *S. aureus* from the lesions is essential to confirm the diagnosis. Gram-positive cocci are also readily demonstrated histologically in tissue sections. *Differential diagnoses* include pasteurellosis, Tyzzer's disease, and listeriosis.

SIGNIFICANCE. Enzootics of neonatal staphylococcal infections may cause a high mortality rate in affected litters. With suppurative respiratory or genital

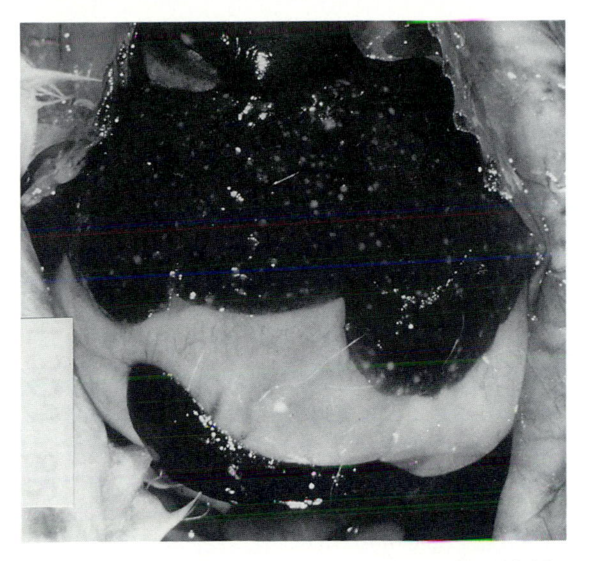

Fig. 6.14. Staphylococcal septicemia in 10-day-old kit. Note the multifocal suppurative lesions in the liver.

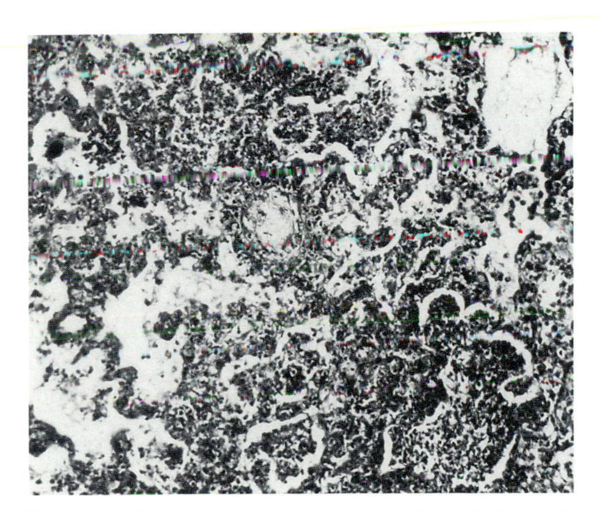

Fig. 6.15. Focal suppurative pneumonia in a fatal case of staphylococcal septicemia in suckling rabbit.

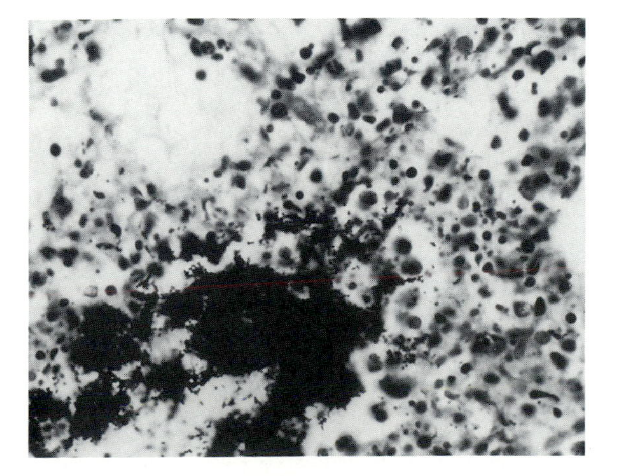

Fig. 6.16. Typical coccoid bacteria associated with lapine staphylococcal pneumonia (Brown and Brenn stain).

tract lesions, or with abscessation suspected to be due to a virulent strain of *S. aureus*, bacterial culture is essential in order to differentiate the disease from other likely causative agents, such as *P. multocida*. In view of the widespread distribution of staphylococci in laboratory animals, characterization of the isolate is necessary in order to determine whether it is a potentially pathogenic strain.

Venereal Spirochetosis

The disease has several synonyms, including "vent disease" and "rabbit syphilis." The causative agent, *Treponema cuniculi*, is a spirochete that has not been successfully grown in artificial media or cell culture. Based on serological surveys, venereal spirochetosis occurs occasionally in conventional facilities (Cunliffe-Beamer and Fox 1981b). However, the disease is seldom detected on cursory examination.

EPIZOOTIOLOGY AND PATHOGENESIS. The venereal route is the most important means of spread, although extragenital contact transmission may also occur. Young animals may develop the disease following contact with an infected dam. However, there is a demonstrated age-related susceptibility. Young rabbits have been shown to be relatively resistant to the infection, either by natural exposure or experimental inoculation (Cunliffe-Beamer and Fox 1981b). There is no evidence of intrauterine transmission. There appears to be a strain-related resistance/susceptibility to clinical disease postexposure. Treponematosis has also been diagnosed in wild rabbits in Britain (Flatt 1974).

PATHOLOGY. Lesions associated with treponemosis may occur in the vulva, prepuce, anal region, muzzle, and periorbital region. Initially, changes are characterized by edema, erythema, and papules at the mucocutaneous junctions. Syphilitic lesions later progress to ulceration and crusting. On microscopic examination, hyperplasia of the epidermis, necrosis of epithelial cells, erosions and ulcerations, with infiltration by plasma cells, macrophages, and heterophils are typical changes (Fig. 6.17). The infection is confined primarily to the epithelium, and other than hyperplasia of the regional lymph nodes, visceral involvement does not normally occur.

DIAGNOSIS. Scrapings from lesions, with wet mount preparation and *dark-field examination* is the recommended method for confirming the diagnosis. Treponemas are recognized by their characteristic spiral shape and corkscrew movement. Demonstration of the spirochetes by silver staining of lesions in histology sections may be used (Fig. 6.18), but this is not as reliable as dark-field examination. *Serology* is a reliable diagnostic procedure. Tests available include the demonstration of reagin antibody (Wasserman antibody) and the fluorescent treponemal antigen test. Rabbits may seroconvert while still infected with the

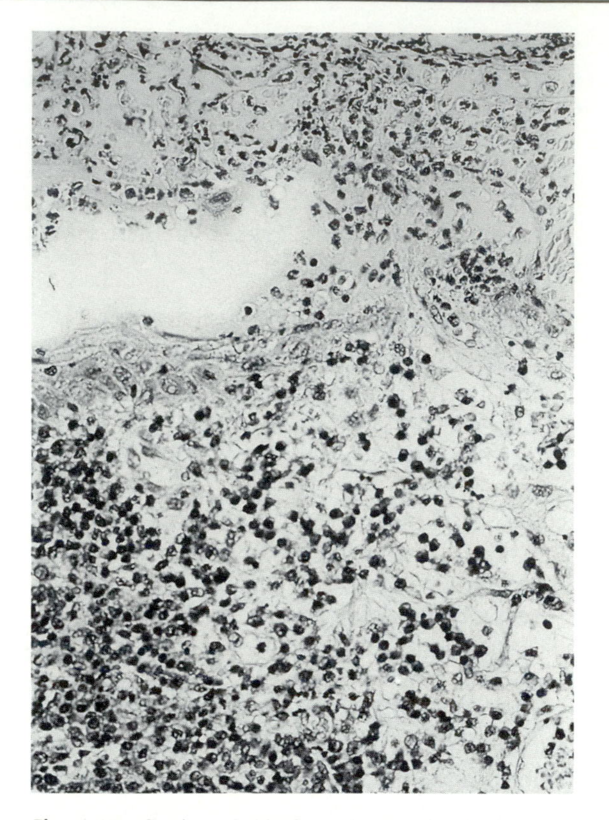

Fig. 6.17. Section of skin from the muzzle of laboratory rabbit with spontaneous treponematosis (rabbit syphilis). There is ulceration of the mucosa with leukocytic infiltration.

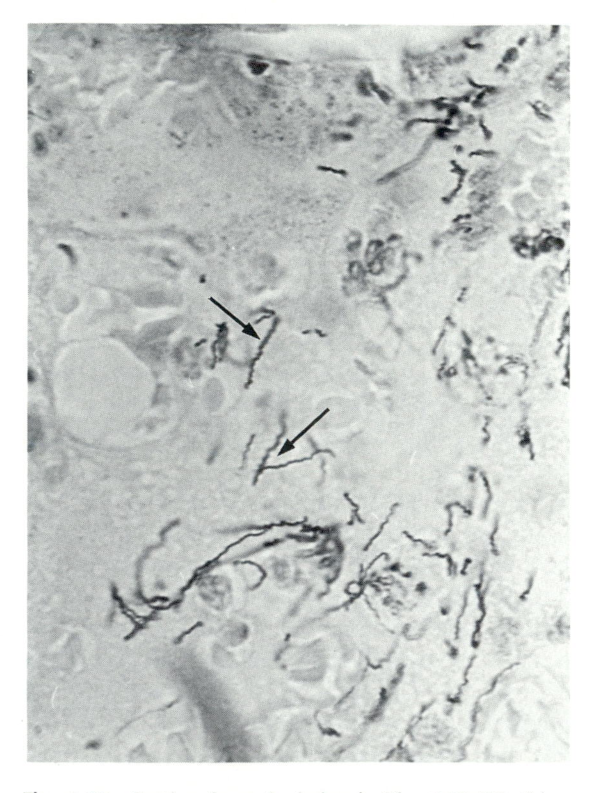

Fig. 6.18. Section from the lesion in Fig. 6.17 (Warthin-Starry stain), illustrating the typical spirochaetes (*arrows*) associated with the lesions.

organism. *Differential diagnoses* include *Pasteurella* infections of the external genitalia and traumatic lesions.

SIGNIFICANCE. Treponematosis is considered to be a self-limiting disease. However, there appears to be increased susceptibility to other infectious agents in rabbits actively infected with *T. cuniculi*. The effect of treponematosis on fertility rates in rabbits has not been determined.

Listeriosis

This bacterial infection is characterized by abortions and sudden deaths, particularly in does in advanced pregnancy. *Listeria monocytogenes* is a small gram-positive, non-spore-forming rod. In 1926, Murray et al. described an acute, fatal disease in young rabbits. Focal hepatitis, ascites, and enlarged mesenteric lymph nodes were typical findings. They observed a marked rise in circulating blood monocytes in both the spontaneous and experimental disease and proposed the name *Bacterium monocytogenes* for the newly recognized organism. The reactive monocytes were observed to phagocytize the organism in vitro with the same avidity as did heterophils (Murray et al. 1926).

EPIZOOTIOLOGY AND PATHOGENESIS. In sporadic outbreaks of the disease, the source of the organism is frequently attributed to contaminated feed or water. Inapparent carriers and shedders also occur. *Listeria monocytogenes* has a specific predilection for the gravid uterus in advanced pregnancy. Adult, nonpregnant does and bucks are usually resistant to the infection. Following oral or conjunctival inoculation of females in advanced pregnancy, abortions, stillbirths, and mortality in the dam usually occur. Pregnancy may be interrupted as early as 24 hr postinoculation. However, inoculation of females by the intravaginal, oral, or conjunctival route either prior to mating or early in pregnancy failed to produce disease (Gray and Killinger 1966). The organism can cross the placental barrier in advanced pregnancy. Uterine infections may persist postkindling and may serve as the source of infection for the next pregnancy. Newborn kits born to infected dams that survive may subsequently develop systemic listeriosis, and stunting and meningoencephalitis are other possible sequelae. On the other hand, young rabbits may shed *Listeria* as an inapparent infection for several weeks postkindling (Gray and Killinger 1966).

PATHOLOGY. Deaths typically occur in does in advanced pregnancy. Straw-colored fluid is frequently present in the peritoneal cavity, occasionally with fibrinous exudate and ecchymoses on the serosal surface of the uterus. Disseminated pale, miliary foci of necrosis on the liver, edema of regional lymph nodes, splenomegaly, and visceral congestion are the usual macroscopic findings. The uterus may contain relatively intact, near-term kits (Fig. 6.19) or fetuses in various stages of decomposition or mummification. In acute cases, the placenta may be edematous and hemorrhagic, but in an infection of longer duration, the placenta is usually thickened, friable, and dull dirty gray, with an irregular surface. Focal miliary hepatic lesions may be evident in stillborn kits. Characteristic microscopic changes seen in adult cases of listeriosis include focal suppurative hepatitis and marked infiltration with heterophils (Fig. 6.20). There may be focal inflammatory lesions in the adrenal cortices, congestion and thromboses in the splenic sinusoids and blood vessels, and acute necrotizing to chronic suppurative metritis and placentitis (Fig. 6.21). In rabbits that die with the acute form of the disease, large numbers of gram-positive bacilli are usually visible, particularly in the placenta. Focal hepatitis and occasionally meningitis have been observed in newborn animals and in kits that succumb within a few days of birth with listeriosis.

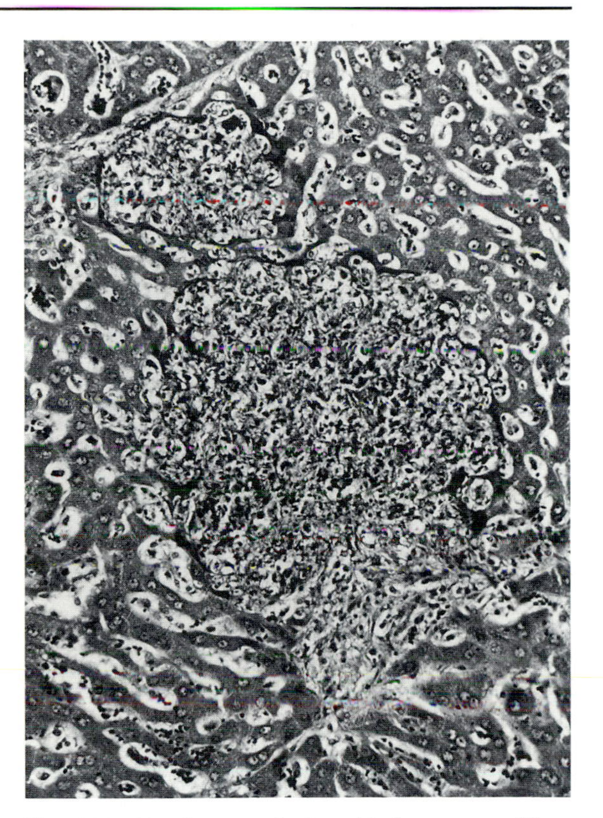

Fig. 6.20. Focal suppurative hepatitis from a case of listeriosis in pregnant doe. There is a clearly delineated lesion, with marked infiltration with heterophils.

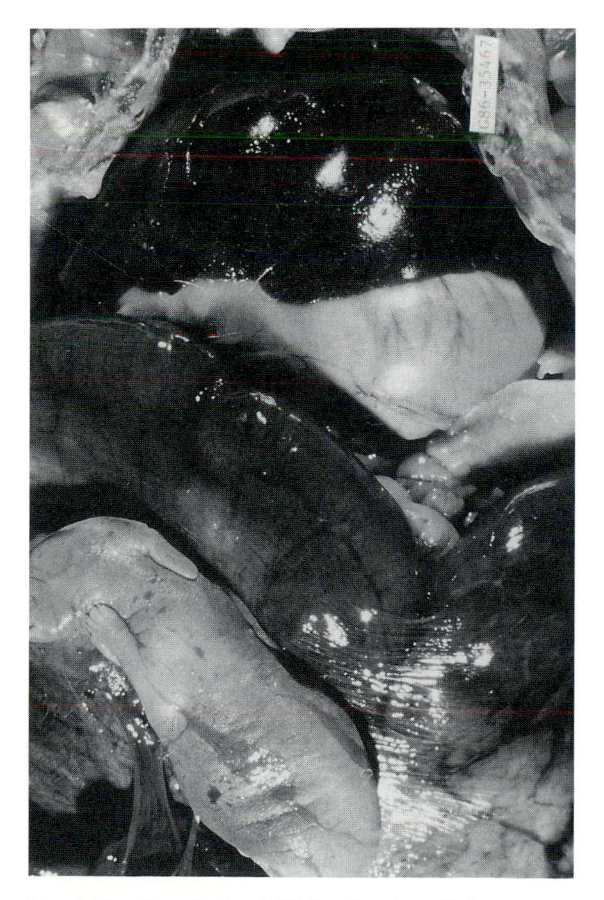

Fig. 6.19. New Zealand White doe that died near-term with acute listeriosis. There are pinpoint-size foci of hepatic necrosis. The kits are intact, with no evidence of maceration. (Courtesy of R.J. Hampson)

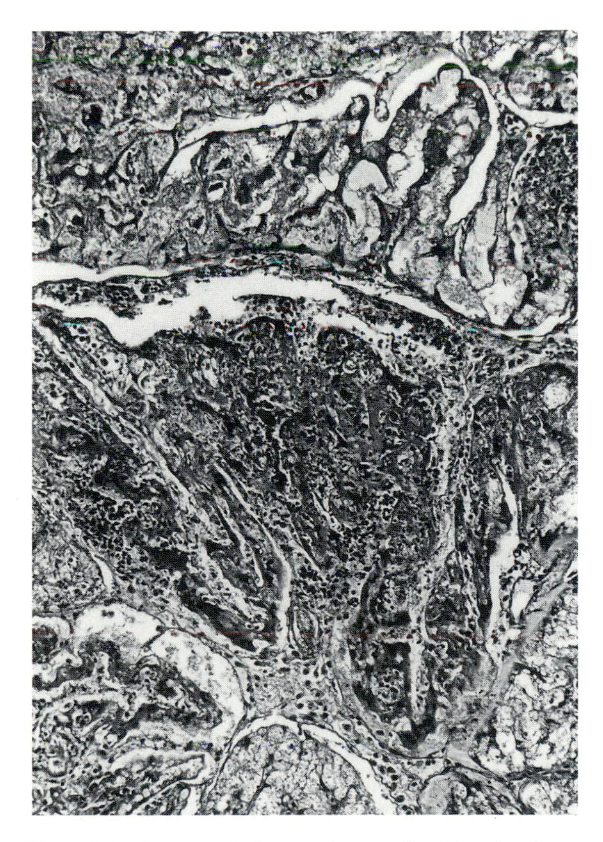

Fig. 6.21. Section of placenta from animal in Fig. 6.20. There is marked placental edema and degeneration of trophoblastic cells, with leukocytic infiltration.

DIAGNOSIS. In acute cases of listeriosis, the organism can usually be readily recovered from the uterine wall, placenta, and fetuses. Blood, liver, and spleen are other likely sources of the organism at necropsy. Recovery of the organism is considered to be a much more satisfactory method to confirm the diagnosis than are serological techniques (Vetesi and Kemenes 1967). *Differential diagnoses*: Diseases causing disseminated foci of hepatic necrosis in the rabbit include Tyzzer's disease, tularemia, and salmonellosis. In perinatal deaths seen in does with acute pasteurellosis and metritis, there may be acute necrotizing uterine lesions, but liver lesions are normally absent.

SIGNIFICANCE. In addition to the mortality associated with lapine listeriosis, the dangers of inadvertent exposure by human contacts should be emphasized.

Salmonellosis

Salmonella infections are relatively rare in domestic rabbits housed in well-managed facilities. Septicemia, diarrhea, abortions, and rapid death have been associated with organisms such as *Salmonella typhimurium* and *S. enteritidis*. Salmonellosis has been identified in specific-pathogen-free rabbits following experimental surgery and irradiation (Newcomer et al. 1983). Pathologic changes associated with lapine salmonellosis include polyserositis, focal hepatic necrosis, splenomegaly, acute enteritis with fibrinous exudation, and suppurative metritis. Laboratory rabbits have been implicated as reservoirs of *S. mbandaka* infection (Newcomer et al. 1983). In view of the public health aspects, and the dangers of interspecies spread, the diagnosis of salmonellosis warrants a thorough investigation and eradication procedures.

Yersiniosis

This disease is characterized by an acute to chronic infection in wild rodents and lagomorphs and in human patients (Flatt 1974). The bacterium, *Yersinia pseudotuberculosis*, in wildlife is usually transmitted by the ingestion of contaminated food or water. Lesions are characterized by focal caseation necrosis of liver, spleen, cecum, and occasionally lymph nodes and reproductive tract. Although occasionally yersiniosis occurs in the wildlife population, it rarely occurs in domestic species.

Bacterial Mastitis

Sometimes referred to as "blue breast," mastitis occurs occasionally in recently kindled and heavily lactating does. The skin overlying the swollen, firm mammary gland(s) has a red to dark blue discoloration. On section, the gland may contain material varying from a fluid consistency to thick purulent exudate. Staphylococci, pasteurellae, or streptococci are the usual pathogens associated with mastitis in this species (Flatt 1974).

Other Bacterial Diseases

Streptococcal septicemia has been reported to occur in young rabbits. Acute diplococcal infections have been observed on rare occasions in domestic rabbits. In isolated cases, *Klebsiella pneumonia* has been associated with acute hemorrhagic bronchopneumonia in domestic rabbits. Suppurative and ulcerative skin lesions have been associated with various infections, including *Corynebacterium pyogenes* and *Fusibacterium necrophorum*. For additional information on uncommon bacterial infections, see Flatt (1974).

Enteritis Complex

The "enteritis complex" is a term used to encompass a disease that commonly occurs in rabbits, particularly in rabbits 5–12 wk of age. Many factors are recognized to play a role in the enteritis complex in domestic rabbits. Frequently, an epizootic (or enzootic) of the disease coincides with recent weaning or changes in feed, environment, or management practices. Important infectious agents now recognized to play a role in this disease complex include coccidia, clostridia *Bacillus piliformis* and *Escherichia coli*, and rotaviruses.

Tyzzer's Disease

The disease was first recognized by Ernest Tyzzer, who described a fatal epizootic in the Japanese waltzing mouse. He noted miliary focal necrosis of the liver and demonstrated bundles of pleomorphic, gram-negative bacilli within the cytoplasm of hepatocytes and in intestinal epithelial cells. He failed to culture the organism but reproduced the disease in mice inoculated with material from fatal cases of the disease. The original article remains a classic description of the disease (Tyzzer 1917). The disease has now been recognized in a variety of laboratory animals, wildlife, and domestic species, including the rabbit (Allen et al. 1965), rat (Jonas et al. 1970), hamster (Nakayama et al. 1975), gerbil (Carter et al. 1969), guinea pig (McLeod et al. 1977), rhesus monkey (Fujiwara 1978), foal (Pulley and Shively 1974), dog (Queresi and Olander 1976), kitten (Kovatch and Zebarth 1973), calf (Webb et al. 1987), and cottontail rabbits (Ganaway et al. 1976).

EPIZOOTIOLOGY AND PATHOGENESIS. The causative agent, *Bacillus piliformis*, is relatively labile in the vegetative phase and cannot be grown in cell-free media. The usual method of culture is in embryonated eggs inoculated by the yolk sac route (Ganaway et al. 1971). The organism may survive for long periods in the spore state and has been shown to remain infectious in contaminated bedding for at least 1 yr (Tyzzer

1917). Antigenic differences have been demonstrated in strains of *B. piliformis* isolated from different species. It has been suggested that the antigenic differences observed may be due to host-associated antigens, not because of the distinctly different host organisms (Fujiwara et al. 1985). It is likely that interspecies infections can occur. Typical lesions have been produced in laboratory animals inoculated orally with isolates from other species (Waggie et al. 1987). The organism is passed in the feces, and the infection usually occurs by ingestion. Transplacental infection has been observed in guinea pigs (Boot and Walvoort 1984) and rats (Fries 1977). There has been speculation, but no proof, that intrauterine transmission may also occur in rabbits (Peeters et al. 1985). Inapparent infections occur, and cortisone treatment has been used to detect subclinically infected animals. There appears to be considerable natural resistance to the organism; frequently corticosteroid treatment is required in order to reproduce the disease consistently. Following oral inoculation in rodents, intestinal and hepatic lesions have been observed by 3 and 4 days, respectively (Waggie et al. 1987). On the other hand, there may be local multiplication in the gut mucosa only, with minimal tissue damage. In mice, B-cell but not T-cell function appears to play an integral part in resistance to the disease (Waggie et al. 1981). Following oral exposure, the usual sequence of events is multiplication of *B. piliformis* in the intestinal mucosa, with tissue damage, dissemination to the liver by the portal circulation with bacteremia, subsequent hepatitis, and on occasion myocarditis (Cutlip et al. 1971; Allen et al. 1965).

Predisposing factors are an important consideration. In rabbits, "stress factors" include shipping, changes in diet, high environmental temperatures, and poor sanitation. Alterations in the gut flora may enhance susceptibility to the disease. In one study, only rabbits treated with sulfaquinoxaline in the drinking water died with Tyzzer's disease (Allen et al. 1965). All ages may be affected during epizootics in domestic rabbits, but young weanlings are most frequently affected. The morbidity may vary from 10% to over 50%. The mortality rate in affected animals is high, and rabbits that survive may be permanently stunted because of residual lesions, particularly stenosis of the gut. The disease is characterized by a sudden outbreak of profuse, watery diarrhea, a short course, and high mortality in affected animals. Rabbits may be found dead with no prior evidence of clinical disease.

PATHOLOGY. On external examination, dehydration and fecal staining in the perineal region are typical findings. There are usually extensive ecchymoses and occasionally fibrinous exudate on the serosal surface of the cecum and colon (Fig. 6.22). The walls of affected areas, particularly the cecum, are markedly thickened and edematous. The cecum and colon contain dirty brown, watery contents, and the mucosal surface is discolored and dull, frequently with an irregular, granular appearance. Fibrinous strands and debris are often adherent to the mucosa. Disseminated pale miliary foci up to 2 mm in diameter are frequently present in the liver (Fig. 6.23). Myocardial lesions, when present, occur as pale, linear streaks, par-

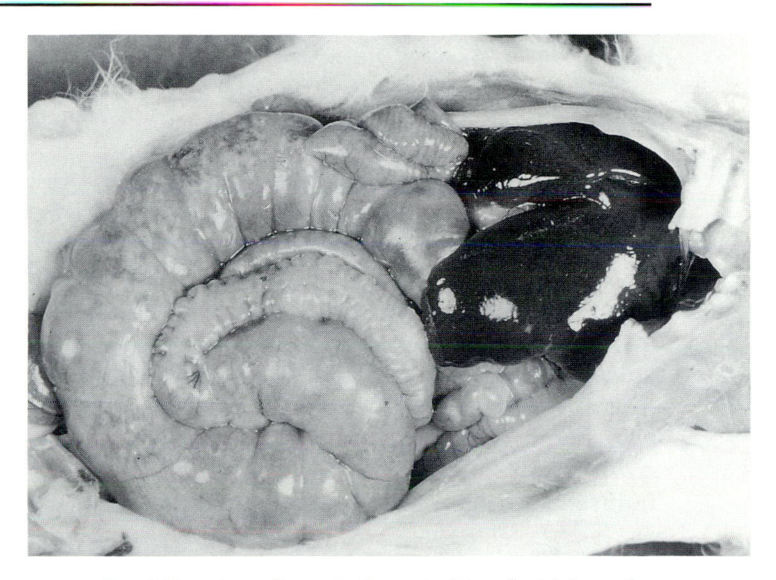

Fig. 6.22. Acute Tyzzer's disease in 12-week-old domestic rabbit. There is marked dilation of the cecum, and fibrinous exudate is adherent to the serosal surface. (Courtesy of R.J. Julian)

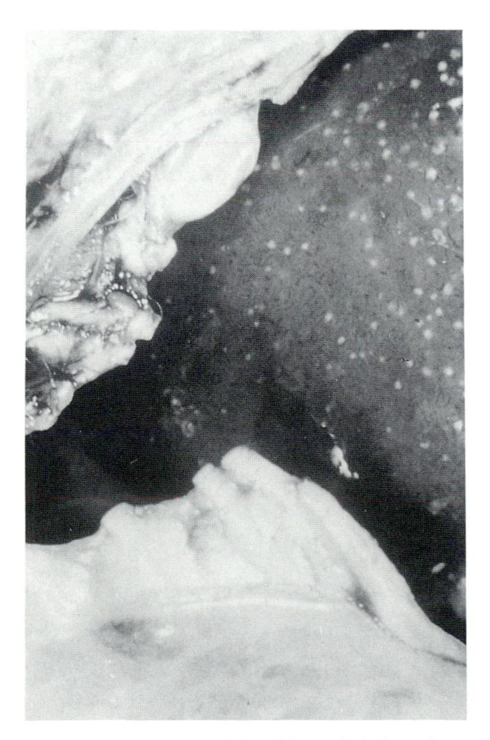

Fig. 6.23. Typical circumscribed focal hepatic lesions in a fatal case of Tyzzer's disease in adult domestic rabbit.

ticularly near the apex of the left ventricle. In affected rabbits that survive, carcasses are thin, usually with identifiable circumferential regions of fibrosis and stenosis in the terminal ileum or large intestine.

Microscopic changes are found consistently in the intestinal tract, usually in the liver, and infrequently in the myocardium. In the intestinal tract, there is focal to segmental necrosis of the mucosa of the cecum, with variable involvement of distal ileum and proximal colon. There is sloughing of enterocytes, and large numbers of bacteria are often present on the surface of the damaged mucosa. Lesions are frequently transmural. There is extensive submucosal edema, necrosis of muscular layers, and concurrent leukocytic infiltration, heterophils predominating (Fig. 6.24). In the liver, focal lesions occur most often adjacent to the periportal areas. Circumscribed areas of coagulation to caseation necrosis contain variable numbers of identifiable heterophils, macrophages, and cell debris. In the myocardium, focal to linear areas of coagulation necrosis may be present, usually accompanied by minimal inflammatory response. In rabbits examined during the acute disease, intracytoplasmic bundles of bacilli are demonstrable in the hepatocytes at the periphery of the focal lesions, in enterocytes, and in the adjacent smooth muscle of the gut. Occasionally bacilli are also visible in myofibers associated with myocardial lesions. Intracytoplasmic, eosinophilic bacilli should be evident in H & E–stained sections. However, paraffin-embedded sections stained with the Warthin-Starry silver method or Giemsa are much better procedures to demonstrate the characteristic bundles of filamentous bacilli associated with the disease (Fig. 6.25). In acute cases of Tyzzer's disease in the rabbit, a careful search may be required in order to locate the organisms.

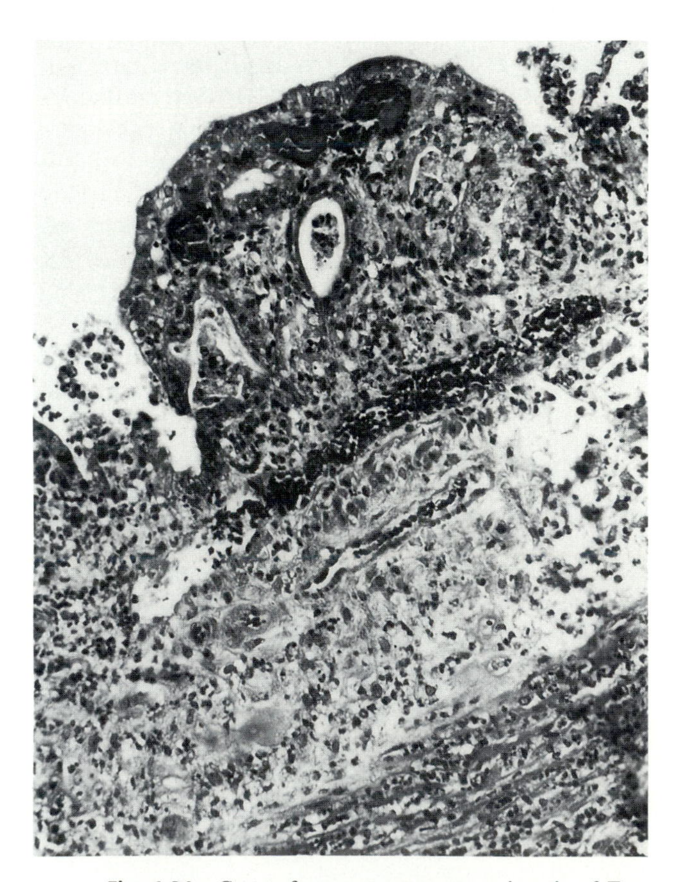

Fig. 6.24. Cecum from a spontaneous epizootic of Tyzzer's disease in a commercial rabbitry. There is a necrotizing transmural typhlitis, with effacement of the mucosal surface and leukocytic infiltration.

Fig. 6.25. Section of liver in lapine Tyzzer's disease. Bacilli (*arrows*) are present in cytoplasm of hepatocytes (Warthin-Starry stain).

Microscopic changes seen in the livers of rabbits that survive the acute stages of the disease include focal fibrosis, with infiltrating macrophages and the presence of multinucleated giant cells and mineralized debris. Focal to segmental fibrosis, with disruption of the architecture, occurs in the large intestine, and occasionally in the myocardium, in surviving animals. Tyzzer's bacilli are not present in lesions examined during the convalescent stages of the disease.

DIAGNOSIS. Giemsa-stained impression smears of the hepatic lesions may reveal the typical beaded filamentous bacilli. The presence of the extensive transmural cecal damage, together with the focal hepatic lesions, should enable the pathologist to differentiate Tyzzer's disease from other infectious diseases, such as listeriosis, neonatal staphylococcosis, other bacterial enteritides, and coccidiosis at necropsy. The demonstration of the typical bacilli in tissue sections are required to confirm the diagnosis. Serological techniques used to detect seropositive animals have included the complement fixation test and the indirect immunofluorescent antibody test. *Differential diagnoses* include listeriosis, staphylococcosis, clostridial infections of the large intestine, and coccidiosis.

SIGNIFICANCE. It is essential that routine bacteriology and fecal flotations be performed to eliminate the possibility of concurrent infections. Elevated intestinal *Escherichia coli* counts and coccidia oocysts have been observed in some outbreaks of Tyzzer's in rabbits (Peeters et al. 1985). The significance of these findings has not been determined. The organism may be harbored as an inapparent infection, which may be activated by manipulations or adverse environmental conditions. In view of speculation regarding interspecies transmission, possible sources of the organism should be investigated; wild rodents have been proposed as one source. The organism is relatively refractory to antibiotics, although tetracycline treatment has been effective in some epizootics of the disease.

Coliform Enteritis. Pathogenic strains of *Escherichia coli* are now considered to be one major cause of enteritis in commercial rabbitries and occasionally in research facilities.

EPIZOOTIOLOGY AND PATHOGENESIS. *E. coli* is normally absent or present in small numbers in the alimentary tract of suckling and weanling rabbits. It has been attributed to the low pH of the stomach, so that the stomach and small intestine are relatively free from bacteria in this species (Smith 1965). Under certain conditions, there may be a marked proliferation of the organisms, and up to 30 million colony-forming units of *E. coli* per gram of feces have been recovered from diarrheic rabbits (Prescott 1978). Several factors may promote this phenomenal growth of *E. coli*. For example, in intestinal coccidiosis, frequently there is a striking concurrent rise in fecal output of *E. coli*. The mucosal damage associated with intestinal coccidiosis may enhance proliferation of *E. coli* and the resorption of endotoxins. A rise in cecal pH occurs in intestinal coccidiosis, and when diets are fed with a high digestive HCl requirement, this may promote the dissociation of volatile cecal fatty acids, which normally exert an antibacterial effect in the gut (Peeters et al. 1984b).

Isolates of *E. coli* recovered from diarrheic rabbits have been categorized as enteropathogenic strains. They cause intestinal disease, do not produce enterotoxins (currently designated as heat-stable or heat-labile), and are not considered to be enteroinvasive. There appear to be significant variations in the pathogenicity of lapine isolates. Strains of low virulence cause problems primarily in rabbitries with poor sanitation and are usually responsive to antibiotic treatment and improved hygiene. On the other hand, highly virulent strains are often refractory to antibiotic therapy. A large number of strains have been isolated from suckling and weanling rabbits with diarrhea, serotyped, and characterized. In general, strains of *E. coli* isolated from naturally occurring diarrheas in suckling or weanling animals produce disease experimentally only in the same age group. For example, strain RDEC-1, isolated from weanling rabbits, attaches only to the enterocytes of weanling rabbits and produces disease in this age group but not in sucklings. This may be due to the absence of the sucrose-isomaltose enzyme complex on the enterocyte brush border of suckling rabbits. The complex develops after weaning and has been shown to permit binding of this strain to enterocytes. In studies of the virulence of strains of *E. coli* isolated from sucklings, the organism attached to the enterocytes in both the large and small intestine. In weaned rabbits with experimental coliform enteritis, bacterial attachment occurred in the ileum, cecum, and colon (Peeters et al. 1984a).

In experimental coliform enteritis in suckling rabbits, mortality occurred within 3 days, while in weaned animals, diarrhea and mortality usually occurred after 1 wk. Isolates of *E. coli* from clinically normal rabbits usually failed to produce detectable disease (Peeters et al. 1984b). In experimental coliform enteritis in susceptible weanlings, the organism initially attaches to the Peyer's patch dome epithelium, and later to the enterocytes of the ileum and large intestine. Intestinal lesions were most extensive at 7–14 days postinoculation, with a corresponding significant rise in fecal *E. coli* (Peeters et al. 1985; Peeters et al. 1984a).

PATHOLOGY. At necropsy, the carcass may be dehydrated, and the perineal region is frequently stained with watery yellow to brown fecal material. The small intestine is usually grossly normal. The cecum and colon frequently are distended with watery yellow to gray-brown contents. There may be serosal ecchymoses, edema of the walls of the cecum and colon, edematous mesenteric lymph nodes, and prominent lymphoid tissue in the Peyer's patches and sacculus rotundus. Microscopically, in sucklings with coliform enteris, large numbers of coccobacilli are usually attached to enterocytes in both the small and large intestine, with polymorphonuclear cell infiltration in the lamina propria. Microscopic changes are normally more extensive in weanlings with the disease. In the small intestine, ileal villi are often blunted, and the lamina propria is edematous, with leukocytic infiltration, heterophils predominating. Enterocytes at the tips of villi are swollen, and bacterial colonies may be attached to these cells, with effacement of the microvillous brush border (Fig. 6.26). In the cecum and colon, there is bacterial attachment and swelling of affected enterocytes, frequently with detachment and

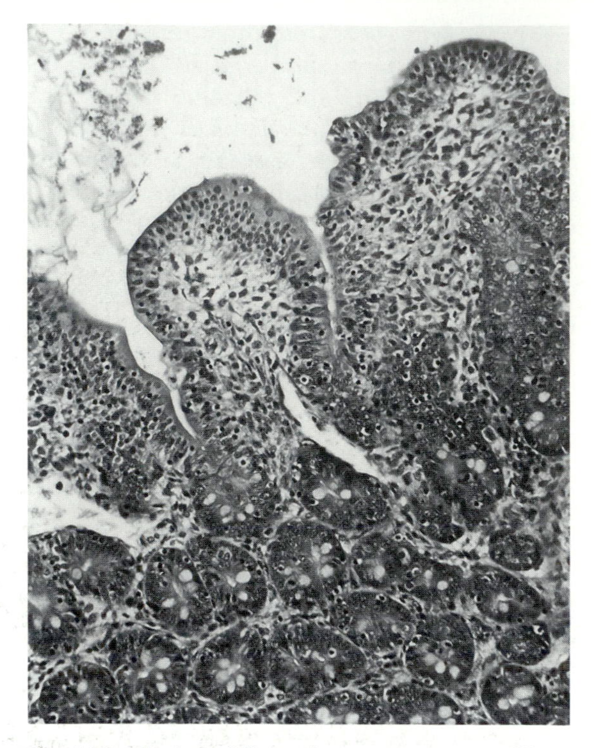

Fig. 6.26. Section of ileum from spontaneous coliform enteritis in weanling rabbit. There is blunting of villi, with flattening and loss of polarity of enterocytes.

ulceration involving primarily the tips of the cecal folds. Nonspecific vacuolation of hepatocytes and renal tubules has also been described (Prescott 1978).

DIAGNOSIS. The age, history, clinical signs, and gross and microscopic findings are useful criteria in making the diagnosis. The characterization of the isolate of *E. coli* is recommended in order to determine whether the strain is likely to be a primary pathogen. Isolates of *E. coli* have been divided into biotypes according to their carbohydrate fermentation patterns. There is a good correlation between biotype and serotype in identified pathogenic strains. *Differential diagnoses* include acute coccidiosis, clostridial enteropathies, viral enteritides, and Tyzzer's disease.

SIGNIFICANCE. Coliform enteritis is an important cause of disease and mortality in commercial rabbitries and occasionally in research facilities. Diagnostic procedures should include screening for other infectious agents. For example, elevated fecal *E. coli* counts frequently occur with intestinal coccidiosis, and it is possible that two or more copathogens may be contributing to the disease. The investigation should include a careful review of possible predisposing factors, such as management and environmental changes, poor sanitation, recent additions to the rabbitry, and recent weaning, that may have had an influence on the disease.

Clostridial Enteropathies/Enterotoxemia. Species of *Clostridi* implicated in the enteritis complex in domestic rabbits include *C. perfringens*, *C. difficile*, and *C. spiroforme*. They are gram-positive bacilli that reside in the gut and grow under anaerobic conditions. Pathogenic strains are capable of producing powerful enterotoxins (or cytotoxins) that can alter normal gut function and produce severe, often fatal enteric disease.

EPIZOOTIOLOGY AND PATHOGENESIS. For several years, *C. perfringens* has been regarded as a likely participant in the enterotoxemia/enteritis complex in domestic rabbits. Type E iota toxin was demonstrated in fatal cases of enterotoxemia and the disease was attributed to *C. perfringens* (Patton et al. 1978). The term "carbohydrate overload" has been associated with the syndrome. The proposed sequence of events was as follows: Rabbits ingesting high-energy feed may fail to degrade and digest the majority of carbohydrates in the small intestine. Significant amounts of carbohydrate may then reach the level of the large intestine, promoting the overgrowth of organisms such as clostridia. This may cause disturbances in the osmolality of intestinal contents, production of enterotoxins, diarrhea, and death. Increasing the dietary fiber has been proposed as one means of reducing mortality due to this complex. It is possible that some previously reported outbreaks of enteropathies associated with *C. perfringens* may have been due to *C. spiroforme*, since antitoxin prepared against *C. perfringens* type E iota toxin will also neutralize similar toxins produced by pathogenic strains of *C. spiroforme* (Carman and Evans 1984). Fatal colitis and enterotoxemia associated with overgrowth of *C. difficile* has occurred following prolonged treatment with penicillin and ampicillin (Rehg and Lu 1981). Subsequent experimental reproduction of the disease following treatment with lincomycin was attributed to *C. difficile* or *C. perfringens* (Rehg and Pakes 1982).

C. spiroforme is now recognized to be the most common clostridial pathogen associated with the enteritis complex in juvenile rabbits. In one survey of diarrheic rabbits, *C. spiroforme* was isolated from over 50% at necropsy, and 90% of the strains isolated were toxigenic. *C. spiroforme* infections commonly occur in rabbitries. Although this organism is not considered to be a normal inhabitant of the alimentary tract in rabbits, it frequently is difficult to produce disease experimentally in healthy rabbits inoculated orally with *C. spiroforme*. The normal gut flora appear to act as a microbial barrier, and a disruption in gut microflora is considered to be an important predisposing factor. Changes in feed, weaning, previous antibiotic treatment, and concurrent infections are examples of events that may permit colonization with *C. spiroforme* and thus trigger an epizootic of enteric disease. In outbreaks of the disease, the organism may be the sole identified pathogen. However, there may be concurrent infections with other pathogens, such as *Escherichia coli*, *B. piliformis*, *Eimeria* spp., rotavirus, and cryptosporidia. In one survey, *C. spiroforme* was considered to be the sole pathogen in 13% of the samples examined, and a copathogen in 33% of samples studied (Peeters et al. 1986). Following the multiplication of pathogenic strains of clostridia in the large intestine, enterotoxins may be produced, resulting in damage to enterocytes, impaired function, profuse diarrhea, subsequent depression, dehydration, and death. In peracute cases, rabbits may be found dead in their cage, with little or no prior evidence of disease. A chronic form may also occur. Anorexia, wasting, and intermittent diarrhea may occur over a period of several days.

PATHOLOGY. In peracute cases, the carcass is usually in good condition, and perineal soiling with diarrheic feces is a variable finding. In subacute to chronic cases, carcasses are frequently thin and dehydrated. Staining of the perineum, belly, and rear legs with watery green to tarry brown feces commonly occurs. Internally, straw-colored fluid may be present in the peritoneal cavity. Extensive ecchymoses are usually present in the cecal serosa, sometimes with involvement of the distal ileum and proximal colon. Epicardial and thymic ecchymoses may also occur. The cecum and adjacent areas are frequently dilated, with watery to mucoid, green to dark brown contents and with gas formation. There may be marked thickening of affected areas, and mucosal changes vary from hemorrhage to ulceration and fibrinous exudation (Fig. 6.27).

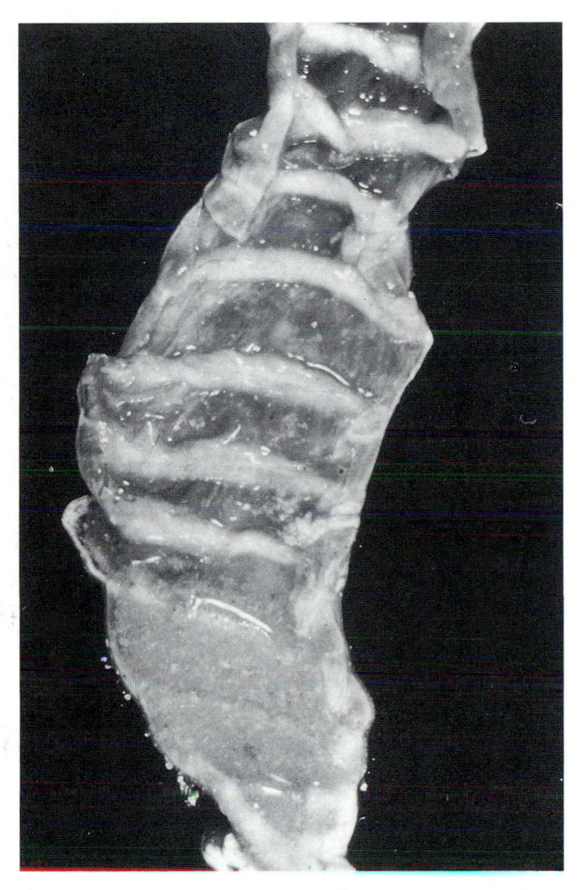

Fig. 6.27. Cecum from a case of spontaneous typhlitis, illustrating changes compatible with clostridial enteropathy. Note the mucosal hemorrhage and the fibrinous exudate on the surface of the mucosa.

Typical microscopic changes present in the cecum of affected animals are those of a necrotizing typhlitis, with irregular denuding of the mucosa, ulceration, fibrinous exudation, and leukocytic infiltration, heterophils predominating (Fig. 6.28). Changes observed in enterocytes vary, including swelling, vacuolation, flattening, denuding, and proliferation. The mucosa and submucosa are congested, frequently with focal hemorrhage, and thrombi may be present in adjacent vessels. Gram-positive bacilli may be present in large numbers on the surface of affected areas of gut mucosa (Fig. 6.29).

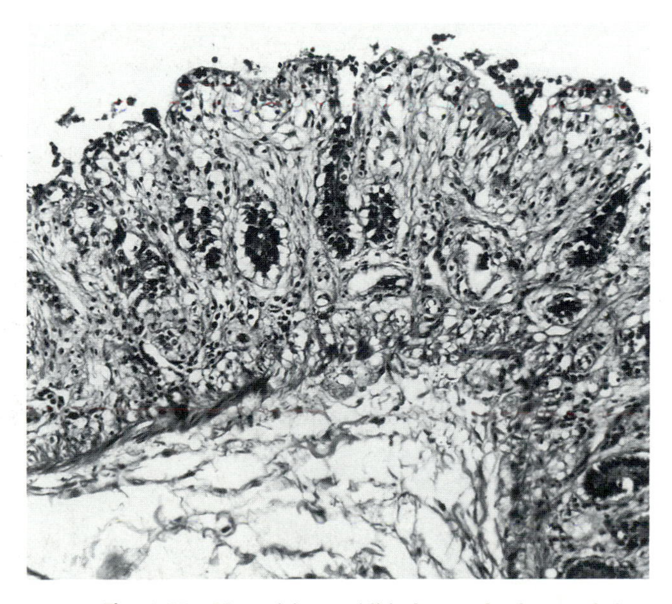

Fig. 6.28. Necrotizing typhlitis from a fatal case of clostridial enteropathy due to *Clostridium spiroforme*. There is degeneration of mucosal cells, sloughing, and disruption of the normal architecture.

DIAGNOSIS. The typical age (usually juveniles) and history of a change in feed, management, or environment may be helpful, particularly if coinciding with an explosive outbreak of diarrhea. Clinical signs may include watery diarrhea, depression, and hypothermia, sometimes with terminal convulsions. Gram-stained cecal smears are helpful. Typical curved and coiled gram-positive organisms are associated with *C. spiroforme* infections (Fig. 6.30). Anaerobic cultures are recommended for positive identification of the or-

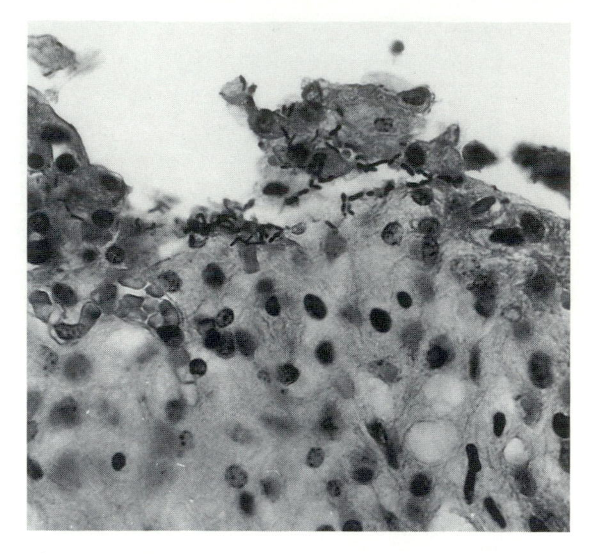

Fig. 6.29. Section of cecum shown in Fig. 6.28 (Brown and Brenn stain). Large numbers of bacilli are adherent to the intestinal mucosa. *C. spiroforme* was isolated in large numbers from the affected areas of gut.

ganism. Several procedures are available for the identification of the clostridial toxin recovered from gut contents or bacterial culture using specific antitoxins. They include skin testing, mouse protection tests, ELISA testing using antibodies to purified toxin antigens, and rocket immunoelectrophoresis. High-speed centrifugation has been used to separate the organism in intestinal contents (Holmes et al. 1988). Aerobic bacterial cultures, fecal flotation, and virology screening are procedures recommended in order to search for possible concurrent infections. *Differential diagnoses* include coccidiosis, *E. coli* infections, and Tyzzer's disease.

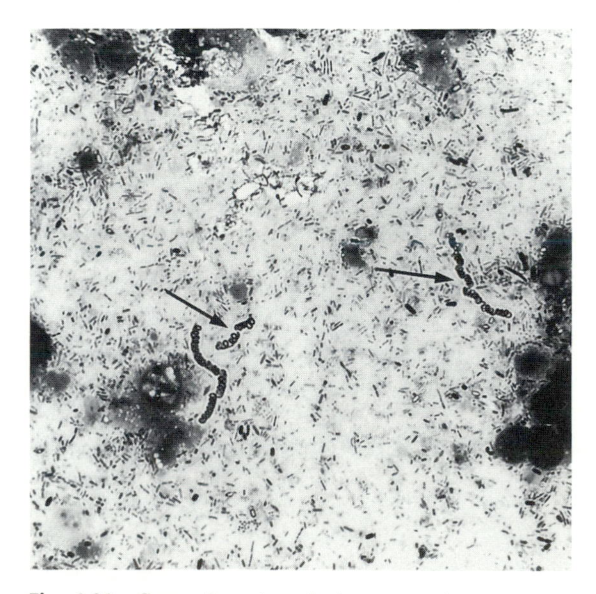

Fig. 6.30. Smear from intestinal contents demonstrating the typical curved and coiled appearance (*arrows*) of *C. spiroforme* (Grain stain).

SIGNIFICANCE. Clostridial infections are recognized as important pathogens in the enteritis complex in this species. A thorough investigation is required in order to identify possible predisposing factors that could upset the normal microbial population of the gut and precipitate an outbreak of the disease. Examples are changes in feed, recent weaning, poor sanitation, adverse environmental conditions, and concurrent infections with other pathogens.

Vibriosis. *Vibrio* organisms have been associated with at least one reported outbreak of enteritis in weanling rabbits. In the cecum, submucosal edema with polymorphonuclear cell infiltration was observed. Enterocytes were flattened and irregular, with focal ulceration. Cecal crypts were hyperplastic, and some crypts were distended with bacteria and cell debris. In Levaditi-stained tissue sections, *Vibrio*-like organisms were demonstrated on the surface and within the cytoplasm of damaged mucosal cells. The organisms in the cecum were confirmed to be *vibrio* by electron microscopy. *Vibrio* were rarely observed in the ceca of controls, and when present, there was no evidence of invasion (Moon et al. 1974). Thus *vibrio* appeared to play a role as a primary pathogen (or copathogen) in these reported cases of acute typhlitis. The prevalence of *Vibrio* infections and the significance of this organism in the enteritis complex has not been determined.

Enterocolitis Associated with *Campylobacter*-like Bacteria. Diarrhea with mortality has been associated with *Campylobacter*-like organisms in suckling, weanling, and young adult domestic rabbits. At necropsy, frequently there were semifluid mucinous contents present, particularly in the colon and rectum. On microscopic examination, varying degrees of involvement were commonly observed in the terminal small intestine, cecum, and colon. The mucosal lesions varied from suppurative and erosive to those that were primarily proliferative in nature. Lesions of the erosive type ranged from focal denuding to segmental loss of enterocytes, with polymorphonuclear cell infiltration in the underlying areas. The proliferative lesions were characterized by multifocal to diffuse hyperplasia of enterocytes lining crypts with mononuclear cell infiltration. Macrophages, and occasionally multinucleated giant cells, were present in the lamina propria of a few animals. *Campylobacter*-like organisms were demonstrated within reactive enterocytes and/or in the luminal exudate of affected animals by light and electron microscopy (Schoeb and Fox 1990).

SIGNIFICANCE. In view of these findings, it is evident that the proliferative changes similar to those observed in some other species that are attributed to

Campylobacter-like organisms also occur in the domestic rabbit. The prevalence of the organism and the incidence of clinical disease due to this bacterium in lagomorphs is currently unknown.

Mucoid Enteropathy. Mucoid enteropathy (ME) is recognized to be a major cause of disease and mortality in young domestic rabbits. Other names for the syndrome have included "mucoid enteritis," "bloat," and "hypoamylasemia." Clinical signs associated with the disease are as follows: anorexia, lethargy, crouched stance, diarrhea, succussion splash, teeth grinding, cecal impaction, and accumulation of large quantities of clear gelatinous mucus in the colon. Cecal impaction and mucus production occur most often in rabbits that live for 7–14 days after the onset of disease. The morbidity is variable but may be high, particularly in rabbits affected during the postweaning period. There is usually a high mortality rate in affected animals regardless of the treatment used. Rabbits 7–10 wk of age are most often affected, but ages ranging from 5 to 20 wk may be involved in outbreaks of ME (Lelkes and Chang 1987).

EPIZOOTIOLOGY AND PATHOGENESIS. For decades, there have been many theories proposed to explain the etiopathogenesis of the disease. Dietary factors have frequently been implicated in ME. This condition was relatively uncommon prior to the feeding of high-energy commercial rations, and rabbits fed a low-fiber diet have been shown to have a higher incidence of the disease than those fed diets high in fiber. It was suggested that dietary components such as lectins or "allergens" may create an unfavorable environment in the cecum, permitting opportunistic bacteria to proliferate (McLeod and Katz 1986). Rabbits on low-fiber diets have lower cecal acetate levels. This short-chain fatty acid is reported to inhibit the growth of potentially harmful opportunistic bacteria in the cecum. With lower acetate levels, "secretagogues" produced by these bacteria may then stimulate the colonic cells to secrete large quantities of mucus (Toofanian and Hamar 1986). Ligation of the cecum or colon will also induce excessive mucus production similar to the naturally occurring disease. In this study, local antibiotic treatment prevented the disease. It was concluded that ME is a toxin-induced secretory disease occurring secondary to constipation and impaction (Toofanian and Targowski 1983). Bacteria such as *E. coli* and clostridia have been isolated from the large intestine in large numbers in some cases of ME. However, inoculation of susceptible rabbits with these isolates seldom produces disease (Lelkes and Chang 1987).

Studies of the cecal microbial flora have revealed that striking changes occur in rabbits with ME. In normal animals, large numbers of ciliated protozoa and large, metachromatically staining bacilli are present in the cecal contents. Rabbits with ME have a dramatic cecal dysbiosis. Large metachromatic bacilli and ciliated protozoa may be present in small numbers or completely absent, and there is a marked rise in coliform bacteria in the cecal contents in ME. Using cannulated animals, cecal inoculation with cecal contents from affected animals resulted in the disease only after acidification of the cecum. Cecal acidification in cannulated animals resulted in a dysbiosis similar to that present in the naturally occurring disease (Lelkes and Chang 1987). The cecum normally provides a relatively constant internal environment; disruption of this internal milieu may precipitate severe disease. Certain rations promote cecal hyperacidosis and subsequent dysbiosis. The resultant changes may lead to the hypersecretion of intestinal fluid; the loss of water, potassium, and bicarbonates; and frequently death. Microbial instability may occur more often in young animals, where homeostatic mechanisms are poorly developed, thus increasing susceptibility to dietary changes associated with weaning.

PATHOLOGY. At necropsy, the stomach may be distended with fluid and gas. The jejunum is frequently distended with translucent, watery fluid, and the cecum is impacted with dried contents and gas. The colon is usually distended with characteristic clear, gelatinous mucus (Fig. 6.31). Microscopically,

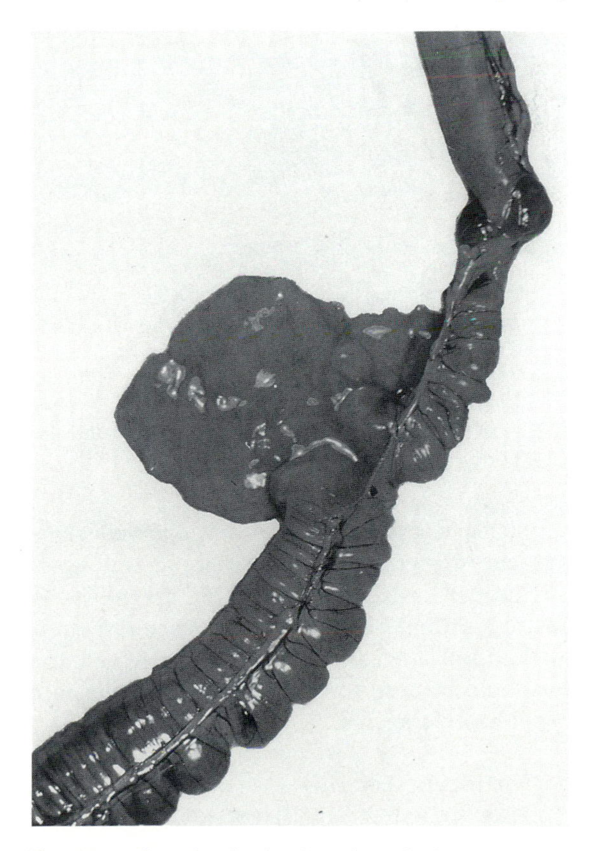

Fig. 6.31. Sacculated colon from juvenile domestic rabbit with mucoid enteropathy. The colon is filled with clear gelatinous material.

there is a striking hyperplasia of goblet cells in the mucosa of the jejunum and ileum, with minimal or no inflammatory response. The prominent goblet cells are readily demonstrated with PAS or alcian blue stains. Lesions are usually minimal to absent in the cecum. In the colon, crypts and the lumen of the gut is distended with mucus and mucous plugs (Fig. 6.32). Hyperplasia of goblet cells in the gallbladder and mild nephrosis have also been described (van Kruinigen and Williams 1972). *Differential diagnoses* include any infectious or management problem that may lead to disruption of the intestinal microflora or function, such as coccidiosis, clostridial infections, hairballs, and constipation of questionable or unknown etiology.

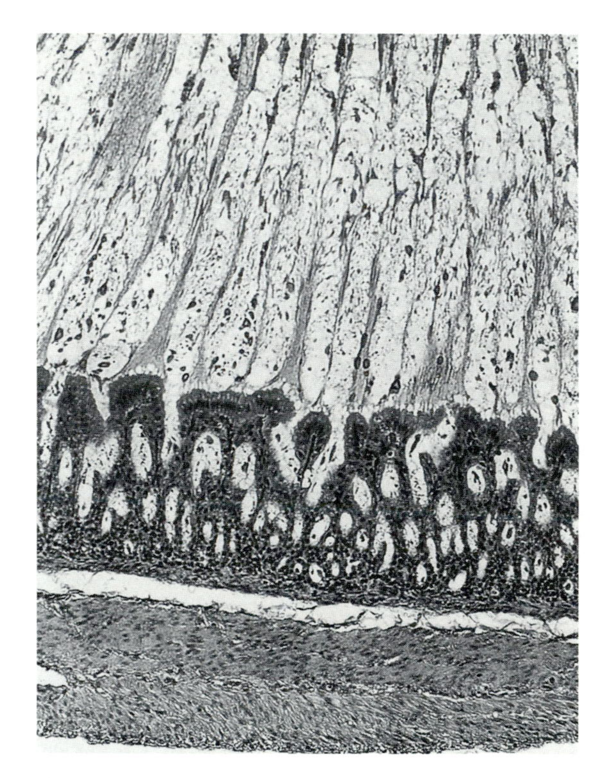

Fig. 6.32. Section from colon of rabbit with mucoid enteropathy. Note the abundant mucus within the lumen and adherent to enterocytes, with minimal inflammatory reaction.

SIGNIFICANCE. The history, clinical signs, and characteristic gross and microscopic findings should provide the basis for a definitive diagnosis. Feeding practices should be investigated. Bacterial culture, fecal flotation, and other procedures deemed appropriate should be used to screen for concurrent infections by pathogenic agents.

Histiocytic Enteritis

EPIZOOTIOLOGY AND PATHOGENESIS. A disease associated with diarrhea and high mortality in young rabbits has been reported from Japan by Umemura and co-workers (1982) and occurs elsewhere. This condition is characterized by the passage of soft to watery feces containing mucus and blood.

PATHOLOGY. At necropsy, there is distension and mucosal thickening of the jejunum and proximal ileum. On microscopic examination, the lamina propria of these affected areas of small intestine is distended with infiltrating inflammatory cells and with distortion of the normal architecture. The cellular infiltrate consists of macrophages, including epithelioid cells and a sprinkling of lymphocytes and plasma cells (Fig. 6.33). The intestinal mucosa overlying the affected areas of jejunum and ileum is relatively normal, with scattered crypt abscesses and a slight increase in goblet cells. Small, slightly curved bacilli have been observed in the cytoplasm of affected enterocytes in toluidine blue–stained 1-μm sections, and in tissues examined by electron microscopy. Organisms in the apical cytoplasm of infected enterocytes resembled organisms seen in enterocytes of hamsters with proliferative ileitis (see Chap. 3) and other species with proliferative bowel disease. In the Japanese study, Ziehl-Neelsen stains for acid-fast bacilli were negative, and no pathogens were recovered on routine bacteriology. A similar condition has been recognized occasionally in young rabbits introduced into research facilities in the United States. Although the lesions are reminiscent of

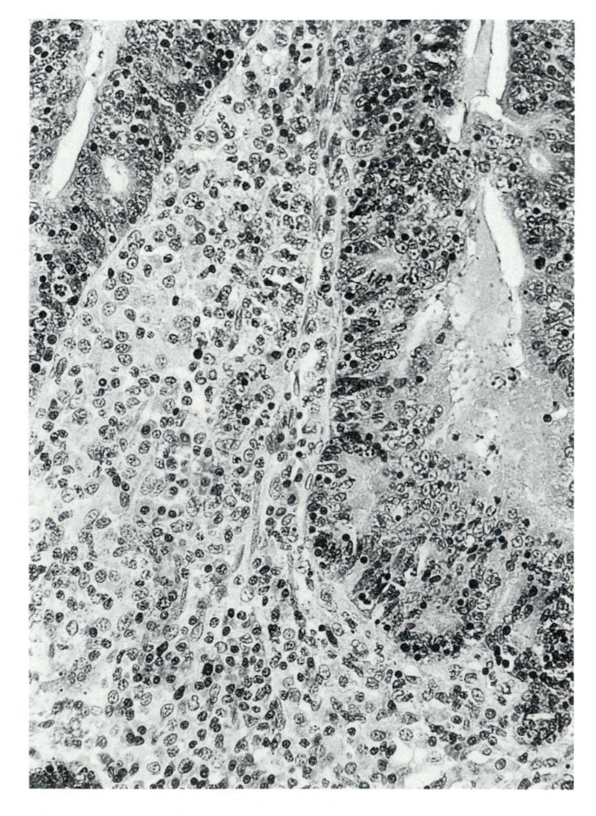

Fig. 6.33. Section of ileum from juvenile rabbit with histiocytic enteritis. There is distortion of the normal mucosal architecture, with marked mononuclear cell infiltration in the lamina propria.

granulomatous enteritides seen in other species, the etiopathogenesis of the condition has not been determined.

Bibliography for Bacterial Infections

Pasteurellosis

Al-Lebban, Z.S., et al. 1988. Rabbit pasteurellosis: Induced disease and vaccination. Am. J. Vet. Res. 49:312–16.

Chengappa, M.M., et al. 1980. A streptomycin-dependent live *Pasteurella multocida* vaccine for the prevention of rabbit pasteurellosis. Lab. Anim. Sci. 30:515–18.

Corbeil, L.B., et al. 1983. Immunity to pasteurellosis in compromised rabbits. Am. J. Vet. Res. 44:845–50.

Dhillon, A.S., and Andrews, D.K. 1982. Abortions, stillbirths, and infant mortality in a commercial rabbitry. J. Appl. Rabbit Res. 5:97–98.

DiGiacomo, R.F., et al. 1989. Atrophic rhinitis in New Zealand rabbits. Lab. Anim. Sci. 37:533.

DiGiacomo, R.F., et al. 1987. Transmission of *Pasteurella multocida* in rabbits. Lab. Anim. Sci. 37:621–23.

DiGiacomo, R.F., et al. 1983. Natural history of infection with *Pasteurella multocida* in rabbits. J. Am. Vet. Med. Assoc. 183:1172–75.

Flatt, R.E. 1974. Bacterial diseases. In *The Biology of the Laboratory Rabbit*, ed. S.H. Weisbroth et al., pp. 194–236. New York: Academic.

Flatt, R.E., et al. 1977. Suppurative otitis media in the rabbit: Prevalence, pathology and microbiology. Lab. Anim. Sci. 27:343–46.

Glass, L.S., and Beasley, J.N. 1990. Infection with and antibody response to *Pasteurella multocida* and *Bordetella bronchiseptica* in immature rabbits. Lab. Anim. Sci. 39:406–10.

Glorioso, J.C., et al. 1982. Adhesion of type A *Pasteurella multocida* to rabbit pharyngeal cells and its possible role in rabbit respiratory tract infections. Infect. Immunol. 35:1103–09.

Holmes, H.T., et al. 1986. Serologic methods for detection of *Pasteurella multocida* infections in nasal culture negative rabbits. Lab. Anim. Sci. 36:640–45.

Holmes, H.T., et al. 1983a. The incidence of vaginal and nasal *Pasteurella multocida* in a commercial rabbitry. J. Appl. Rabbit Res. 6:95–96.

———. 1983b. Pasteurella contaminated water valves: Its incidence and implications. J. Appl. Rabbit Res. 6:123–24.

Lu, Y-S., and Pakes, S.P. 1981. Protection of rabbits against experimental pasteurellosis by a streptomycin-dependent *Pasteurella multocida* serotype 3A live mutant vaccine. Infect. Immunol. 34:1018–24.

Manning, P.J., et al. 1987. A dot-immunobinding assay for the serodiagnosis of *Pasteurella multocida* infection in laboratory rabbits. Lab. Anim. Sci. 37:615–20.

Nielsen, J.P. 1989. Personal communication. National Veterinary Laboratory, Copenhagen, Denmark.

Percy, D.H., et al. 1988. Incidence of *Pasteurella* and *Bordetella* infections in fryer rabbits: An abattoir survey. J. Appl. Rabbit Res. 11:245–46.

Percy, D.H., et al. 1984. Characterization of *Pasteurella* isolated from rabbits in Canada. Can. J. Comp. Med. 48:36–41.

Ringler, D.H., et al. 1985. Protection of rabbits against experimental pasteurellosis by vaccination with a potassium thiocyanate extract of *Pasteurella multocida*. Infect. Immunol. 49:498–504.

Scharf, R.A., et al. 1981. A modified barrier system for the maintenance of *Pasteurella*-free rabbits. Lab. Anim. Sci. 31:513–15.

Watson, W.T., et al. 1975. Experimental respiratory infections with *Pasteurella multocida* and *Bordetella bronchiseptica* in rabbits. Lab. Anim. Sci. 25:459–64.

Webster, L.T. 1925. Epidemiological studies on respiratory infections of the rabbit. Pneumonia associated with *Bacterium leptisepticum*. J. Exp. Med. 43:555–72.

Zaoutis, T.E., et al. 1991. Screening rabbit colonies for antibodies to *Pasteurella multocida* by an ELISA. Lab. Anim. Sci. 41:419–22.

Bordetella bronchiseptica Infections

Bemis, D.A., and Wilson, S.A. 1985. Influence of potential virulence determinants on *Bordetella bronchiseptica*–induced ciliostasis. Infect. Immunol. 50:35–42.

Feinstein, R.E., and Rehbinder, C. 1988. Health monitoring of purpose bred laboratory rabbits in Sweden: Major findings. Scand. J. Lab. Anim. Sci. 15:49–67.

Matsuyama, T., and Taking, T. 1980. Scanning electron microscopic studies of *Bordetella bronchiseptica* on the rabbit tracheal mucosa. J. Med. Microbiol. 13:159–61.

Percy, D.H., et al. 1988. Incidence of *Pasteurella* and *Bordetella* infections in fryer rabbits: An abattoir survey. J. Appl. Rabbit Res. 11:245–46.

Watson, W.T., et al. 1975. Experimental respiratory infections with *Pasteurella multocida* and *Bordetella bronchiseptica* in rabbits. Lab. Anim. Sci. 25:459–64.

Respiratory Infections due to CAR Bacillus

Kurisu, K., et al. 1990. Cilia-associated respiratory bacillus infection in rabbits. Lab. Anim. Sci. 40:413–15.

Staphylococcal Infections

Hagen, K.W. 1963. Disseminated staphylococcal infection in young domestic rabbits. J. Am. Med. Assoc. 142:1421–22.

Okerman, L., et al. 1984. Cutaneous staphylococcosis in rabbits. Vet. Rec. 114:313–15.

Osebald, J.W., and Gray, D.M. 1960. Disseminated staphylococcal infection in wild jack rabbits. J. Infect. Dis. 106:91–94.

Renquist, D., and Soave, O. 1969. Staphylococcal pneumonia in a laboratory rabbit: An epidemiological follow-up study. J. Am. Med. Assoc. 155:1221–23.

Snyder, S.B., et al. 1976. Disseminated staphylococcal disease in the laboratory rabbit (*Oryctolagus cuniculus*). Lab. Anim. Sci. 26:86–88.

Venereal Spirochetosis

Cunliffe-Beamer, T.L., and Fox, R.R. 1981a. Venereal spirochaetosis of rabbits: Description and diagnosis. Lab. Anim. Sci. 31:366–71.

———. 1981b. Venereal spirochaetosis of rabbits: Epizootiology. Lab. Anim. Sci. 31:372–78.

Flatt 1974 (*see* General Bibliography).

Listeriosis

Gray, M.L., and Killinger, A.H. 1966. *Listeria moncytogenes* and listeriosis. Bacteriol. Rev. 30:309–82.

Murray, E.G.D., et al. 1926. A disease of rabbits characterized by a large mononuclear leucocytosis caused by a hitherto undescribed bacillus *Bacterium monocytogenes* (n.sp.). J. Pathol. Bacteriol. 40:407–39.

Vetesi, F., and Kemenes, F. 1967. Studies on listeriosis in pregnant rabbits. Acta Vet. Acad. Sci. Hung. 17:27–38.

Watson, G.L., and Evans, M.G. 1985. Listeriosis in a rabbit. Vet. Pathol. 22:191–93.

Salmonellosis

Newcomer, C.E., et al. 1983. The laboratory rabbit as reservoirs of *Salmonella mbandaka*. J. Infect. Dis. 147:365.

Yersiniosis, Bacterial Mastitis, and Other Bacterial Diseases

Flatt 1974 (*see* General Bibliography).

Tyzzer's Disease

Allen, A.M., et al. 1965. Tyzzer's disease syndrome in laboratory rabbits. Am. J. Pathol. 46:859–82.

Boot, R., and Walvoort, H.C. 1984. Vertical transmission of *Bacillus piliformis* infection (Tyzzer's disease) in a guinea pig: Case report. Lab. Anim. 18:195–99.

Carter, G.R., et al. 1969. Natural Tyzzer's disease in Mongolian gerbils (*Meriones unguiculatus*). Lab. Anim. Care 19:648–51.

Cutlip, R.C., et al. 1971. An epizootic of Tyzzer's disease in rabbits. Lab. Anim. Sci. 21:356–61.

Fries, A.S. 1979. Studies on Tyzzer's disease: Transplacental transmission of *B. piliformis* in rats. Lab. Anim. 13:43–46.

______. 1977. Studies on Tyzzer's disease: Application of immunofluorescence for detection of *Bacillus piliformis* and for demonstration and determination of antibodies to it in sera from mice and rabbits. Lab. Anim. 11:69–73.

Fujiwara, K. 1978. Tyzzer's disease. Jpn. J. Exp. Med. 48:467–80.

Fujiwara, K., et al. 1985. Antigenic relatedness of Tyzzer's disease occurring in Japan and other regions. Jpn. J. Vet. Sci. 47:9–16.

Ganaway, J.R., et al. 1976. Tyzzer's disease in free-living cottontail rabbits (*Sylvilagus floridanus*) in Maryland. J. Wildl. Dis. 12:545–49.

Ganaway, J.R., et al. 1971. Tyzzer's disease. Am. J. Pathol. 64:717–32.

Jonas, A.M., et al. 1970. Tyzzer's disease in the rat: Its possible relationship with megaloileitis. Arch. Pathol. 90:516–28.

Kovatch, R.M., and Zebarth, G. 1973. Naturally occurring Tyzzer's disease in a cat. J. Am. Vet. Med. Assoc. 162:136–38.

McLeod, C.G., et al. 1977. Intestinal Tyzzer's disease and spirochetosis in a guinea pig. Vet. Pathol. 14:229–35.

Nakayama, M., et al. 1975. Transmissible gastroenteritis in hamsters caused by Tyzzer's organism. Jpn. J. Exp. Med. 45:33–41.

Peeters, J.E., et al. 1985. Naturally-occurring Tyzzer's disease (*Bacillus piliformis* infection) in commercial rabbits: A clinical and pathological study. Ann. Rech. Vet. 16:69–79.

Pulley, L.T., and Shively, J.N. 1974. Tyzzer's disease in a foal. Light and electron-microscopic observation. Vet. Pathol. 11:203–11.

Quereshi, S.R., et al. 1976. Tyzzer's disease in a dog. J. Am. Vet. Med. Assoc. 168:602–4.

Tyzzer, E.E. 1917. A fatal disease of the Japanese waltzing mouse caused by a spore-bearing bacillus (*B. piliformis* n.sp.). J. Med. Res. 37:307–38.

Waggie, K.S., et al. 1987. Lesions of experimentally induced Tyzzer's disease in Syrian hamsters, guinea pigs, mice and rats. Lab. Anim. 21:155–60.

Waggie, K.S., et al. 1981. A study of mouse strain susceptibility to *B. piliformis* (Tyzzer's disease): The association of B-cell function and resistance. Lab. Anim. Sci. 31:139–42.

Webb, D.M., et al. 1987. *Bacillus piliformis* infection (Tyzzer's disease) in a calf. J. Am. Vet. Med. Assoc. 191:431–34.

Coliform Enteritis

Moon, H.W., et al. 1983. Attaching and effacing of rabbit and human enteropathogenic *Escherichia coli* in pig and rabbit intestines. Infect. Immunol. 41:1340–51.

Peeters, J.E., et al. 1988a. A selective citrate-sorbose medium for screening certain enteropathogenic attaching and effacing *Escherichia coli* in weaned rabbits. Tijdschrift Diergeneesk. 57:264–70.

Peeters, J.E., et al. 1988b. Biotype, serotype, and pathogenicity of attaching and effacing enteropathogenic *Escherichia coli* strains isolated from diarrheic commercial rabbits. Infect. Immunol. 56:1442–48.

Peeters, J.E., et al. 1985. Scanning and transmission electron microscopy of attaching effacing *Escherichia coli* in weanling rabbits. Vet. Pathol. 22:54–59.

Peeters, J.E., et al. 1984a. Experimental *Escherichia coli* enteropathy in weanling rabbits: Clinical manifestations and pathological findings. J. Comp. Pathol. 94:521–28.

Peeters, J.E., et al. 1984b. Pathogenic properties of *Escherichia coli* strains isolated from diarrheic commercial rabbits. J. Clin. Microbiol. 20:34–39.

Prescott, J.F. 1978. *Escherichia coli* and diarrhea in the rabbit. Vet. Pathol. 15:237–48.

______. 1977. Rabbit cecal flora in health and disease. Ph.D. diss., University of Cambridge.

Smith, H.W. 1965. Observations on the flora of the alimentary tract of animals and factors affecting its composition. J. Pathol. Bacteriol. 89:95–122.

Clostridial Enteropathies/Enterotoxemia

Carman, R.J., and Borriello, S.P. 1984. Infectious nature of *Clostridium spiroforme*-mediated rabbit enterotoxaemia. Vet. Microbiol. 9:497–502.

Carman, R.J., and Evans, R.H. 1984. Experimental and spontaneous clostridial enteropathies of laboratory and free living lagomorphs. Lab. Anim. Sci. 34:443–52.

Cheeke, P.R., and Patton, N.M. 1978. Effect of alfalfa and dietary fiber on the growth performance of weanling rabbits. Lab. Anim. Sci. 28:167–72.

Holmes, H.T., et al. 1988. Isolation of *Clostridium spiroforme* from rabbits. Lab. Anim. Sci. 39:167–68.

Patton, N.M., et al. 1978. Enterotoxemia in rabbits. Lab. Anim. Sci. 28:536–40.

Peeters, J.E., et al. 1986. Significance of *Clostridium spiroforme* in the enteritis-complex of commercial rabbits. Vet. Microbiol. 12:25–31.

Rehg, J.E., and Lu, Y-S. 1981. *Clostridium difficile* colitis in a rabbit following antibiotic therapy for pasteurellosis. J. Am. Vet. Med. Assoc. 179:1296–97.

Rehg, J.E., and Pakes, S.P. 1982. Implication of *Clostridium difficile* and *Clostridium perfringens* iota toxins in experimental lincomycin-associated colitis of rabbits. Lab. Anim. Sci. 32:253–57.

Vibriosis

Moon, H.W., et al. 1974. Intraepithelial vibrio associated with acute typhlitis of young rabbits. Vet. Pathol. 11:313–26.

Enterocolitis Associated with *Campylobacter*-like Bacteria

Schoeb, T.R., and Fox, J.G. 1990. Enterocolitis associated with intraepithelial *Campylobacter*-like bacteria in rabbits (*Oryctolagus cuniculus*). Vet. Pathol. 27:73–80.

Mucoid Enteropathy

Lelkes, L., and Chang, C-L. 1987. Microbial dysbiosis in rabbit mucoid enteropathy. Lab. Anim. Sci. 37:757–64.

McLeod, C.G., and Katz, W. 1986. Toxic components in commercial rabbit feeds and their role in mucoid enteritis. S. Afr. J. Sci. 82:375–79.

Toofanian, F., and Hamar, D.W. 1986. Cecal short-chain fatty acids in experimental rabbit mucoid enteropathy. Am. J. Vet. Res. 47:2423–25.

Toofanian, F., and Targowski, S. 1983. Experimental production of rabbit mucoid enteritis. Am. J. Vet. Res. 44:705–8.

van Kruiningen, H.J., and Williams, C.B. 1972. Mucoid enteritis in rabbits: Comparison to cholera and cystic fibrosis. Vet. Pathol. 9:53–77.

Histiocytic Enteritis

Umemura, T., et al. 1982. Histiocytic enteritis of rabbits. Vet. Pathol. 19:326–29.

General Bibliography

Flatt, R.E. 1974. Bacterial diseases. In *The Biology of the Laboratory Rabbit*, ed. S.H. Weisbroth et al., pp. 194–236. New York: Academic.

Harkness, J.E., and Wagner, J.E. 1989. *The Biology and Medicine of Rabbits and Rodents*. Philadelphia: Lea and Febiger.

Kraus, A.L., et al. 1984. Biology and diseases of rabbits. In *Laboratory Animal Medicine*, ed. J.G. Fox et al., pp. 207–40. New York: Academic.

Percy, D.H., et al. 1993. The enteritis complex in domestic rabbits: A field study. Can. Vet. J. 34: 95–102.

MYCOTIC INFECTIONS

Dermatophytosis ("Ringworm")

Clinical cases of dermatophytosis are relatively uncommon in domestic lagomorphs. When present, lesions are usually located around the head and ears, sometimes with secondary spread to the paws. Affected areas are typically raised, circumscribed, and erythematous, with crusted surface and hair loss (Fig. 6.34). *Trichophyton mentagrophytes* is most frequently involved (Banks and Clarkson 1967), but *Microsporum canis* infections have also been recognized in rabbits (Vogtsberger et al. 1986). Microscopic examination of skin scrapings from the periphery of lesions cleared in 10% KOH should reveal the typical arthrospores. Examination of tissue sections for the characteristic fungi and culture on the appropriate media are both useful diagnostic procedures in confirming the diagnosis. On histopathology, characteristic changes include hyperkeratosis, epidermal hyperplasia, and folliculitis, with mononuclear and polymorphonuclear cell infiltration. Stains such as the methenamine silver and PAS-staining procedures are used to best demonstrate the typical arthrospores investing infected hair shafts (Fig. 6.35). *Differential diagnoses* include idiopathic "molt," hair loss in does during nest building, and "barbering" seen occasionally in group-housed juvenile rabbits.

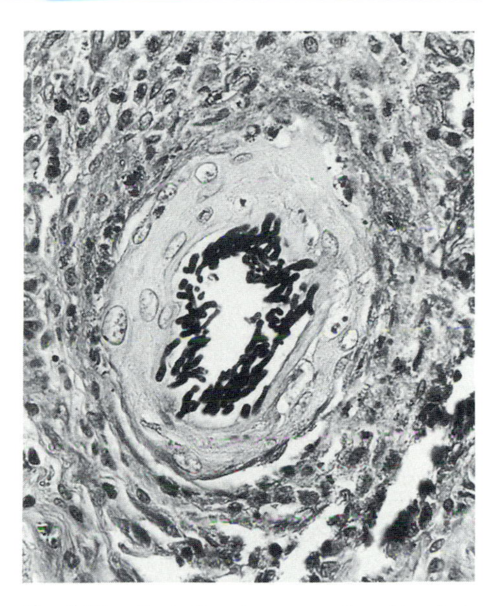

Fig. 6.35. Typical arthrospores investing hair follicles in rabbit (PAS stain).

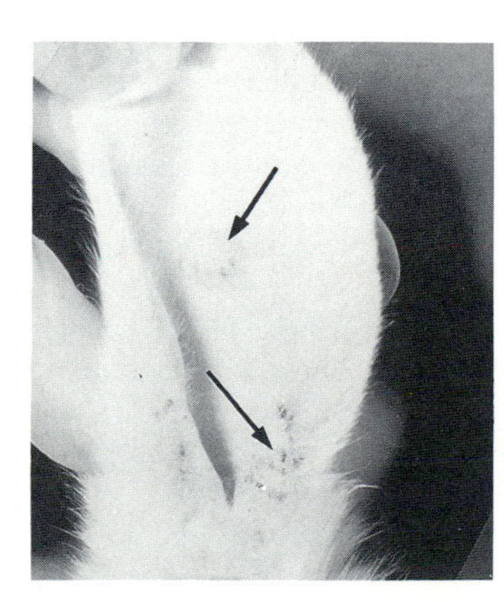

Fig. 6.34. Raised, circumscribed scaling lesions with reddened periphery (*arrows*) in a spontaneous case of dermatophytosis due to *Trichophyton mentagrophytes* infection.

SIGNIFICANCE. The diagnosis of dermatophytosis then requires a thorough investigation to determine the possible sources of the infection. The disease is readily transmitted to susceptible human contacts; thus careful screening, culling, and slaughter are recommended. If animals are to be treated, oral griseofulvin has been used with some success (Hagen 1969). Rabbits may occasionally harbor pathogenic dermatophytes as an inapparent infection.

Aspergillus Infection

Pulmonary granulomas are occasionally encountered in rabbits at necropsy. They consist of circumscribed inflammatory lesions with a central area of coagulation necrosis with mononuclear inflammatory cell response. Typical septate hyphae are evident, particularly with PAS or methenamine silver stains (Fig. 6.36).

Bibliography for Mycotic Infections

Banks, K.L., and Clarkson, T.B. 1967. Naturally occurring dermatomycosis in the rabbit. J. Am. Med. Assoc. 151:926–29.

Hagen, K.W. 1969. Ringworm in domestic rabbits: Oral treatment with griseofulvin. Lab. Anim. Care 19:635–38.

Vogtsberger, L.M., et al. 1986. Spontaneous dermatomycosis due to *Microsporum canis* in rabbits. Lab. Anim. Sci. 36:294–97.

PARASITIC DISEASES

Endoparasitic Infections

Protozoal Infections

Intestinal Coccidiosis. Coccidiosis is a widespread, major disease problem in commercial rabbitries and in research facilities. Of the species of *Eimeria*

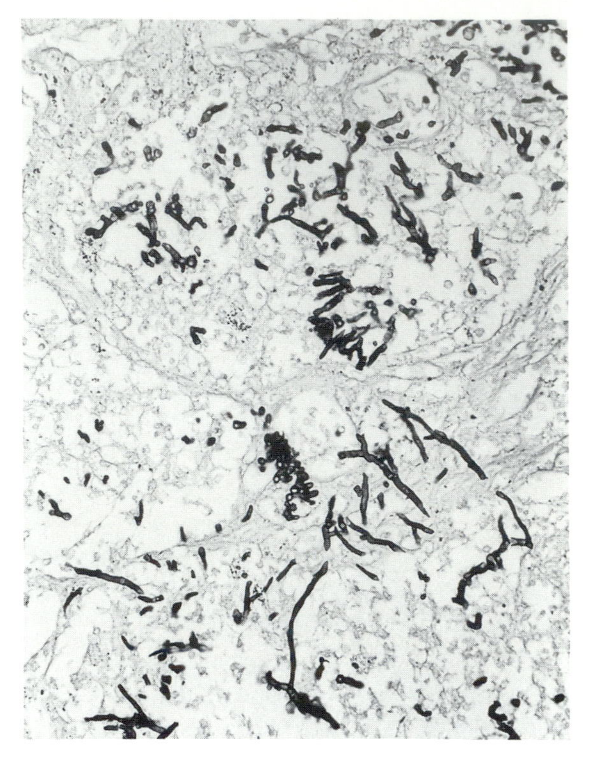

Fig. 6.36. Section of granuloma from lung of domestic rabbit (PAS stain), illustrating hyphae associated with focal pulmonary aspergillosis.

associated with intestinal coccidiosis in the rabbit *E. intestinalis* and *E. flavescens* are considered the most pathogenic; *E. magna*, *E. irresidua,* and *E. piriformis* moderately pathogenic; and *E. perforans*, *E. neoleporis*, and *E. media* the least pathogenic (Varga 1982). In one survey in England, *E. magna*, *E. media*, and *E. perforans* were most frequently identified. Over 65% of the animals tested were carrying two to four species of *Eimeria* (Catchpole and Norton 1979). In one study of experimental coccidiosis, 5-wk-old coccidia-free rabbits were inoculated with *E. intestinalis*, a relatively pathogenic species. Rabbits inoculated with 100 oocysts or less were asymptomatic. In animals receiving 10,000 or more sporulated oocysts, the mortality rate was around 50%. Hemodilution, hypokalemia, and a marked rise in *E. coli* bacterial counts occurred during the course of the disease. Enterocyte destruction and villous atrophy was evident at 7–10 days postinoculation, with repair in survivors evident by 2 wk postinoculation (Peeters et al. 1984).

LIFE CYCLE. Following passage in the feces, oocysts require 1 or more days to sporulate at room temperature before they are infective. When ingested, sporulated oocysts release sporozoites, which invade enterocytes and multiply by schizogony. Depending on the species of *Eimeria*, one or more asexual cycles then occur, followed by gametogony and oocyst passage in the feces. The prepatent period is from 5 to 12 days, depending on the species. There may be a phenomenal number of progeny from a single ingested oocyst. One oocyst of *E. magna* may produce over 25,000,000 oocysts in a susceptible host (Varga 1982).

EPIZOOTIOLOGY AND PATHOGENESIS. Rabbits most frequently develop clinical disease during the weaning period. The most damaging stage in the life cycle is the sexual cycle, where there may be extensive destruction of enterocytes and cells in the lamina propria in affected sections of the gut. Because oocysts require sporulation at room temperature before they are infective, reingestion of the "night feces" (cecotrophy) does not appear to play a role in the dissemination of the disease. In well-managed operations where coccidiostats are not used, control is dependent on rigorous sanitation practices. Exposure to relatively small numbers of oocysts should result in a subclinical infection, with appropriate immune response. However, immunity to one species of *Eimeria* is unlikely to provide good protection against other species. In many commercial operations, the feed is routinely medicated with anticoccidials to control the disease. However, this should not be considered an acceptable substitute for rigid sanitation practices.

PATHOLOGY. At necropsy, the perineal region and belly are frequently smeared with watery dark green to brown feces. The animal may be thin and dehydrated, depending on the duration of the disease. The cecum and colon contain dark green to brown watery, foul-smelling material. The mucosa of affected areas of the gut is congested and edematous, occasionally with hemorrhagic areas.

The location of microscopic changes varies with the species of *Eimeria* involved, but they are normally concentrated in the caudal half of the small intestine and in the cecum. In acute coccidiosis, there is destruction of enterocytes, villous atrophy in affected areas of small intestine, and marked leukocytic infiltration, heterophils predominating. Gametocytes and oocysts are frequently evident in the intestinal mucosa in affected areas (Fig. 6.37).

DIAGNOSIS. Fecal flotations, mucosal scrapings, and microscopic examination for oocysts is a standard diagnostic procedure. An approximate oocyst count and oocyst speciations are recommended, particularly in view of the recognized variation in pathogenicity among species of *Eimeria*. In acute cases of coccidiosis, oocysts may not be present in the feces but will be evident in sections of the appropriate areas of small and large intestine. Bacteriology cultures are essential, since there frequently is a significant rise in the bacterial count, especially coliforms. These bacteria may play an important role as opportunistic infections in clinical outbreaks of coccidiosis. *Differential diagnoses* include coliform enteritis, Tyzzer's disease, clostridial enteropathies, viral enteritides, and mucoid enteropathy.

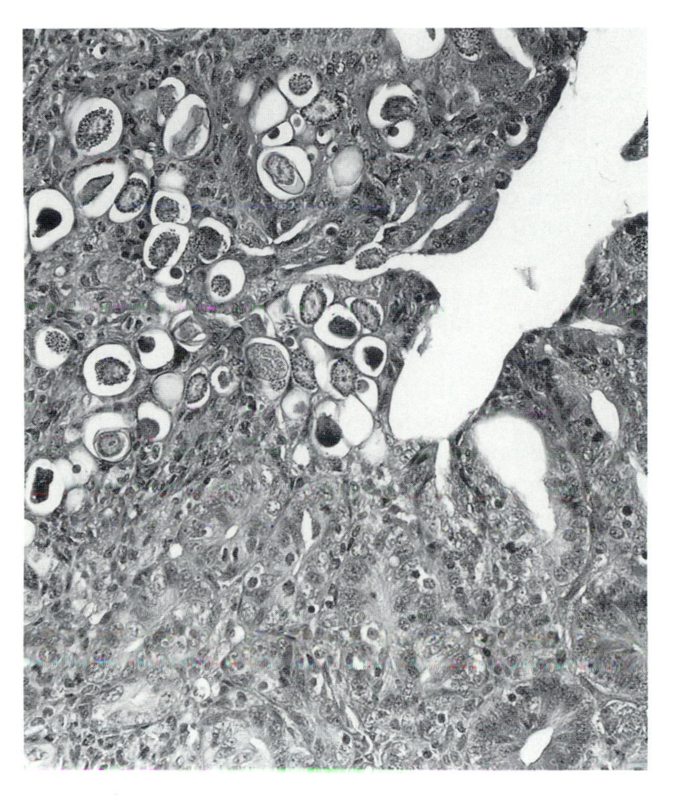

Fig. 6.37. Cecum from rabbit approximately 10 wk of age with coccidiosis. Large numbers of micro- and macrogametocytes are present in the cecal mucosa. There is marked infiltration with leukocytes in the lamina propria, with the disruption of the normal architecture.

SIGNIFICANCE. Coccidiosis is an important cause of clinical (or subclinical) disease in domestic rabbits, frequently causing weight loss and mortality. Events such as changes in management practices, feed changes, and experimental procedures may be sufficient to precipitate a clinical outbreak of the disease.

Hepatic Coccidiosis

EPIZOOTIOLOGY AND PATHOGENESIS. *Eimeria stiedae* infestations occur in both domestic and wild rabbits and represent an important cause of poor weight gains, disease, and mortality in commercial rabbitries. Following the ingestion of sporulated oocysts, sporozoites invade the duodenal mucosa and migrate to the lamina propria prior to systemic migration. Sporozoites have been demonstrated in the regional mesenteric lymph nodes within 12 hr postexposure and in the liver by 48 hr. Organisms have been reported to migrate to the liver in mononuclear cells via lymphatics (Horton 1967). However, viable sporozoites have also been demonstrated in the peripheral blood and bone marrow in *E. stiedae*–inoculated rabbits, and the hematogenous route has been proposed as means of migration to the liver (Owen 1970). After migration, sporozoites invade the epithelial cells of the bile ducts and schizogony begins. Following the gametogenous stage, oocysts are formed, released into the bile ducts, and passed to the intestine. The prepatent period is approximately 15–18 days. Oocysts may be shed in the feces for up to 7 or more wk postexposure. Oocysts are normally resistant to environmental change; thus contaminated premises and fomites may be a source of infective sporulated oocysts for several months. *E. stiedae* infestations may be manifest either as clinical or subclinical disease.

Weanling rabbits are most often affected. Frequently a significant number of livers collected from fryer rabbits in abattoirs are condemned because of hepatic coccidiosis. A dose-related effect has been observed in experimentally infected animals. In young rabbits inoculated orally with varying numbers of oocysts (100 to 100,000 per animal), mortality rates in animals that received either 10,000 or 100,000 oocysts were 40% or 80%, respectively. No fatalities occurred at lower dosages (Barriga and Arnoni 1979). Significant variations in liver enzymes and blood chemistry have been observed during the course of the disease. Four stages have been proposed: (1) the initial stage of metabolic dysfunction that coincides with hepatocyte damage during schizogony; (2) the cholestatic stage, with elevated transaminases and serum bilirubin; (3) the stage of metabolic dysfunction, characterized by hypoglycemia and hypoproteinemia; and (4) the period of immunodepression in heavily infested animals resulting in an inability to curtail the production of oocysts in the biliary system (Barriga and Arnoni 1981).

PATHOLOGY. At necropsy, affected animals are frequently thin and pot-bellied and lack body fat reserves. There may be dark brown to green soiling in the perineal region. Ascites is a variable finding. Depending on the degree of liver involvement, there may be hepatomegaly and in severe cases icterus. In the liver, there are variable numbers of raised, bosselated, yellow to pearl gray circumscribed lesions 0.5–2 cm in diameter scattered throughout the parenchyma of the liver. On cut surface, lesions contain fluid green to inspissated, dark green to tan material (Fig. 6.38). The gallbladder is thickened and contains viscid green bile and debris.

Microscopically, there is marked dilation of bile ducts, extensive periportal fibrosis, and mixed inflammatory cell infiltration in the periportal regions. In affected bile ducts, there is hyperplasia of epithelium, with papillary projections lined by reactive epithelial cells overlying collagenous tissue stroma. Infiltrating periductal inflammatory cells include lymphocytes, macrophages, and a sprinkling of polymorphs. Large numbers of gametocytes and oocysts are usually present in parasitized ducts (Fig. 6.39). In lesions of some duration, organisms may be sparse to absent in bile ducts, with prominent periportal fibrosis.

Fig. 6.38. Cut surface of liver from a case of hepatic coccidiosis (*Eimeria stiedae*). Bile ducts are dilated and filled with inspissated material.

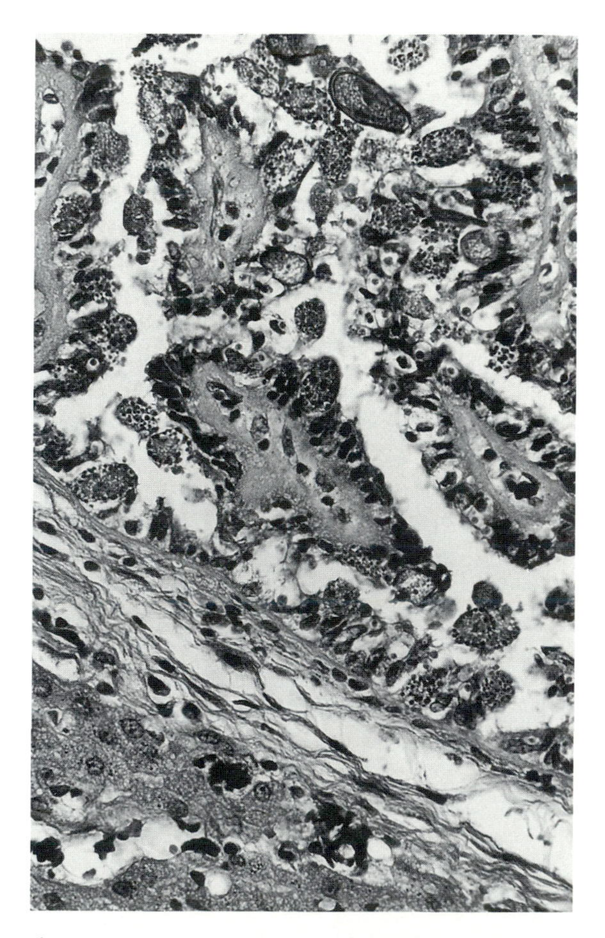

Fig. 6.39. Histological section, illustrating hepatic coccidiosis. There is a proliferative cholangitis with periportal fibrosis. Gametocytes and oocysts are present in bile duct epithelium and in the lumen of dilated ducts.

DIAGNOSIS. The diagnosis may be confirmed at necropsy by wet mount preparations. Oocysts are usually readily observed in aspirates from the gallbladder or in impression smears of sectioned lesions. The characteristic proliferative biliary changes and organisms seen histologically are pathognomonic of the disease.

SIGNIFICANCE. Aside from clinical disease and mortality associated with hepatic coccidiosis, growth rates may be compromised significantly in affected rabbits. The changes in serum chemistry seen during the acute and convalescent stages of the disease indicate that significant metabolic aberrations do occur. There is some evidence that there may be an impaired immune response in rabbits heavily infected with *E. stiedae* (Barriga and Arnoni 1981). Improved sanitation is a major consideration in the effective control of the disease.

Encephalitozoonosis (Nosematosis). *Encephalitozoon cuniculi* is an obligate intracellular microsporidian parasite that affects a variety of mammalian hosts, most commonly the domestic rabbit. The organism is characterized by the presence of a coiled polar filament in the mature spore stage. Taxonomists have disagreed on the appropriate classification for the organism. In the past, it was classified in the genus *Encephalitozoon*. This was later changed to *Nosema*. However, by electron microscopy, the characteristic diplokarya seen during the developmental cycle in protozoa of the *Nosema* genus were not present in this parasite, thus the reversion to the genus *Encephalitozoon*. Following the extrusion of the sporoplasm from the spore coat, the sporoplasm may then invade a susceptible host cell. Penetration may be due to the mechanical forces exerted by the extruded polar filament or due to an active migratory process by the sporoplasm. Following entry into the cell, multiplication occurs in association with a cytoplasmic vacuole. Sporoblasts develop into mature spores and finally the cell ruptures, releasing organisms that can then repeat the cycle (Pakes 1974).

EPIZOOTIOLOGY AND PATHOGENESIS. "Infectious motor paralysis" attributed to a protozoan parasite was first reported in laboratory rabbits by Wright and Craighead (1922). Encephalitozoonosis is now recognized to be widespread in domestic rabbits. The incidence of seropositive animals in some conventional colonies may be relatively high, and seropositive animals have also been detected in specific-pathogen-free rabbits (Bywater and Kellet 1978)). Rabbits seropositive for *E. cuniculi* have not been observed in wild rabbits (Cox and Ross 1980); thus the disease may be confined to domesticated lagomorphs. The organism has a wide host range. Susceptible species include the mouse (Innes et al. 1962), guinea pig (Moffat and Schiefer 1973), squirrel monkey (Anver et al. 1972), cat (van Rensburg and Du Plessis 1971), and dog (Plowright 1952). In general, encephalitozoonosis appears to be a more severe disease in species such as dogs and monkeys. In the rabbit, the disease is usually a subclinical infection, and renal lesions are frequently detected as an incidental finding. In surveys on the incidence of diseased kidneys attributed to *E. cuniculi* at slaughter or necropsy, positive cases have varied from approximately 5% to over 25% (Flatt and Jackson 1970; Robinson 1954). Occasionally nervous signs, with mortality, occur in young New Zealand White rabbits with heavy infestations. Dwarf rabbits appear to be especially susceptible to encephalitozoonosis, with clinical signs. Torticollis and other neurological manifestations attributed to *E. cuniculi* were observed in a series of pet dwarf rabbits. Renal lesions were also observed (Kunstyr and Nauman 1985).

PATHOGENESIS. The usual source of the infection is considered to be spores shed in the urine from rabbits actively infected with the disease. Transplacental infection has been reported to occur, although there is disagreement on this issue (Owen 1980; Hunt et al. 1972). Rabbits are readily infected experimentally by the oral or respiratory route, and invasion by inhalation has been identified as a possible portal of entry under field conditions (Cox et al. 1979). Following ingestion/oral inoculation, it appears unlikely that multiplication occurs in the intestinal mucosa. The spores appear to pass via infected mononuclear cells into the systemic circulation. Initially, target organs are those of high blood flow, such as lung, liver, and kidney. In rabbits inoculated orally with *E. cuniculi* and examined at 31 days postinoculation (pi), moderate to marked lesions were demonstrated primarily in the lung, liver, and kidney, and occasionally in the myocardium. No lesions were present in the central nervous system at 1 mo pi. At 3 mo pi, moderate to severe lesions were evident histologically in the kidney, and minimal in the lung, liver, and heart. Lesions were evident in the brain at this stage postexposure (Cox et al. 1979). Serum titers may be detectable by 3–4 wk pi and reach high titers by 6–9 wk pi. Spores have been seen in the urine at 1 mo pi and may be excreted in large numbers up to 2 mo pi. Only small numbers are excreted thereafter. Shedding of spores is essentially terminated by 3 mo pi. Spores survive for less than 1 wk at 4°C but may remain viable for at least 6 wk at 22°C.

PATHOLOGY. At necropsy, affected animals are usually in good flesh, and frequently lesions seen macroscopically are regarded as an incidental finding. Lesions are usually confined to the kidney and appear as focal, irregular, depressed pale areas 1–100 mm in diameter. In severely affected kidneys, lesions frequently coalesce with adjacent foci (Figs. 6.40 and 6.41). On cut surface, indistinct, linear, pale gray-white areas may extend into the underlying cortex.

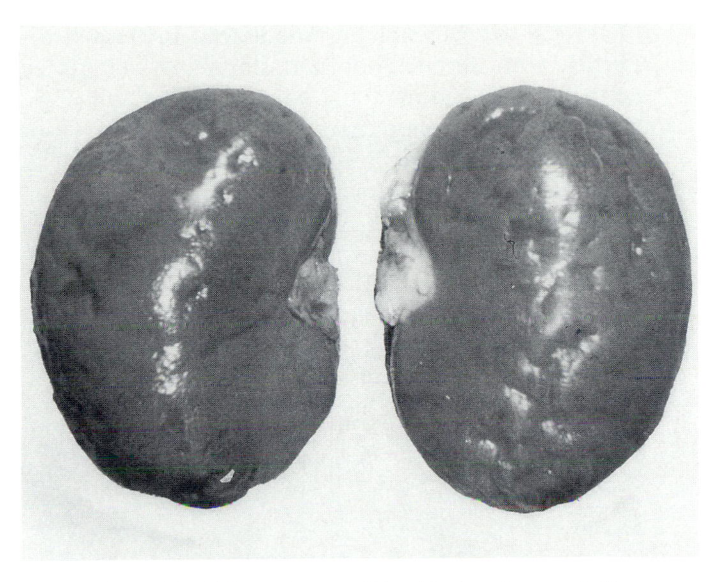

Fig. 6.40. Kidneys from rabbit with chronic encephalitozoonosis. There are multiple irregular, pitted areas on the surface of the cortices.

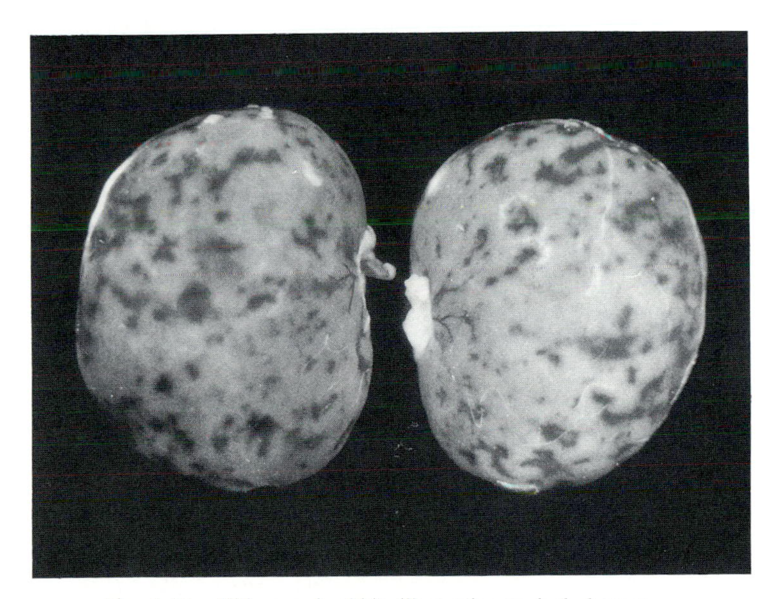

Fig. 6.41. Kidneys of rabbit, illustrating typical changes associated with recent infection with *Encephalitozoon cuniculi*. There are multiple dark red, depressed, irregular foci present in the cortex.

On histopathology, granulomatous lesions are evident in the interstitium of the lung, kidney, and liver by 1 mo postexposure. In the lung, focal to diffuse interstitial pneumonitis, with mononuclear cell infiltration, may occur (Fig. 6.42). Hepatic lesions are characterized by a focal granulomatous inflammatory response (Fig. 6.43), with periportal lymphocytic infiltration. Focal lymphocytic infiltrates may also occur in the myocardium. In the kidney, early lesions consist of a focal to segmental granulomatous interstitial nephritis, with degeneration and sloughing of affected epithelial cells and mononuclear cell infiltration (Figs. 6.44 and 6.45). Lesions may be present at all levels of the renal tubule, usually with minimal involvement of the glomeruli. Using tissue Gram stains (e.g., Brown and Brenn), the spores are evident as ovoid, gram-positive organisms approximately 1.5 x 2.5 μm in size. Staining procedures using carbol fuchsin will stain the organisms a distinct purple color. Spores may be present within epithelial cells, in macrophages, in in-

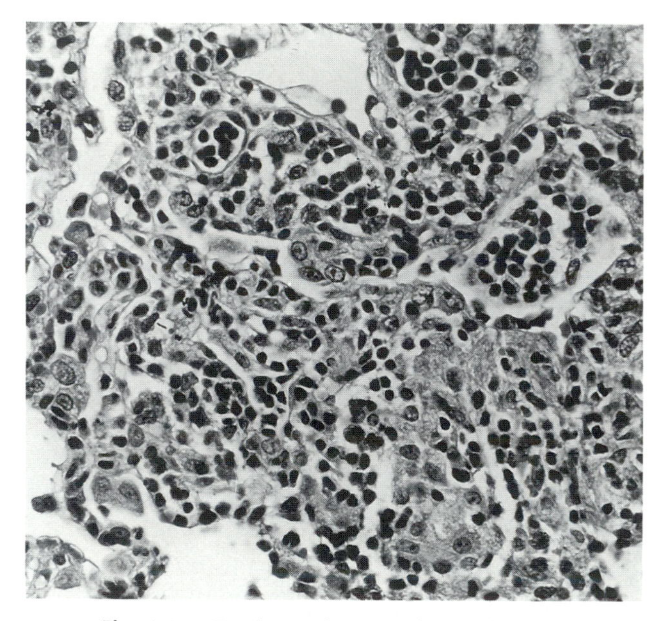

Fig. 6.42. Focal granulomatous interstitial pneumonitis, typical transient lesions associated with recent *Encephalitozoon* infection.

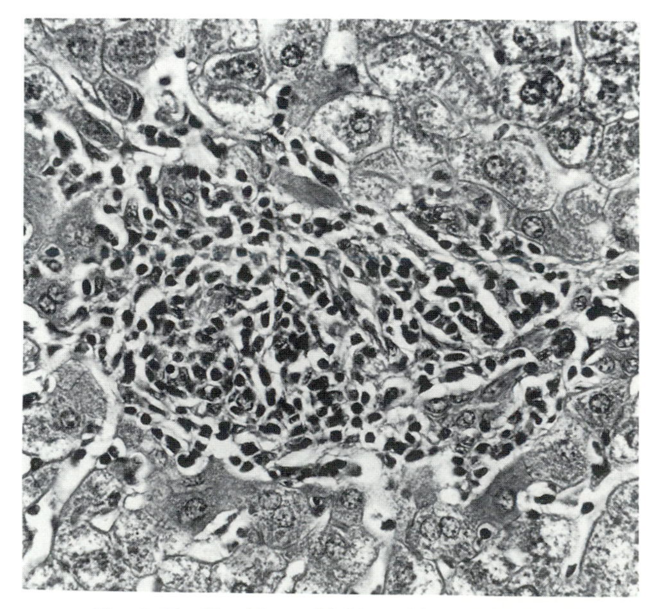

Fig. 6.43. Focal interstitial hepatitis associated with early *Encephalitozoon* infection. There is marked periportal infiltration with mononuclear cells.

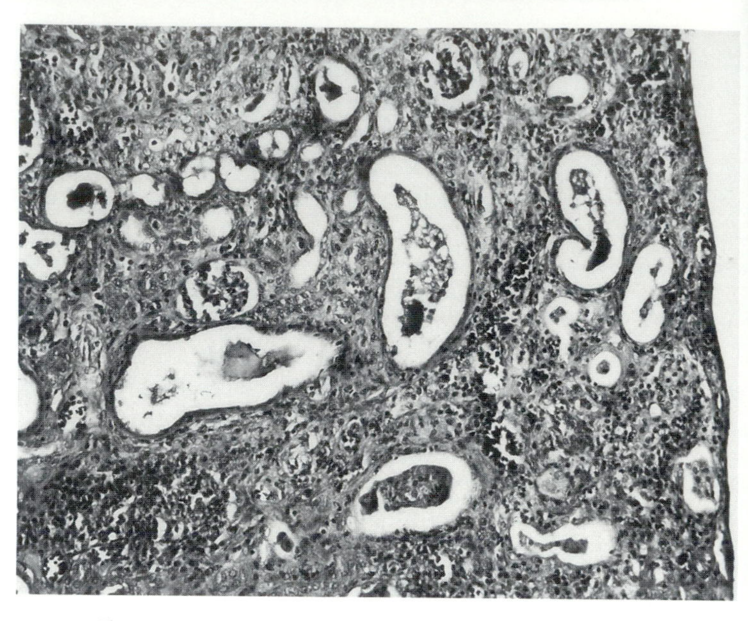

Fig. 6.44. Kidney from case of lapine encephalitozoonosis. There is a marked granulomatous interstitial nephritis, with mononuclear cell infiltration.

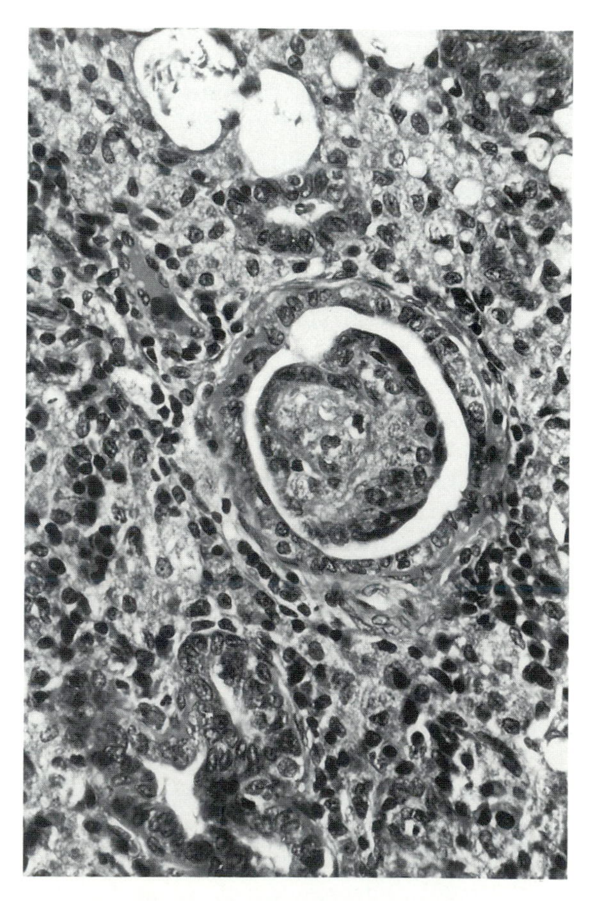

Fig. 6.45. Higher magnification of Fig. 6.44, illustrating renal encephalitozoonosis in New Zealand White rabbit. There is degeneration of tubules, with mononuclear cell infiltration. Cellular debris is present in scattered tubules.

flammatory foci, or free within collecting tubules. At 1–2 mo postexposure, organisms are usually readily demonstrated in the kidney. In renal lesions of longer duration, interstitial fibrosis, collapse of the parenchyma, and mononuclear cell infiltration are typical changes. The organism is usually eliminated by this stage of the disease.

In the central nervous system, lesions normally do not occur until at least 30 days postexposure. Changes are those of a focal nonsuppurative granulomatous meningoencephalomyelitis, with astrogliosis and perivascular lymphocytic infiltration (Fig. 6.46). Using appropriate stains, organisms may be evident as collections of spores within parasitized astroglial cells or as scattered organisms within granulomatous inflammatory foci (Fig. 6.47). Characteristic lesions may also be present in the central nervous system in the absence of identifiable organisms. In one report, *E. cuniculi* infection was also associated with cataractous change in the eye of a rabbit (Ashton et al. 1976). Several cases of uveitis with cataractous change in dwarf rabbits have been seen. Typical organisms are demonstrable within the affected lens stroma (Wolfer et al. 1992). Such cases are likely due to congenital infections with *E. cuniculi*.

DIAGNOSIS. The identification of characteristic lesions and the demonstration of the organisms in tissue sections are the standard diagnostic procedures used to confirm the diagnosis. The organisms can be readily differentiated from other protozoal infections,

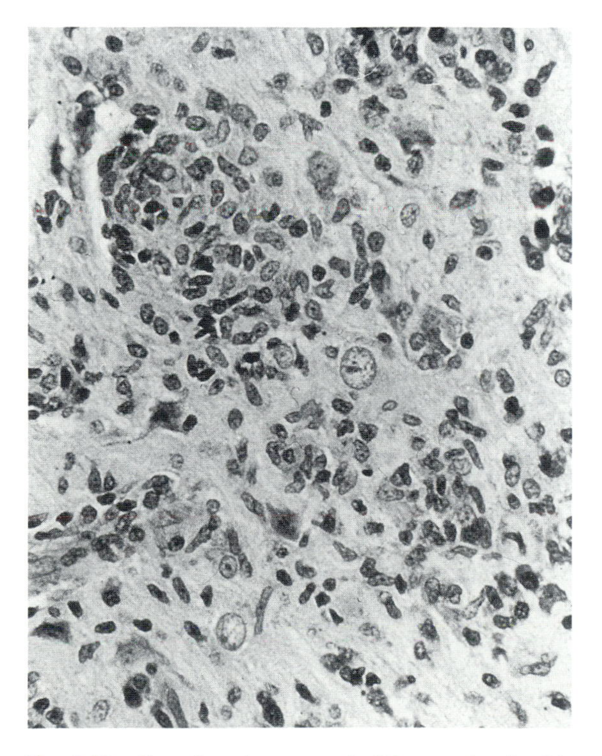

Fig. 6.46. Granulomatous encephalitis associated with chronic *Encephalitozoon* infection.

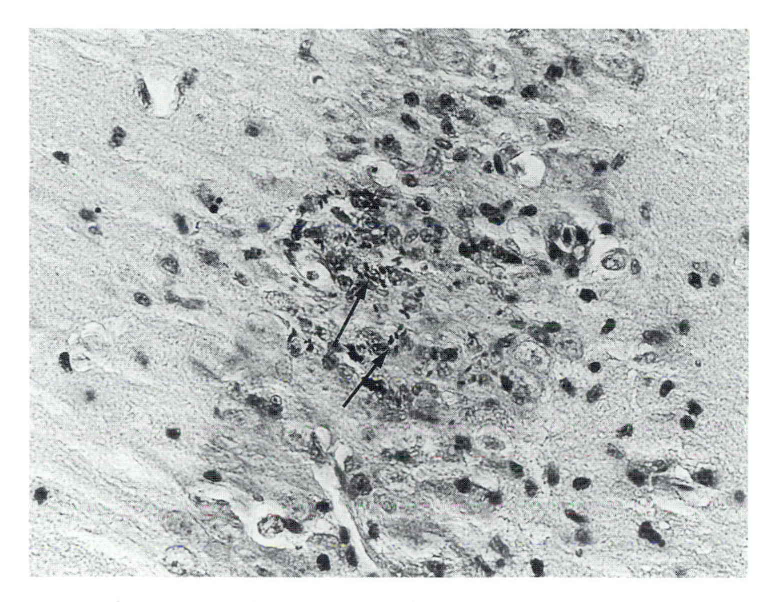

Fig. 6.47. Typical ovoid organisms are scattered in the neuropil (*arrows*) associated with encephalitozoonosis (Brown and Brenn stain).

such as toxoplasmosis, by the nature of the inflammatory response and staining properties of the organism. *Toxoplasma* organisms are gram-negative and do not stain with carbol fuchsin stains. In addition, Serology tests are available to identify animals that have been exposed to the organism. Serology tests currently used include the modified India ink immunoreaction test, indirect immunofluorescence microscopy, and a dot ELISA test (Beckwith et al. 1988). An intradermal skin test has also been used to detect infected rabbits (Pakes et al. 1972). The serology tests have been the most widely used. Accurate tests have been made on single drops of blood and from tissues and fluids collected at necropsy (Waller et al. 1980).

CONTROL Regular serological testing will readily identify infected animals. Since seroconversion precedes renal shedding, infected animals can be identified before they are excreting organisms (Cox et al. 1979; Bywater and Kellett 1978).

SIGNIFICANCE. Although usually a subclinical infection, occasionally neurological disease and renal insufficiency may occur in heavily infested animals. In addition, it is likely that growth weights and feed conversion are compromised during the course of the disease. The granulomatous lesions in the target tissues such as kidney and brain are a frequent source of confusion and frustration for researchers doing histological evaluations of tissues during the course of an experiment. There is also evidence that alterations in the immune response may occur during the course of the disease (Kunstyr et al. 1986). In addition, a variable response to implanted biomaterials has been observed in rabbits infected with *E. cuniculi* (Ansbacher et al. 1988). Currently there is no convincing evidence that encephalitozoonis is a zoonosis (Bywater 1979). Serological surveys at 4-wk intervals and culling of seropositive rabbits may be used to eliminate the organism from infected colonies (Bywater and Kellet 1978). Some strains of dwarf rabbits appear to be particularly likely to develop severe disease postexposure to *E. cuniculi*.

Cryptosporidiosis. Cryptosporidia have been identified in a variety of species, including calves, lambs, foals, and guinea pigs. An organism referred to as *Cryptosporidium cuniculus* has been identified in the small intestine of rabbits (Rehg et al. 1979). Animals are usually asymptomatic, and the organism may be demonstrated as an incidental finding. Microscopically, occasionally the villi of the terminal small intestine are shortened and blunted. Round to ovoid bodies are present on the brush border of epithelial cells. Changes on enterocytes are minimal and consist of elongation or shortening of microvilli adjacent to attachment sites (Inman and Takeuchi 1979).

SIGNIFICANCE. Based on current information, cryptosporidia rarely occur as a primary pathogen in enteritis in rabbits. The role of *C. cuniculus* as a copathogen has not been fully studied. The organism can induce significant shortening of intestinal villi during the course of an infection and thus could produce a potential problem when interpreting intestinal changes during experimental studies.

Toxoplasmosis

Antibodies to *Toxoplasma gondii* have been detected in the sera of rabbits in the United States, but clinical disease rarely occurs. In one reported outbreak of the disease, anorexia, pyrexia, and neurological disorders were the usual presenting signs. At necropsy, there were multiple foci of necrosis with a granulomatous inflammatory response present in the lung, liver, and spleen. Both tachyzoites and tissue cysts were associated with the lesions (Leland et al. 1992).

Helminth Infections: Pinworms (Oxyuriosis)

EPIZOOTIOLOGY AND PATHOGENESIS. Of the Oxyuridae, *Passalurus ambiguus* is a common parasite of domestic rabbits, and frequently the typical eggs, slightly flattened on one side, are observed on fecal flotation in asymptomatic rabbits. Adult worms are located in the cecum and other areas of the large intestine.

DIAGNOSIS. The identification of the typical pinworm eggs on fecal flotation and/or the adult worms in the large intestine will confirm the diagnosis.

SIGNIFICANCE. Infestations with *Passalurus* are considered to be relatively harmless by most parasitologists, although clinicians frequently elect to treat affected animals. In one report, impaired weight gains, poor breeding performance, and occasionally death were attributed to heavy infestations with pinworms (Duwel and Brech 1981).

Other Helminth Infections

In domestic rabbits, Trichostrongylidae include gastrointestinal helminths of the genera *Nematodirus* and *Trichostrongylus*. They are not considered to be a problem under normal circumstances. Of the cestodes, infestations with adult tapeworms is rare in the domestic rabbit. However, lagomorphs may serve as the intermediate host for *Taenia pisiformis*, a parasite whose definitive host is primarily the canine species. In one survey of slaughtered rabbits, approximately 7% had lesions consistent with hepatic cysticercosis (Flatt and Campbell 1974). Raised, light tan, focal to linear, solitary or multiple lesions up to 3 mm in diameter are present on the surface of the liver. Microscopically, lesions often consist of a necrotic center containing cell debris, inflammatory cells, and remnants of the parasite. At the periphery, there is fibrosis, with multinucleated giant cell formation, epitheliod cells, polymorphs, and mononuclear cells. In older lesions, mineralization and fibrous tissue proliferation occur. For additional information on endoparasites in this species, see Wescott (1974).

***Baylisascaris* Infection.** Infestation with *Baylisascaris procyonis* commonly occurs in the natural host, the raccoon. There appears to be a relatively amiable host-parasite relationship in the raccoon. However, when an unnatural host such as a rabbit (or a human being) accidentally ingests infective eggs, frequently a devastating cerebrospinal disorder results (Kazacos and Kazacos 1988; Kazacos et al. 1983). Typical neurological signs include torticollis, ataxia, circling, opisthotonus, and recumbancy. If not euthanized, animals usually die as a result of the unremitting nervous signs.

EPIZOOTIOLOGY AND PATHOGENESIS. Hay or bedding contaminated with raccoon feces containing *B. procyonis* eggs is the usual source of the parasite. Following passage in raccoon feces, embryonation requires approximately 30 days before the eggs are infective. Eggs will remain infective for at least a year under appropriate environmental conditions (Kazacos and Kazacos 1988). Following the accidental ingestion of embryonated eggs, the larvae are released in the intestine and undergo aggressive somatic and pulmonary migration. The severe damage, particularly in the central nervous system, is attributed to the rapid growth of the larvae and the ability to migrate into the brain, resulting in extensive trauma in target areas. In addition, metabolic wastes and enzymes elaborated by the parasite may evoke a vigorous inflammatory response.

PATHOLOGY. At necropsy, multiple, circumscribed, raised white nodules up to 1.5 mm in diameter are often present in the subepicardial and subendocardial regions of the liver and the serosal surface of the liver. Microscopic examination of the visceral lesions reveals focal granulomas, with mononuclear cells and polymorphs infiltrating the area. Remnants of the parasite are usually present within the lesions. In the central nervous system, lesions are most often present in the gray and white matter in the brain stem and cerebellar regions, but the cerebrum, including the hippocampus, may be involved. Sites of parasitic migration are characterized by extensive malacia and astrogliosis. Large numbers of Gitter cells and gemistocytic astrocytes may be present in lesions interpreted to be of several days duration (Fig. 6.48). Infiltrating inflammatory cells include lymphocytes, macrophages, eosinophils, and heterophils. Within the neuropil adjacent to the lesions, frequently there are nematode larvae with characteristic excretory columns and lateral alae (Fig. 6.49).

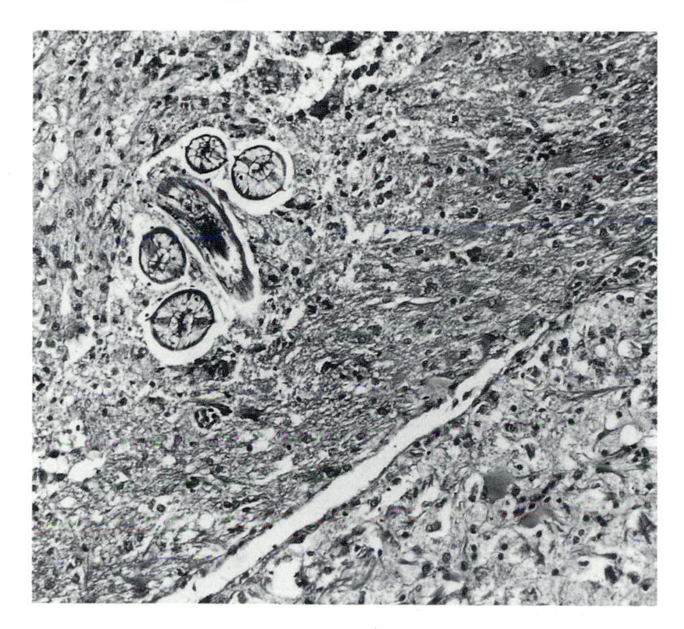

Fig. 6.49. Cerebral *Baylisascaris* infestation in New Zealand White rabbit. Note the focus of malacia with larvae present within lesion.

DIAGNOSIS. The lesions should be correlated with a possible source of *Baylisascaris* eggs. Larvae may be removed from the brain either by the Baermann method or by artificial digestion for positive identification.

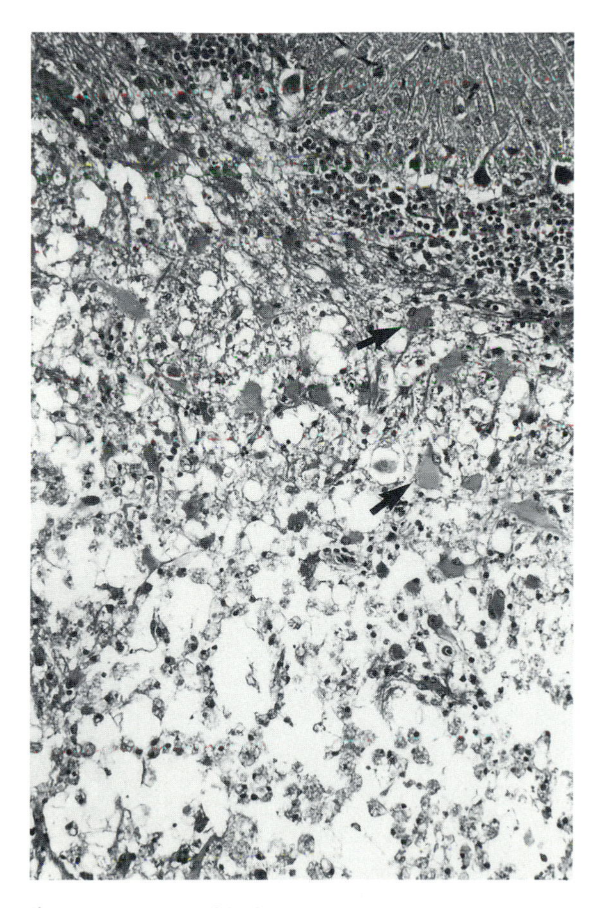

Fig. 6.48. Focus of leukomalacia in the cerebellum associated with cerebral *Baylisascaris procyonis* infection in domestic rabbit. There are prominent gemisocytic astrocytes (*arrows*) within the lesion. (Courtesy of R.J. Hampson)

SIGNIFICANCE. *Baylisascaris* infection represents one possible cause of neurological disease in rabbits. Zoonotic aspects should be emphasized. However, rabbits are a "dead end" host and thus cannot serve as a source of infective eggs for human contacts.

Ectoparasitic Infections

Mite Infection (Acariasis): *Psoroptes cuniculi*

Psoroptes cuniculi (ear mite) infection is the most common and costliest disease due to an ectoparasite in domestic rabbits. The mites are obligate, nonburrowing parasites that chew and pierce the epidermal layers of the external ear, evoking a marked inflammatory response.

EPIZOOTIOLOGY AND PATHOGENESIS. The mite normally spends its entire life span in the external ear of the rabbit. The life cycle (egg to egg) is usually completed in around 3 wk. Up to 10,000 mites may be present in a severely infested ear (Kraus 1974).

PATHOLOGY. In heavily parasitized ears, foul-smelling branlike crusts may fill the external ear canal and extend up the ear (Fig. 6.50). The ear is often thickened and edematous. The mites can be easily demonstrated in wet mount preparations from the ear (Fig. 6.51). Frequent local applications of oily preparations containing an acaricide have been used for routine treatment. Ear mite infestations have also been treated effectively with ivermectin (Wright and Riner 1985).

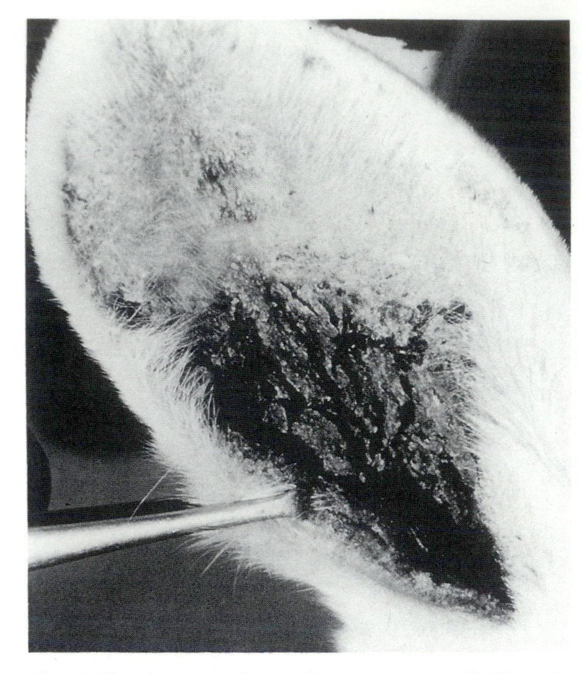

Fig. 6.50. Acariasis due to *Psoroptes cuniculi*. Note the crusting and oil debris present in the external ear canal.

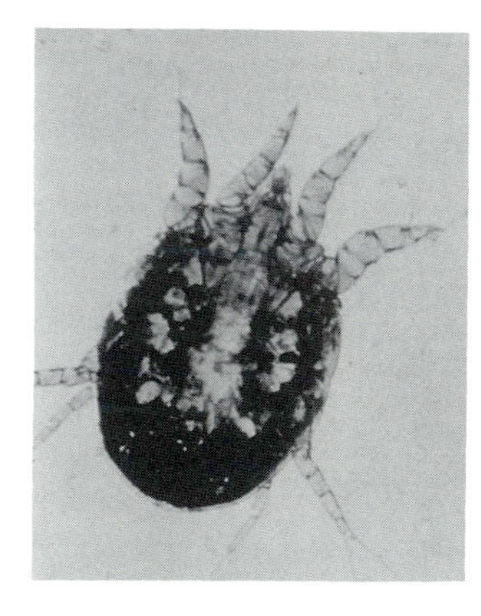

Fig. 6.51. Typical *Psoroptes* mite in a wet mount preparation collected from the affected ear.

Other Ectoparasitic Infections

Sarcoptic mange mites (*Notoedres cati*, *Sarcoptes scabei*) have been associated with alopecia and dermatitis involving the face, nose, lips, and external genitalia (Lin et al. 1984). Pruritis is common, and self-mutilation may occur. Cheyletid fur mites (*Cheyletiella parasitovorax*) may be present without producing detectable disease. When lesions are present, they are usually located on the dorsal trunk and scapular areas. They consist of areas of scaliness with crusting and variable degrees of hair loss (Fig. 6.52). Pruritis is usually absent. There may be a high prevalence of fur mites in commercial rabbitries. For additional information on ectoparasites, see Kraus (1974).

Bibliography for Parasitic Diseases

Coccidiosis

Barriga, O.O., and Arnoni, J.V. 1981. Pathophysiology of hepatic coccidiosis in rabbits. Vet. Parasitol. 8:201–10.

______. 1979. *Eimeria stiedae*: Oocyst output and hepatic function of rabbits with graded infections. Exp. Parasitol. 48:407–14.

Catchpole, J., and Norton, C.C. 1979. The species of *Eimeria* in rabbits for meat production in Britain. Parasitology 79:249–57.

Horton, R.J. 1967. The route of migration of *Eimeria stiedae* (Lindemann, 1865) sporozoites between the duodenum and bile ducts of the rabbit. Parasitology 57:9–17.

Owen, D. 1970. Life cycle of *Eimeria stiedae*. Nature 227:304.

Peeters, J.E., et al. 1984. Clinical and pathological changes after *Eimeria intestinalis* infection in rabbits. Zentralbl. Veterinaermed. [B] 31:9–24.

Rutherford, R.L. 1943. The life cycle of four intestinal coccidia of the domestic rabbit. J. Parasitol. 29:10–32.

Varga, I. 1982. Large-scale management systems and parasite populations: Coccidia in rabbits. Vet. Parasitol. 11:69–84.

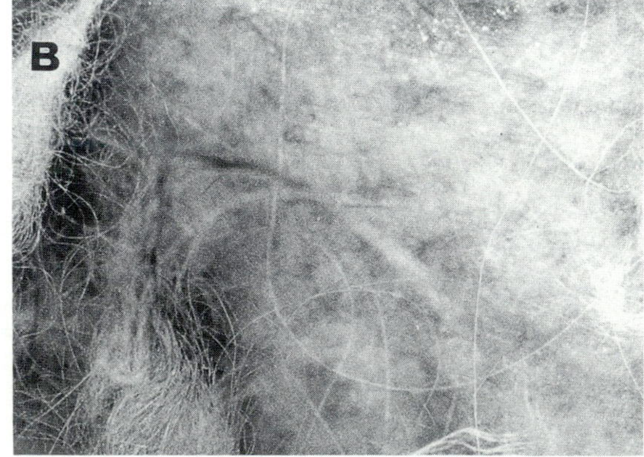

Fig. 6.52. (*A*) Dwarf rabbit with a history of doing poorly. Note the unkempt appearance of hair coat, *Cheyletiella* infestation. (*B*) Inguinal region, illustrating the scaling lesions with hair loss.

Encephalitozoonosis

Ansbacher, L., et al. 1988. The influence of *Encephalitozoon cuniculi* on neural tissue responses to implanted biomaterials in the rabbit. Lab. Anim. Sci. 38:689–95.

Anver, M.R., et al. 1972. Congenital encephalitozoonosis in a squirrel monkey. Vet. Pathol. 9:475–80.

Ashton, N., et al. 1976. Encephalitozoonosis (nosematosis) causing bilateral cataract in a rabbit. Br. J. Ophthalmol. 60:618–31.

Beckwith, C., et al. 1988. Dot enzyme-immunoabsorbent assay (dot ELISA) for antibodies to *Encepalitozoon cuniculi*. Lab. Anim. Sci. 38:573–76.

Bywater, J.E.C. 1979. Encephalitozoonosis a zoonosis? Lab. Anim. 13:149–51.

Bywater, J.E.C., and Kellett, B.S. 1978. The eradication of *Encephalitozoon cuniculi* from a specific pathogen-free rabbit colony. Lab. Anim. Sci. 28:402–4.

Cox, J.C. 1977. Altered immune responsiveness associated with *Encephalitozoon cuniculi* infection in rabbits. Infect. Immunol. 15:392–95.

Cox, J.C., and Ross, J. 1980. A serological survey of *Encephalitozoon cuniculi* infection in the wild rabbit in England and Scotland. Res. Vet. Sci. 28:396.

Cox, J.C., et al. 1979. An investigation of the route and progression of *Encephalitozoon cuniculi* infection in adult rabbits. J. Protozool. 26L:260–65.

Cox, J.C., et al. 1972. Presumptive diagnosis of *Nosema cuniculi* in rabbits by immunofluorescence. Res. Vet. Sci. 13:595–97.

Flatt, R.E., and Jackson, S.J. 1970. Renal nosematosis in young rabbits. Vet. Pathol. 7:492–97.

Hunt, R.D., et al. 1972. Encephalitozoonosis: Evidence for vertical transmission. J. Infect. Dis. 126:212–14.

Innes, J.R.M., et al. 1962. Occult encephalitozoonosis of the central nervous system of mice. J. Neuropathol. Exp. Neurol. 21:519–33.

Kunstyr, I., and Naumann, S. 1985. Head tilt in rabbits caused by pasteurellosis and encephalitozoonosis. Lab. Anim. 19:208–13.

Kunstyr I., et al. 1986. Humoral antibody response of rabbits to experimental infection with *Encephalitozoon cuniculi*. Vet. Parasitol. 21:223–32.

Lyngset, A. 1980. A survey of serum antibodies to *Encephalitozoon cuniculi* in breeding rabbits and their young. Lab. Anim. Sci. 30:558–61.

Moffat, R.E., and Schiefer, B. 1973. Microsporidiosis (encephalitozoonosis) in the guinea pig. Lab. Anim. Sci. 23:282–83.

Owen, D.G. 1980. Investigation into the transplacental transmission of *Encephalitozoon cuniculi* in rabbits. Lab. Anim. 14:35–38.

Pakes 1974 (*see* General Bibliography).

Pakes, S.P., et al. 1972. A diagnostic skin test for encephalitozoonosis (nosematosis) in rabbits. Lab. Anim. Sci. 22:870–77.

Pattison, M., et al. 1971. An outbreak of encephalomyelitis in broiler rabbits caused by *Nosema cuniculi*. Vet. Rec. 88:404–5.

Plowright, W. 1952. An encephalitis-nephritis syndrome in the dog probably due to congenital *Encephalitozoon* infection. J. Comp. Pathol. 62:83–92.

Robinson, J.J. Common infectious disease of laboratory rabbits questionably attributed to *Encephalitozoon cuniculi*. Arch. Pathol. 58:71–84.

van Rensburg, I.B.J.; and Du Plessis, J.L. 1971. Nosematosis in a cat: A case report. J. S. Afr. Med. Assoc. 42:327–31.

Waller, T., et al. 1980. Immunological diagnosis of encephalitozoonosis from postmortem specimens. Vet. Immunol. Immunopathol. 1:353–60.

Wolfer, J. 1992. Spontaneous lens capsule rupture in the rabbit. Vet. Pathol. 29:449.

Wright, J.H., and Craighead, E.M. 1992. Infectious motor paralysis in young rabbits. J. Exp. Med. 36:135–40.

Cryptosporidiosis

Inman, L.R., and Takeuchi, A. 1979. Spontaneous cryptosporidiosis in an adult female rabbit. Vet. Pathol. 16:89–95.

Rehg, J.E., et al. 1979. *Cryptosporidium cuniculus* in the rabbit (*Oryctolagus cuniculus*). Lab. Anim. Sci. 29:656–60.

Toxoplasmosis

Dubey, J.P. 1988. *Toxoplasmosis of Animals and Man*. Boca Raton, Fla.: CRC.

Leland, M.M., et al. 1992. Clinical toxoplasmosis in domestic rabbits. Lab. Anim. Sci. 42:318–19.

Helminth Infections

Dade, A.W., et al. 1975. An epizootic of cerebral nematodiasis in rabbits due to *Ascaris columnaris*. Lab. Anim. Sci. 25:65–69.

Duwel, D., and Brech, K. 1981. Control of oxyuriasis in rabbits by fenbendazole. Lab. Anim. 15:101–5.

Flatt, R.E., and Campbell, W.W. 1974. Cysticercosis in rabbits: Incidence and lesions of the naturally occurring disease in young domestic rabbits. Lab. Anim. Sci. 24:914–18.

Kazacos, K.R., and Kazacos, E.A. 1988. Diagnostic exercise: Neuromuscular condition in rabbits. Lab. Anim. Sci. 38:187–89.

Kazacos, K.R., et al. 1983. Fatal cerebrospinal disease caused by *Baylisascaris procyonis* in domestic rabbits. J. Am. Vet. Med. Assoc. 183:967–71.

Wescott, R.B. 1974. Helminth parasites. In *The Biology of the Laboratory Rabbit*, ed. S.H. Weisbroth et al., pp. 317–29. New York: Academic.

Ectoparasitic Infections

Flatt, R.E., and Weimers, J. 1976. A survey of fur mites in domestic rabbits. Lab. Anim. Sci. 26:758–61.

Kraus, A.L. 1974. Arthropod parasites. In *The Biology of the Laboratory Rabbit*, ed. S.H. Weisbroth et al., pp. 287–315. New York: Academic.

Lin, S.L., et al. 1984. Diagnostic exercise (sarcoptic mange). Lab. Anim. Sci. 34:353–55.

Wright, F.C., and Riner, J.C. 1985. Comparative efficacy of injection routes and doses of ivermectin against *Psoroptes* in rabbits. Am. J. Vet. Res. 46:752–54.

General Bibliography

Pakes, S.P. 1974. Protozoal diseases. In *The Biology of the Laboratory Rabbit*, ed. S.H. Weisbroth et al., pp. 263–86. New York: Academic.

MISCELLANEOUS DISORDERS

Ulcerative Dermatitis ("Sore Hocks")

The affected area of skin typically involves a circumscribed region of varying size on the plantar aspect of the metatarsal bones. Lesions consist of a circumscribed, ulcerated area covered by granulation tissue and necrotic debris (Fig. 6.53). Purulent exudate may be adherent to the lesions. The problem is most commonly seen in heavy, mature adults. Poor sanitation, trauma from poor-quality, wire-bottom cages, and hereditary predisposition are examples of factors that may influence the incidence of the disease. *Staphylococcus aureus* is the most frequent isolate from lesions.

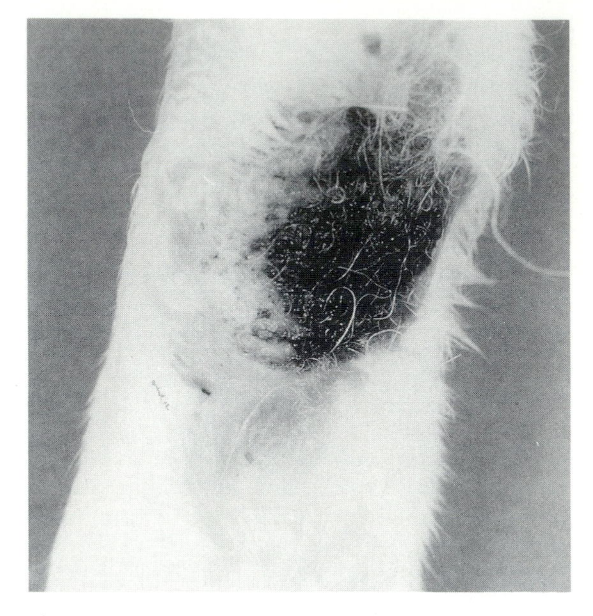

Fig. 6.53. Pododermatitis sore hock in adult domestic rabbit.

Vertebral Fracture

Posterior paralysis due to vertebral fracture or dislocation occurs all too often in laboratory rabbits. The axial and appendicular skeletons of domestic rabbits are relatively fragile in proportion to their muscle mass. Thus an unsupported, sudden movement of the hindlimbs may exert sufficient leverage on the lumbosacral junction to cause a vertebral fracture. Depending on the duration of the problem prior to euthanasia and necropsy, the hindquarters may be soiled with urine and fecal material consistent with incontinence. The site of the fracture (or luxation) is usually the lumbosacral region (L7) (Jones et al. 1982). There may be extensive hemorrhage in the underlying psoas muscles. Changes vary from luxation to multiple fractures of the affected vertebra, with extensive damage to the lumbosacral spinal cord.

Hair Chewing (Barbering)

Occasionally hair loss due to hair chewing (barbering) occurs. Patchy alopecia may be present on the face and back, with no evidence of a concurrent dermatitis (Fig. 6.54). Hair chewing most commonly occurs in young, group-housed rabbits. Skin scrapings for microscopic examination and fungal culture are recommended in order to eliminate the possibility of dermatophyte infection. Boredom and low-roughage diets have been implicated as contributing factors to this condition.

Gastric Trichobezoar (Hairball)

Hairballs are frequently present as an incidental finding at necropsy in rabbits that have died from other causes. Depending on the size and location, however, anorexia, wasting, and occasionally death may occur in severely affected animals. In one study of clinically affected animals, hematology and blood chemistry values were within normal limits (Wagner et al. 1974). Animals are usually alert, with reduced feed and water consumption. The antemortem diagnosis is usually confirmed by palpation and sometimes by contrast radiography. Predisposing factors implicated include excessive grooming and hair chewing due to boredom and insufficient roughage in the diet to promote passage of hair through the alimentary tract. Mineral oil, oral treatment with substances containing papain (e.g., fresh pineapple juice), and feeding supplemental roughage have been used as methods to treat the disorder.

Deaths attributed to hairballs do occur. At necropsy, animals may be in fair to poor condition.

Fig. 6.54. Typical hair chewing (barbering) pattern in juvenile rabbit.

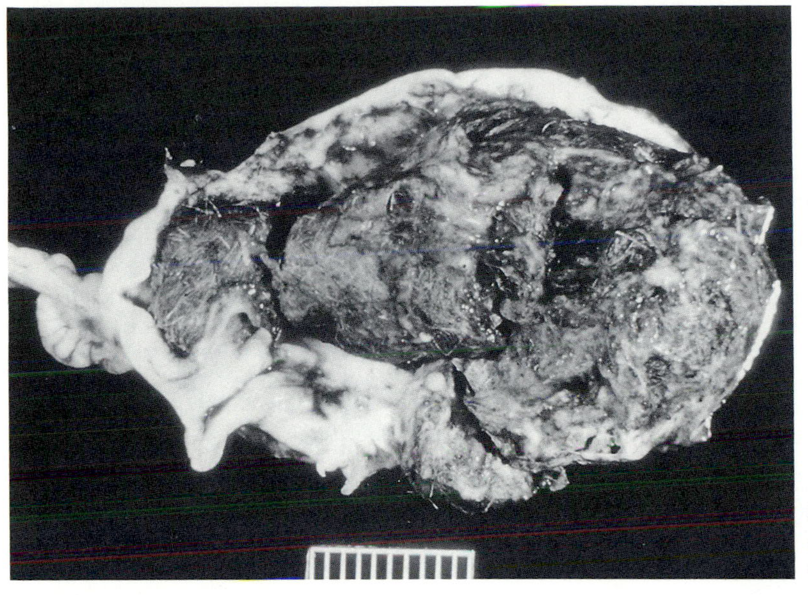

Fig. 6.55. Trichobezoar (hairball) and stomach from adult rabbit. The animal was alert but anorectic and was euthanized.

Large trichobezoars usually fill the stomach, extending into the pyloric region (Fig. 6.55). The intestinal tract usually contains scanty ingesta. Hepatic lipidosis is a characteristic finding. Gastric rupture and peritonitis may occur.

BIBLIOGRAPHY FOR MISCELLANEOUS DISORDERS

Jones, T., et al. 1982. Diagnostic exercise (vertebral fracture). Lab. Anim. Sci. 32:489–90.

Lee, K.P., et al. 1978. Acute peritonitis in the rabbit (*Oryctolagus cuniculi*) resulting from gastric trichobezoar. Lab. Anim. Sci. 28:202–4.

Wagner, J.E., et al. 1974. Spontaneous deaths in rabbits resulting from gastric trichobezoars. Lab. Anim. Sci. 24:826–30.

NUTRITIONAL AND METABOLIC DISEASES

Vitamin E Deficiency

With the current rigid quality-control standards for commercial feed production, confirmed nutritional problems are relatively rare. In addition, synthetic diets are prepared routinely by some researchers for specialized projects; thus there is also the possibility of human error in these formulations. There are several reports of nutritional muscular dystrophy due to vitamin E deficiency (Yamimi and Stein 1989; Ringler and Abrams 1970). In addition to stiffness and muscle weakness, neonatal mortality and infertility are manifestations of vitamin E deficiency in domestic rabbits. At necropsy pale, mineralized streaks may be present in musculature, such as the diaphragm, paravertebral regions, and hindlimbs. Similar lesions have been seen grossly in choline- or potassium-deficient rabbits (Hunt and Harrington 1974). Typical changes seen microscopically are hyaline degeneration of affected myofibers and clumping and mineralization of the sarcoplasm (Fig. 6.56). Macrophages may be present in reactive areas. Collapse of sarcolemmal sheaths and interstitial fibrosis frequently occur in lesions of some duration.

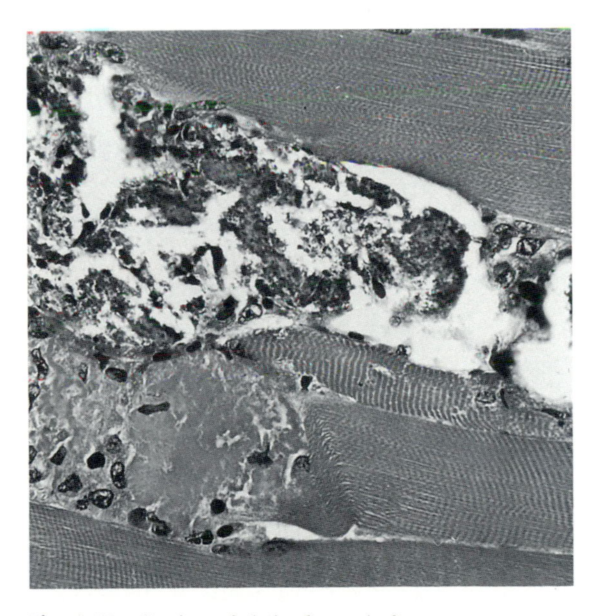

Fig. 6.56. Section of skeletal muscle from a spontaneous case of nutritional myopathy due to vitamin E–selenium deficiency. Note the degeneration of myofibers, with concurrent mineralization.

Hypervitaminosis D

Hypervitaminosis D has been reported to occur occasionally in domestic rabbits fed an improperly formulated diet (Zimmerman et al. 1990; Stevenson et al. 1976). Histologic changes include medial degeneration and mineralization of major arteries. Mineralization of the glomerular tufts, basement membranes, and tubules of the kidney may also occur. In the long bones, there is deposition of basophilic material (presumably osteoblasts) on the periosteal and endosteal surfaces, medullary trabeculae, and haversian systems (Fig. 6.57). For additional information on nutritional requirements and disorders, see Hunt and Harrington (1974).

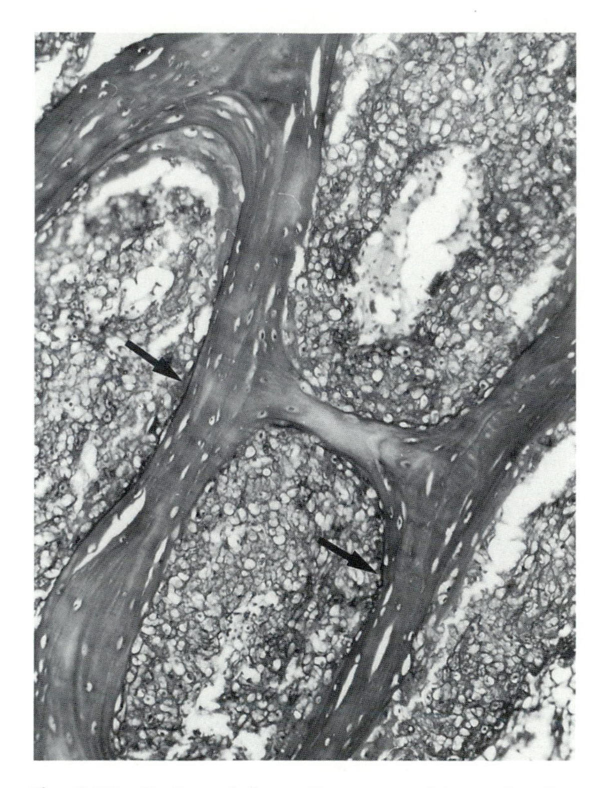

Fig. 6.57. Section of femur from case of hypervitaminosis D. There is deposition of basophilic material (*arrows*) on the endosteal surfaces. (Courtesy of N.C. Palmer)

Disorders Associated with Hypo- or Hypervitaminosis A

The clinical manifestations of vitamin A deficiency or toxicity are similar in domestic rabbits and are characterized by poor conception rates, congenital anomalies, fetal resorptions, abortion, and weak, thin kits. Congenital defects associated with hypervitaminosis A include microencephaly, hydrocephalus, and cleft palate.

Pregnancy Toxemia

This is a poorly characterized and understood condition that occurs in does, usually during the last week in pregnancy. Obesity and fasting are important predisposing factors (Harkness and Wagner 1989). The disease is characterized by low morbidity and high mortality. Primiparous does are especially at risk, and metabolic toxemia also may occur on occasion in obese, "stressed," nonpregnant rabbits. Mobilization of fat deposits for energy may result in metabolic acidosis and ketosis, with resultant anorexia, depression, and death. At necropsy animals are usually obese, with fatty infiltration of the liver, kidney, and adrenal glands.

Bibliography for Nutritional and Betabolic Disorders

Cheeke, P.R. 1987. Vitamins. In *Rabbit Feeding and Nutrition,* by P.R. Cheeke, pp. 136–53. New York: Academic.

DiGiacomo, R.F., et al. 1992. Hypervitaminosis A and reproductive disorders in rabbits. Lab. Anim. Sci. 42:250–54.

Flatt, R.E., et al. 1974. Metabolic, traumatic, and miscellaneous diseases of rabbits. In *The Biology of the Laboratory Rabbit*, ed. S.H. Weisbroth et al., pp. 435–51. New York: Academic.

Greene, H.S.N. 1937. Toxemia of pregnancy in the rabbit. Clinical manifestations and pathology. J. Exp. Med. 65:809–32.

Harkness, J.E., and Wagner, J.E. 1989. *The Biology and Medicine of Rodents and Rabbits*. Philadelphia: Lea and Febiger, pp. 147–48.

Hunt, C.E., and Harrington, D.D. 1974. Nutrition and nutritional diseases. In *The Biology of the Laboratory Rabbit*, ed. S.H. Weisbroth et al., pp. 403–33. New York: Academic.

Ringler, D.H., and Abrams, G.D. 1971. Laboratory diagnosis of vitamin E deficiency in rabbits fed a faulty commercial ration. Lab. Anim. Sci. 21:383–88.

______. 1970. Nutritional muscular dystrophy and neonatal mortality in a rabbit breeding colony. J. Am. Vet. Med. Assoc. 157:1928–34.

Stevenson, R.G., et al. 1976. Hypervitaminosis D in rabbits. Can. Vet. J. 17:54–57.

Yamimi, B., and Stein, S. 1989. Abortion, stillbirth, neonatal death, and nutritional myodegeneration in a rabbit breeding colony. J. Am. Vet. Med. Assoc. 194:561–62.

Zimmerman, T.E., et al. 1990. Soft tissue mineralization in rabbits fed a diet containing excess vitamin D. Lab. Anim. Sci. 40:212–14.

HEREDITARY DISORDERS

Congenital Glaucoma (Buphthalmia)

This condition occurs most frequently in New Zealand White rabbits. Buphthalmia is characterized clinically by enlargement of one or both eyes, with subsequent corneal opacity. Abnormalities may occur within the first few weeks of life, but usually they are first evident by 3–5 mo of age. The primary defect has been identified as an absence or underdevelopment of the outflow channels, with incomplete cleavage of the iridocorneal angles (Tesluk et al. 1982). Thus with impaired drainage of aqueous humor from the anterior chamber, the increased intraocular pressure results in megaloglobus, increased corneal diameter, and protrusion of the corneal contours (Fig. 6.58). The sclera is relatively immature at this stage and thus expands to accommodate the increased volume of aqueous humor within the globe. The defect is inherited as an autosomal recessive, with incomplete penetrance (Hanna et al. 1962).

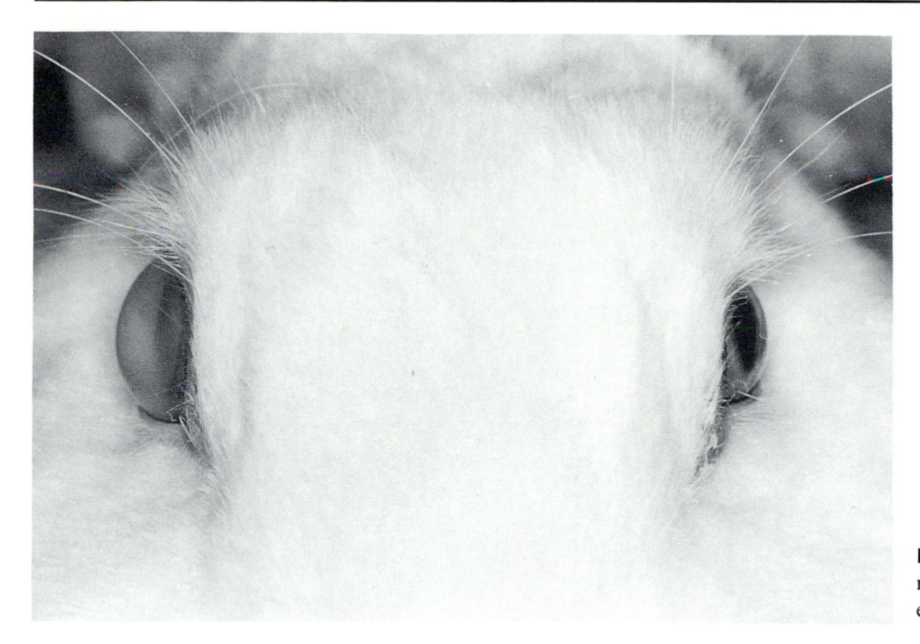

Fig. 6.58. Unilateral buphthalmia. There is marked coning of the cornea of the affected eye.

Malocclusion

In lagomorphs, normal occlusion brings the lower incisor teeth into apposition against the upper secondary incisors ("peg teeth") located behind the large upper incisors. In rabbits with malocclusion, usually the mandible is abnormally long relative to the maxilla. The overshot lower jaw results in misalignment, failure of the incisors to wear normally, and impaired mastication. In domestic rabbits, combined lengths of growth of both the upper and lower incisors has been shown to be over 20 cm/yr (Shadle 1936). Thus malocclusion will cause overgrowth of the incisors in a relatively short period of time (Fig. 6.59). The defect appears to be inherited as an autosomal recessive. Overgrowth of the premolar and molar teeth has also been recognized to occur in rabbits. For additional information on other inherited conditions, including behavioral, neuromuscular, and skeletal abnormalities, see Lindsey and Fox (1974).

Endometrial Venous Aneurysms

Multiple endometrial venous aneurysms have been associated with persistent urogenital bleeding (Bray et al. 1992). At necropsy, there are multiple blood-filled endometrial varices that consist of dilated, thin-walled veins (Fig. 6.60). These varices ap-

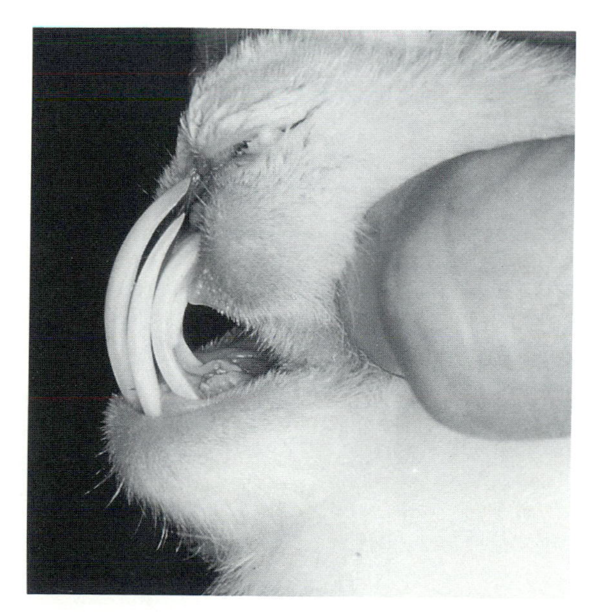

Fig. 6.59. Malocclusion in an adult New Zealand White rabbit. Note the concurrent overgrowth of the accessory incisor teeth.

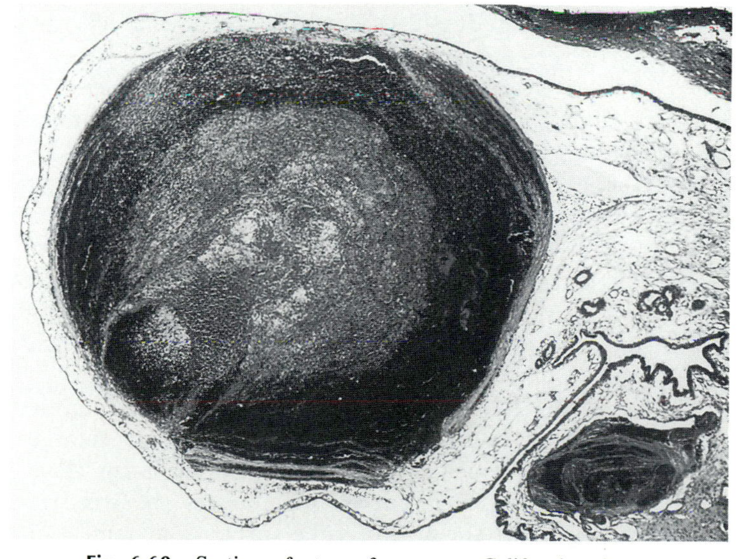

Fig. 6.60. Section of uterus from a pet Californian doe with a history of intermittent vaginal bleeding and hematuria. Animal became anemic and was euthanized *in extremis.* Note the marked dilation of the endometrial vessels (endometrial venous aneurysms) with thrombus formation. (Courtesy of M. Savic)

parently rupture and bleed periodically into the uterine lumen, with subsequent hematuria. They have been observed in nonpregnant multiparous does. The aneurisms are considered to be a congenital defect, and there is no evidence that predisposing factors, such as trauma or bleeding disorders, play a role in the disease.

BIBLIOGRAPHY FOR HEREDITARY DISORDERS

Bray, M.V., et al. 1992. Endometrial venous aneurysms in three New Zealand White rabbits. Lab. Anim. Sci. 42:360–62.

Fox, R.R., and Crary, D.D. 1971. Mandibular prognathism in the rabbit: Genetic studies. J. Hered. 62:23-27.

Hanna, B.L., et al. 1962. Recessive buphthalmos in the rabbit. Genetics 47:519–29.

Lindsey, J.R., and Fox, R.R. 1974. Inherited diseases and variations. In *The Biology of the Laboratory Rabbit*, ed. S.H. Weisbroth et al., pp. 377–401. New York: Academic.

Shadle, A.R. 1936. The attrition and extrusive growth of the four major incisor teeth of domestic rabbits. J. Mammal. 17:15–21.

Tesluk, G.C., et al. 1982. A clinical and pathological study of inherited glaucoma in New Zealand White rabbits. Lab. Anim. 16:234–39.

Zeman, W.V., and Fielder, F.G. 1969. Dental malocclusion and overgrowth in rabbits. J. Am. Vet. Med.Assoc. 155:1115–19.

NEOPLASMS

Uterine Adenocarcinoma

This is the most commonly encountered spontaneous neoplasm occurring in *Oryctolagus cuniculi*. The relatively low incidence of this tumor seen in most commercial rabbitries and research facilities is due to the fact that these animals are usually relatively young. In one study, the incidence of uterine adenocarcinoma in does 2–3 yr of age was around 4%, and in does 5–6 years of age, around 80%. A variety of breeds were affected (Greene 1958). Thus there is a striking increase in the incidence of uterine tumors with increasing age.

PATHOGENESIS. The carcinogenic effects of estrogens have been implicated in the evolution of uterine cancer in this species. However, the evidence is conflicting. In a more recent study, the administration of estrogens to female Dutch rabbits actually reduced the incidence of endometrial adenocarcinomas (Baba and Von Haam 1972).

PATHOLOGY. On gross examination, the tumors appear as nodular, frequently multicentric enlargements and usually involve both uterine horns (Figs. 6.61 and 6.62). On cut surface, masses are firm, frequently with a cauliflowerlike surface and central ulcerations. Serosal implantation and metastases to the lung and liver may occur. Typical microscopic changes are those of an adenocarcinoma, with invasion of the underlying layers forming acinar and tubular structures (Fig. 6.63). In rapidly growing tumors, necrotic

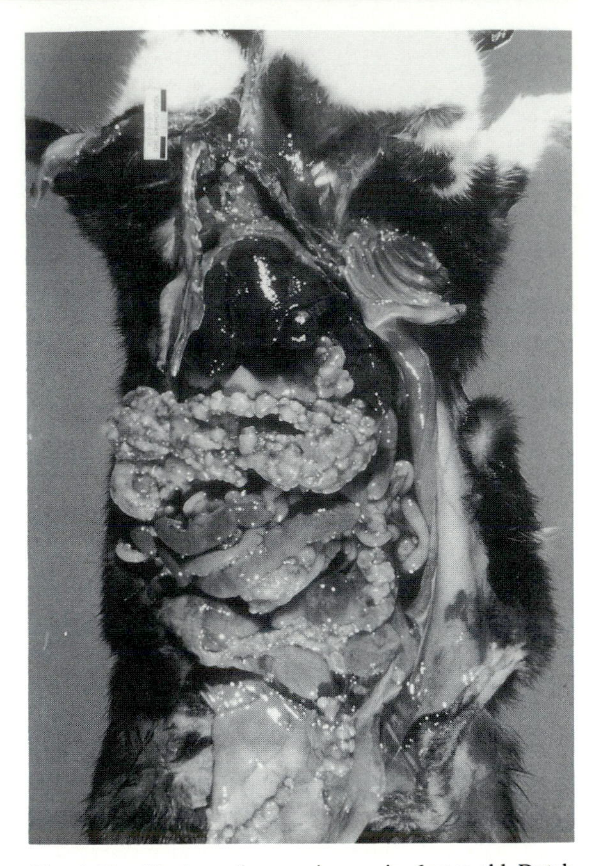

Fig. 6.61. Uterine adenocarcinoma in 6-year-old Dutch Belted doe. There are multiple implants in the peritoneal cavity.

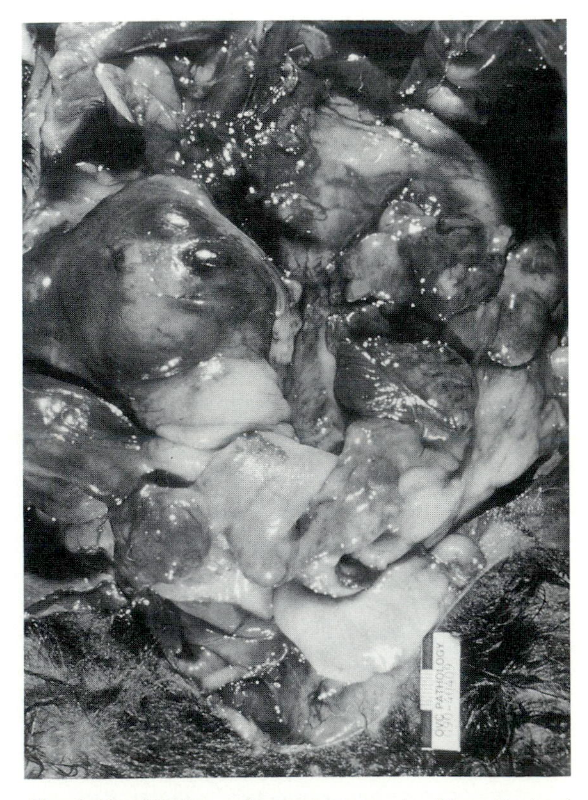

Fig. 6.62. Uterine adenocarcinoma in New Zealand White doe, illustrating the multiple nodules present in both uterine horns.

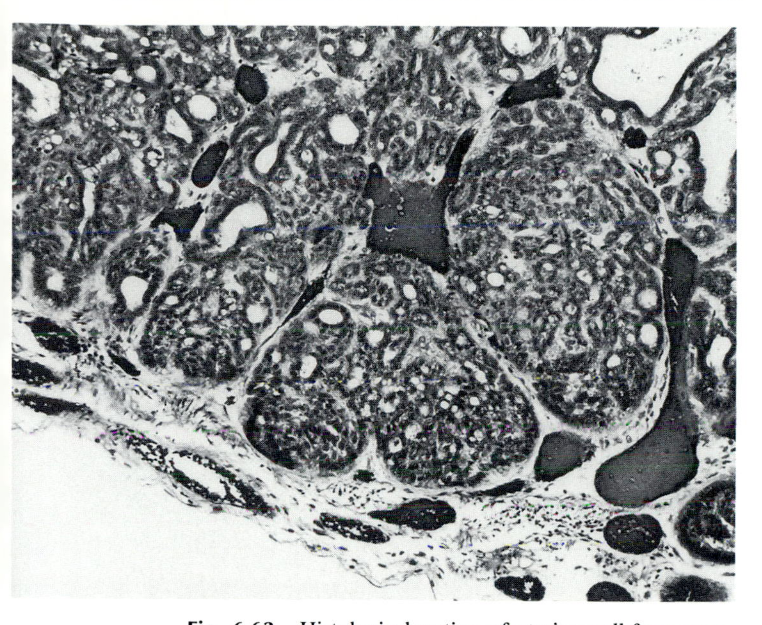

Fig. 6.63. Histological section of uterine wall from spontaneous uterine adenocarcinoma in aged New Zealand White doe. The neoplastic process extends into the myometrium and to the serosal surface of the uterus. Note the acinar structures lined by anaplastic epithelial cells.

areas are frequently observed. Metastases and tumor implants usually are similar to the primary neoplasm, often with a prominent stromal component.

Lymphosarcoma

Lymphoreticular neoplasms are the most common malignancy encountered in juvenile and young adult domestic rabbits. Anemia, depressed hematocrit, and terminally, elevated BUN are typical changes seen clinically. Affected animals are usually aleukemic, but leukemia occasionally occurs. Although lymphoproliferative disorders have been associated with *Herpesvirus sylvilagus* infections in the cottontail rabbit, there is no convincing evidence that viral agents are associated with the disease in domestic rabbits. An autosomal recessive gene has been implicated as an important factor in susceptibility to the disease (Fox et al. 1970).

PATHOLOGY. At necropsy, the typical findings are as follows: The kidneys are enlarged and pale gray to tan, with irregular cortical outlines. On cut surface, the changes are usually confined to the renal cortices (Fig. 6.64). The liver is enlarged, pale, and swollen. Splenomegaly, enlarged lymph nodes, and prominent intestinal lymphoid tissue are typical features seen grossly. The wall of the stomach may be markedly thickened, with irregular surface plaques and mucosal ulceration. On histopathology, there are diffuse infiltrates of lymphoblastic cells in the interstitial regions of the renal cortex, with distortion of the normal architecture and relative sparing of the glomeruli (Fig. 6.65). In the liver, there are periportal to diffuse sinus-

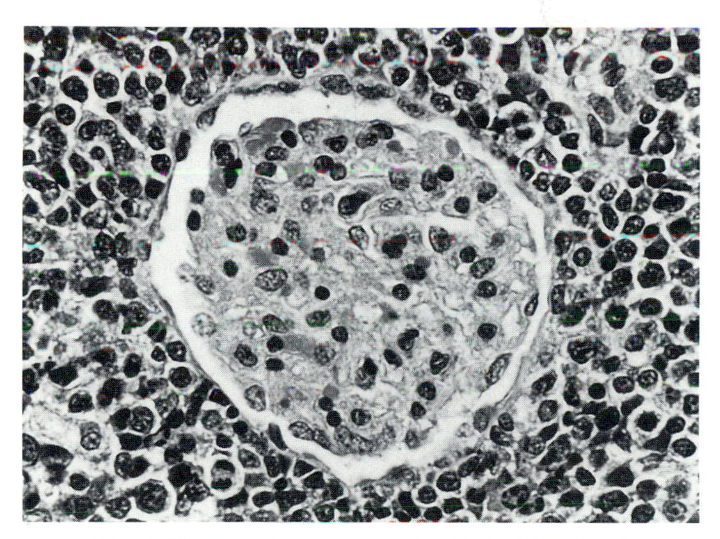

Fig. 6.65. Lymphosarcoma in New Zealand White rabbit. This section of kidney illustrates marked neoplastic lymphocytic infiltrate in the interstitium adjacent to a glomerulus.

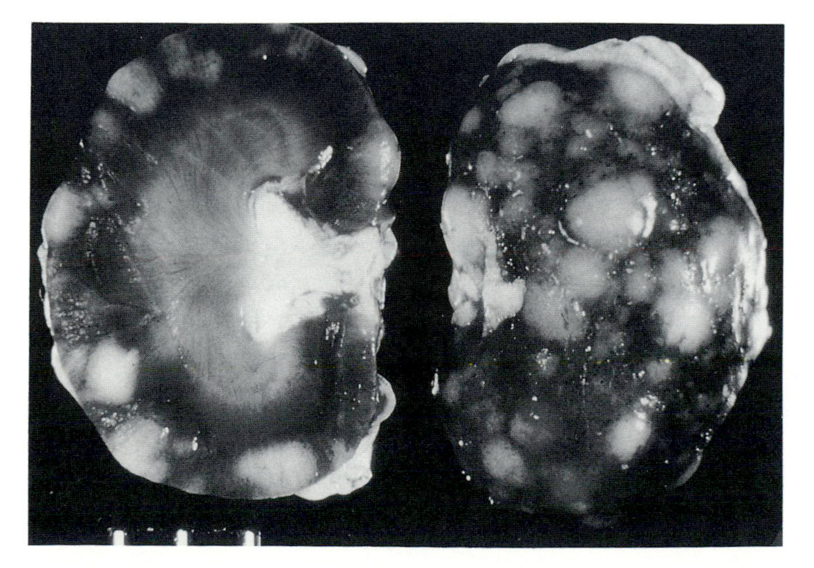

Fig. 6.64. Kidneys from young domestic rabbit with spontaneous lymphosarcoma. There are multiple raised, pale, nodular masses on the cortices, a frequent finding in lapine lymphoma.

oidal infiltrates of neoplastic cells. Diffuse infiltration occurs in the spleen, and infiltrates are also usually present in lymph nodes and in lymphoid tissue of the alimentary tract and stomach. Neoplastic cells may also be present in the uveal tract, adrenal gland, and ovary. Neoplastic lymphocytes range from 5 to 35 μm in diameter, with prominent nucleoli. There is usually a high mitotic index.

Other Neoplasms

Bile duct adenoma, bile duct carcinoma, osteosarcoma, embryonal nephroma, leiomyoma, leiomyosarcoma, squamous cell carcinoma, thymoma, mammary adenocarcinoma (Fig. 6.66), and papillomatosis are other spontaneous tumors that occur in this species (Weisbroth 1974). For additional information on virus-associated papillomatosis, see Viral Infections.

Bibliography for Neoplasms

Anderson, W.I , et al. 1990. Bilateral testicular seminoma in a New Zealand White rabbit (*Oryctolagus cuniculus*). Lab. Anim. Sci. 40:420–21.

Baba, N., and Von Haam, E. 1972. Animal model: Spontaneous adenocarcinoma in aged rabbits. Am. J. Pathol. 68:653–56.

Baba, N., et al. 1970. Nonspecific phosphatases of rabbit endometrial carcinoma. Arch. Pathol. 90:65–71.

Fox, R.R., et al. 1970. Lymphosarcoma in the rabbit: Genetics and pathology. J. Natl. Cancer Inst. 45:719–30.

Greene, H.S.N. 1958. Adenocarcinoma of the uterine fundus in the rabbit. Ann. N.Y. Acad. Sci. 75:535–42.

Greene, H.S.N., and Newton, R.I. 1948. Evolution of cancer of the uterine fundus in the rabbit. Cancer 1:88–99.

Toth, L.A., et al. 1990. Lymphocytic leukemia and lymphosarcoma in a rabbit. J. Am. Vet. Med. Assoc. 197: 627–29.

Weisbroth, S.H. 1974. Neoplastic diseases. In *The Biology of the Laboratory Rabbit*, ed. S.H. Weisbroth, et al., pp. 331–75. New York: Academic.

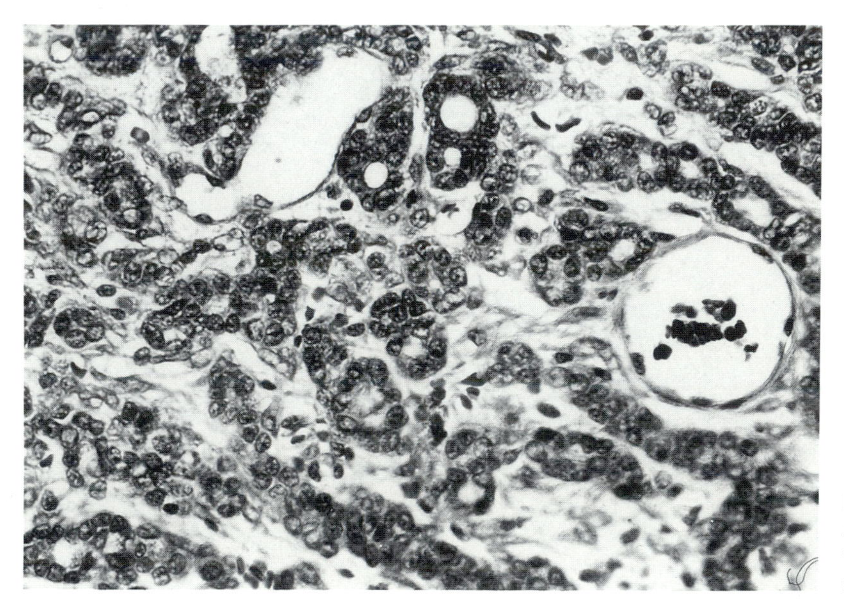

Fig. 6.66. Mammary adenocarcinoma in an aged New Zealand White doe. There was marked local infiltration with multiple metastases to the lung. There was no uterine involvement. (Courtesy of M.L. Brash)

INDEX

GERBIL

GUINEA PIG

HAMSTER

MOUSE

RABBIT

RAT